Lecture Notes in Computer Science 1025

Edited by G. Goos, J. Hartmanis and J. van Leeuwen

Advisory Board: W. Brauer   D. Gries   J. Stoer

Springer
Berlin
Heidelberg
New York
Barcelona
Budapest
Hong Kong
London
Milan
Paris
Santa Clara
Singapore
Tokyo

Colin Boyd (Ed.)

Cryptography and Coding

5th IMA Conference
Cirencester, UK, December 18-20, 1995
Proceedings

 Springer

Series Editors

Gerhard Goos
Universität Karlsruhe
Vincenz-Priessnitz-Straße 3, D-76128 Karlsruhe, Germany

Juris Hartmanis
Department of Computer Science, Cornell University
4130 Upson Hall, Ithaca, NY 14853, USA

Jan van Leeuwen
Department of Computer Science,Utrecht University
Padualaan 14, 3584 CH Utrecht, The Netherlands

Volume Editor

Colin Boyd
The Manchester School of Engineering
Oxford Road, M13 9PL, Manchester, UK

Cataloging-in-Publication data applied for

Die Deutsche Bibliothek - CIP-Einheitsaufnahme

Cryptography and coding : 5th IMA conference, Cirencester,
UK, December 18 - 20, 1995 ; proceedings / Colin Boyd (ed.). -
Berlin ; Heidelberg ; New York ; Barcelona ; Budapest ; Hong
Kong ; London ; Milan ; Paris ; Santa Clara ; Singapore ;
Tokyo : Springer, 1995
  (Lecture notes in computer science ; Vol. 1025)
  ISBN 3-540-60693-9
NE: Boyd, Colin [Hrsg.]; GT

CR Subject Classification (1991): E.3-4, G.2.1, C.2, J.1

1991 Mathematics Subject Classification: 11T71, 68P25, 94A60, 94Bxx

ISBN 3-540-60693-9 Springer-Verlag Berlin Heidelberg New York

© Springer-Verlag Berlin Heidelberg 1995
Printed in Germany

Typesetting: Camera-ready by author
SPIN 10512350   06/3142 – 5 4 3 2 1 0   Printed on acid-free paper

Preface

The first IMA Conference on Cryptography and Coding took place in December 1986. The second conference had to wait another three years, but since December 1989 the series has become bi-annual. The topics of cryptography and coding are inextricably linked; indeed the modern theories of both have their roots in the seminal work of Shannon. This conference is perhaps unique in concentrating on both areas and provides a valuable opportunity to explore the fruitful relationships between the two; many of the papers in this volume are concerned with the overlap.

This time there was a record of 48 papers submitted for inclusion. These were from an international authorship composed as follows: UK (27 submissions), France (4), Japan (2), Norway (2), Russia (2), Spain (2), Australia, Belgium, Germany, Italy, Malta, South Africa, Switzerland, USA, Yugoslavia. I would like to thank the authors of all papers, both those whose work is included in these Proceedings, and those whose work could not be accommodated. Without their months of research and painful writing up there would be no conference. As well as contributed papers we have been fortunate to enlist six eminent researchers to talk on particularly relevant topics of their choice.

The record number of submitted papers put an additional strain on the committee members. I am very grateful to them all for their work in assessing the papers in a short time and for freely giving me the benefit of their experience and support in a variety of ways. They are: Mike Darnell (University of Leeds), Paddy Farrell (University of Manchester), Mick Ganley (Racal Airtech) John Gordon (Concept Laboratories), Chris Mitchell (Royal Holloway), Fred Piper (Royal Holloway), Michael Walker (Vodaphone). I would also like to thank Pamela Bye, IMA Conference Officer, who dealt with all correspondence with the authors and was always ready to give advice and assistance.

The papers in this volume are presented in the order that they are intended to appear in the conference programme. As has become traditional at this conference, papers are not divided into related groups but are 'randomly' mixed.

Colin Boyd
Manchester, October 1995

Contents

A Broadcast Key Distribution Scheme Based on Block Designs

Valeri Korjik*, Michael Ivkov*, Yuri Merinovich*,
Alexander Barg**, and Henk C.A. van Tilborg **

*Department of Communication Theory,
St. Petersburg University of Telecommunications, 191065 St. Petersburg, Russia,
bymey@iec.spb.su
**Department of Mathematics and Computing Science,
Eindhoven University of Technology, 5600 MB, Eindhoven, The Netherlands,
abarg@win.tue.nl and henkvt@win.tue.nl

Abstract. A key distribution scheme for broadcast encryption is proposed. It is based on block designs. In this scheme a centre provides each user (receiver) of the system with a set of keys. If at a later stage, some users are no longer entitled to the messages, they should also no longer be able to decrypt them. This should even be the case if these illegitimate users can form a coalition and exchange the keys that they have obtained before (provided that the size of this coalition does not exceed some value).

By means of block designs a tradeoff can be made between the size of the largest admissible coalition and the total length of the keys that each user has to store. The proposed system is unconditionally secure and seems better suited for large coalitions than existing schemes.

Key words : key distribution, broadcast encryption, s-resilience, connectivity, fractional covering, block design.

1 Introduction

Consider the key distribution problem for broadcast encryption. In such a scheme, a centre transmits encrypted messages to a set $\mathcal{V}$ of users. The size v of this set can be very large. For the encryption, the centre uses a key K_0, which should be known to the users of the system. At a later stage, some users may no longer be entitled to the messages (for instance, if their subscription has expired). Let C, $|C| = \sigma$, be the set of illegitimate users. To continue broadcasting, the centre has to replace K_0 by another key, K_1, known only to the users in $\mathcal{V} - C$. We assume that this situation can iterate, i.e., that some additional users can become unauthorized and join the set C. Then K_1 has to be replaced by K_2, and so on. If some users renew their subscription, they can be made authorized again after the next replacement of the master key. It is logical to assume that the only channel available for the transmission of the keys is the broadcast channel itself since otherwise the centre has to maintain a host of private channels toward each

Design Choices and Security Implications in Implementing Diffie-Hellman Key Agreement

Paul C. van Oorschot

Bell-Northern Research, Ottawa, Canada

Abstract. 20 years ago, Diffie and Hellman conceived the idea of exponential key exchange, now commonly known as Diffie-Hellman key agreement. Today, the long-predicted wide-scale commercial deployment of cryptographic technology is finally occurring. As designers seek to provide implementations of public-key cryptography which offer adequate security while minimizing the computational requirements thereof, various options are available regarding parameter selection and other design choices. We examine a subset of options intended to reduce the computational costs of Diffie-Hellman key agreement protocols, and their related security implications.

user. Therefore, we assume that during the installation of the system, each user is supplied with a collection of keys, which can be used to broadcast master keys K_i. We allow the centre to use any of the private keys of user u for this purpose. It may happen that to broadcast a new master key to all remaining authorized users (to the set $\mathcal{V} - C$) the centre will need to transmit a considerable amount of messages, each destinated to a certain user or a group of users in $\mathcal{V} - C$.

We call the procedure described the *dynamical key management*. Obviously, if σ is too large, the key management can become impossible since some legitimate users can no longer be reached by the centre, and the scheme becomes disconnected. The maximal value s for which the key management is still possible for all C of size $\sigma \leq s$, is called the *resilience* of the scheme.

There are two trivial approaches to this problem. Under the first approach, every user u has his own unique key k_u, which is used by the centre to transmit to this user the key K_j that the broadcaster is going to use. Note that this approach requires the transmission of $v - \sigma$ keys before the broadcasting can start.

Under the second approach, with each $(v - \sigma)$-subset U of the set of v users, $0 \leq \sigma \leq s$, corresponds a key k_U, which is given to the members of U. For every subset C of size at most s, the key $k_{\mathcal{V} \setminus C}$ corresponding to the complement of the subset can be used to broadcast to the remaining users. In this system no extra round of key transmission is necessary (it is a so-called "zero message" scheme), but each user has to store $\sum_{i=0}^{s} \binom{v}{i}$ keys, which is quite impractical.

To reduce the number of private keys and messages transmitted, we assume that private keys are communicated to subsets of users rather than to individual users. Each subset can be reached during a single transmission round, thus reducing the number of rounds.

In [1], solutions are described that provide tradeoff between the number of messages before transmission and the number of keys that each user has to store. Here, such a scheme is proposed based on particular properties of block designs. In contrast to [1], we assume that all illegitimate users are known to the centre. On the other hand, our approach is unconditionally secure since it involves no "public-key-type" reasoning. The scheme suggested below compares favorably to the results of [1] when $s \log s$ grows faster than $v^{1/4}$.

2 The Key Distribution Scheme

As discussed above, a key distribution scheme is basically an incidence structure.

Definition 1.1 A *key distribution scheme* for a set of users $\mathcal{V}$ of size v is a collection $\mathcal{B}$, $|\mathcal{B}| = b$, of subsets of $\mathcal{V}$.

To every $B \in \mathcal{B}$ we associate a key k_B, which is given by the centre to all the users contained in B. The key management can be arranged as follows. Let $C \subset \mathcal{V}$ be a coalition of illegitimate users known to the centre. Then all keys k_B for which B contains at least one element from the coalition are compromised and should be removed. The remaining keys should be used in the most effective way to communicate a new transmission key to all users outside the coalition. For this to be possible it is necessary and sufficient that

$$\bigcup_{B \in \mathcal{B}, B \cap C = \emptyset} B = \mathcal{V} - C. \tag{1}$$

We are free to derive the scheme from any suitable incidence structure (hypergraph). Equation (1) reduces the construction problem of key distribution schemes to coverings in subgraphs that are obtained from the original hypergraph by deleting the compromised vertices (keys).

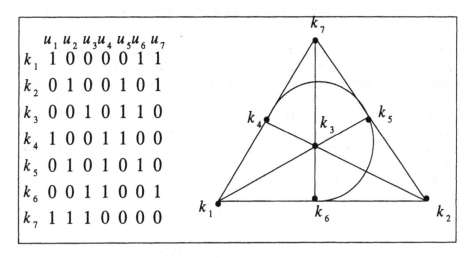

Fig. 1. An incidence structure with 7 points and 7 blocks

Example 1. Figure 1 shows a key distribution scheme with 7 users and 7 private keys. Key k_1 is shared by users u_1, u_6, and u_7, etc. It can be represented in two ways, namely, as a binary incidence matrix M whose columns are the points and rows the blocks of the structure, and as as a hypergraph with vertices numbered by the rows and edges by the columns of M.

Definition 1.2 Let $\mathcal{D} = (\mathcal{V}, \mathcal{B})$ be an incidence structure and $C \subseteq \mathcal{V}$. Then the *residual structure* $\mathcal{D}_C$ of $\mathcal{D}$ with respect to C consists of the restriction of all the blocks $B \in \mathcal{B}$ disjoint from C to the set $\mathcal{V} - C$.

As said above, the performance of the key distribution scheme is determined by (a) its resilience, (b) the number of messages that the centre has to transmit to maintain the dynamical key management, and (c) the number of private keys each user has to store. We deem the first two parameters to be more important since they determine the performance of the system as a whole.

Definition 1.3 Let s be the maximum number such that Eq. (1) holds for every $C \subset \mathcal{V}$ with $0 \leq |C| \leq s$. Then the key distribution scheme based on the incidence structure $\mathcal{D}$ is called *s-resilient*.

Thus, if $|C| > s$, the key management is generally not possible since some users cannot be reached and the residual structure is not connected. For $|C| = \sigma \leq s$, we are interested in the minimum number of messages that the centre has to transmit to successfully handle any coalition C of size σ. In a certain abuse of terminology, we call this parameter the connectivity.

Definition 1.4 The *connectivity* $\chi_\sigma(\mathcal{D})$ is the maximal, over all subsets $C \subset \mathcal{V}, |C| = \sigma$, size of the minimal vertex covering of the residual structure $\mathcal{D}_C$.

Example 1 (continued). Suppose users u_1 and u_2 become unauthorized. Then the residual structure consists of the remaining 5 users and keys k_3 and k_6. This structure still allows to reach all remaining users. The scheme is 2-resilient, and $\chi_0(\mathcal{D}) = \chi_1(\mathcal{D}) = 3, \chi_2(\mathcal{D}) = 2$.

Obviously, any estimate on the size of coverings in hypergraphs gives a bound on connectivity in the corresponding schemes. For general hypergraphs, we quote a classical covering theorem discovered in [3, 4], see also [5]. Let us view the incidence system $\mathcal{D} = (\mathcal{V}, \mathcal{B})$ as a hypergraph $\mathcal{H}(V, E)$, where $V = \mathcal{B}$ and $E = \mathcal{V}$. The residual structure $\mathcal{D}_C$ with respect to a given subset C of size s yields a subgraph $H \subset \mathcal{H}$ with $E(H) = E - C$ and $V(H) = V - X$ obtained by deleting all edges in C and the subset $X \subset V$ formed by vertices incident with these edges.

We wish to estimate the minimum number of subsets in the residual structure $\mathcal{D}_C$ covering all its elements, i.e., the minimal size of the vertex covering of H. Let $D(H) = \max_{v \in V(H)} \deg v$. Let $\tau : V(H) \to \mathbf{R}_+$ be a function such that

$$\sum_{p \in E} \tau(p) \geq 1 \quad \text{for all } E \in E(H) \tag{2}$$

(any function with such properties is called a fractional covering of H). Finally, let

$$\tau^*(H) = \min \sum_{p \in V(H)} \tau(p).$$

Theorem 2.1. [3, 4]. *The greedy algorithm outputs a vertex covering S of H of size*

$$|S| \leq (1 + \log D(H))\tau^*(H). \qquad \square$$

In this generality, this theorem is not very useful for the construction of key distribution schemes. To obtain more concrete results, let us restrict the class of incidence structures $\mathcal{D} = (\mathcal{V}, \mathcal{B})$ considered. The required property is that for every pair (u, C), $u \in \mathcal{V}, C \subset \mathcal{V}, |C| \leq s, u \notin C$, there exists a $B \in \mathcal{B}, B \cap C = \emptyset$, such that $u \in B$. Let $\mathcal{B}_u = \{B \in \mathcal{B}|u \in B\}$, $\mathcal{B}_{uc} = \{B \in \mathcal{B}_u|c \in B\}$. Thus, we are concerned with the fact that

$$|\mathcal{B}_u| > \left| \bigcup_{c \in C} \mathcal{B}_{uc} \right|.$$

In particular, this is guaranteed if

$$|\mathcal{B}_u| > \sum_{c \in C} |\mathcal{B}_{uc}|. \tag{3}$$

An example of incidence structures that give information about the numbers $|\mathcal{B}_{uc}|$ is 2-designs. Let us recall some of their properties used below (see, e.g., [2]). A $2 - (v, k, \lambda)$ design is a finite set $\mathcal{V}$, $|\mathcal{V}| = v$, whose elements are also called points, and a collection $\mathcal{B}$ of its k-subsets (blocks) such that every pair of elements of $\mathcal{V}$ is contained in λ blocks. We assume that $0 \leq \lambda \leq k \leq v - 1$. Every point of $\mathcal{V}$ occurs in one and the same number of blocks r. Let $|\mathcal{B}| = b$. Counting ones in the incidence matrix of the design in two ways shows that

$$bk = vr, \qquad\qquad b \geq v, \quad k \leq r, \tag{4}$$
$$\lambda(v - 1) = r(k - 1). \tag{5}$$

If $b = v$ and, thus, $k = r$, the design is called symmetric. For symmetric designs, (5) implies

$$v = 1 + \frac{k(k - 1)}{\lambda}. \tag{6}$$

A symmetric design with the parameters $2 - (k^2 - k + 1, k, 1)$ is called a finite projective plane of order $d = k - 1$ with points (lines) associated with points (blocks) of the design. From the above definition we see that every two points are incident with a unique line and every two lines meet at a unique point.

Proposition 2.2. *The key distribution scheme based on a $2 - (v, k, \lambda)$ design is s-resilient if*

$$r > s\lambda. \tag{7}$$

For projective planes, this inequality is also a necessary condition.

Proof. Inequality (7) is a reformulation of (3) for the case of $|\mathcal{B}_{uc}| = \lambda$. The second part is obvious since the argument preceding (3) becomes exact. □

Remark 1. Thus, the scheme based on a projective plane of order d is d-resilient.

Remark 2. It is also possible to use in our construction general $t - (v, k, \lambda_t)$ designs, i.e., incidence structures in which every t-subset of $\mathcal{V}$ is contained in λ_t blocks. In this case, every $t - 1, t - 2, \ldots, 2$-subset of $\mathcal{V}$ is contained in one and the same number of blocks, $\lambda_{t-1}, \lambda_{t-2}, \ldots, \lambda_2$, respectively. This numbers are easily calculated from the parameters of the t-design. Then (7) should be replaced by $r \geq \min\{\lambda_2 s, \lambda_3 \lceil s/2 \rceil, \ldots, \lambda_t \lceil s/(t - 1) \rceil\}$, which, generally, can give a better estimate for s. However, since this does not give better asymptotical results for the connectivity and since for large t, t-designs are rare, we prefer to stay with the less general case.

We now turn our attention to the implementation complexity (connectivity) of our schemes. To shorten notation, denote $\chi_\sigma(\mathcal{D})$ by $\chi_{\sigma,\lambda}$ and write χ_σ for $\chi_{\sigma,1}$. First we deal with the case $\sigma = 0$, i.e. the situation when all users are still legitimate.

Proposition 2.3. *The connectivity $\chi_{0,\lambda}$ of an s-resilient key distribution scheme based on a $2 - (v, k, \lambda)$ design is at most r.*
For schemes based on projective planes, $\chi_0 = r$.

Proof. The union of all r blocks intersecting a fixed point of the design contains all points of the design. For projective planes, this is the smallest possible size of a covering since every two lines meet at exactly one point, and, thus, taking $r - 1$ lines, we would cover not more than $1 + (r - 1)(r - 1) < v$ points. □

Let us now apply Theorem 2.1 to derive a general bound on the connectivity of the schemes suggested. Suppose the scheme is s-resilient and the number of unauthorized users is $|C| = \sigma$.

Proposition 2.4. *The connectivity χ_σ of the s-resilient key distribution scheme based on a $2 - (v, k, \lambda)$ design*

$$\chi_{\sigma,\lambda} \leq \frac{1}{r - \lambda\sigma}(1 + \log k)(v - \sigma), \quad \text{for any } \sigma < r/\lambda. \tag{8}$$

In particular, for schemes based on projective planes of order $k - 1$,

$$\chi_\sigma \leq \frac{1}{k - \sigma}(1 + \log k)(v - \sigma) \quad \text{for any } \sigma < k. \tag{9}$$

Proof. Let us derive (8) from Theorem 2.1. Take the residual incidence structure obtained from the 2-design and construct from it hypergraph H as above. Clearly, $D(H) = k$ since every remaining block contains k points. For every $p \in V(H)$, set

$$\tau(p) = \frac{1}{r - \lambda\sigma}.$$

This choice satisfies (2) since $r - \lambda\sigma$ is the minimum possible number of vertices incident with an edge in H. Therefore, by Theorem 2.1 and the definition of τ^*,

$$\chi_{\sigma,\lambda} \leq (1 + \log D(H))\tau^*(H) \leq (1 + \log k)|V(H)|\tau(p),$$

which proves (8).
Inequality (9) follows immediately because for projective planes $r = k$ and $\lambda = 1$.

□

Let us study the asymptotic behavior of this construction. Let $v \to \infty$, then in order to keep $\tau(p)$ in the proof well-defined we must assume that σ is bounded away from s, i.e., $\sigma = s/c$ for some real $c > \lambda s/r$. Then (8) becomes

$$\chi_{\sigma,\lambda} \leq \frac{c}{cr - \lambda s}(1 + \log k)(v - \frac{s}{c}) < \frac{c}{c - 1}\frac{1}{r}(1 + \log k)(v - \frac{s}{c}).$$

For projective planes, when $k = r$ and $v = r(r - 1) + 1$, we get

$$\chi_\sigma < \frac{c}{c - 1}r(1 + \log r). \tag{10}$$

This will be used in Sect. 3 for the asymptotic comparison of the suggested scheme with Fiat–Naor's one.

Observe that the class of schemes derived from block designs is suited for the dynamical key management. Indeed, if in addition to a group of illegitimate users, some other ones become unauthorized, then the scheme still allows proper key distribution as long as the overall number of unauthorized users is within the resilience limit of the system. The centre has to construct a covering of the new residual structure, which is feasible since the greedy algorithm has the complexity of order at most $O(vb^2)$.

For small values of σ, one can easily obtain better bounds than those of Proposition 2.4. For instance, consider projective planes of order $k-1$. One has

$$\chi_1 = 2k - 3, \quad \chi_i = 2k - 4, \ 2 \leq i \leq 5. \tag{11}$$

This is seen by counting the points covered by two intersecting pencils of lines and treating separately points on the lines incident with the points in C (we refer to Fig. 2, which shows the case of $\sigma = 3$). Generalizing this argument, one obtains the following proposition.

Proposition 2.5. *Let $v \to \infty$, $\sigma = $ const. Then the connectivity of the scheme based on a projective plane with v points is $O(\sqrt{v})$.*

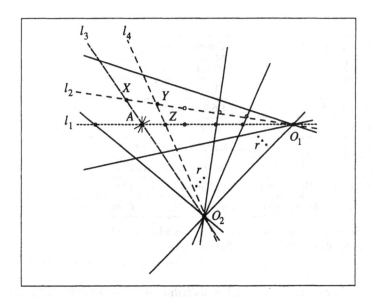

Fig. 2. The plane with the deleted subset $C = \{X, Y, Z\}$ and all the lines incident with it. The covering is formed by all the lines through O_1 and O_2 except $\ell_1, \ldots, \ell_4$ plus a line through A.

Proof. Let $|C| = \text{const}$ and consider two intersecting pencils of lines, $\mathcal{P}_1$ and $\mathcal{P}_2$, as in Fig. 2. The set $\mathcal{L}$ of lines in $\mathcal{P}_1 \cup \mathcal{P}_2$ that intersect C, cannot be used for the covering. Clearly, the number of points outside C not covered by $\mathcal{P}_1 \cup \mathcal{P}_2 - \mathcal{L}$ is at most $|C|^2$. To cover them, we use one line per point. Though this is not the most economical way of constructing the covering, this is sufficient for our claim since the total number of lines used is bounded by $O(k) = O(\sqrt{v})$. □

An important parameter of key distribution schemes is the number of private keys. Let us give simple bounds to their number in terms of the size of C.

Proposition 2.6. *The number b_σ of lines in the residual structure of a projective plane of order $k - 1$ with respect to a set C of σ points satisfies*

$$(k - \sigma)(k - 1) \leq b_\sigma \leq k(k - 1 - \sigma) + \binom{\sigma}{2} + 1$$

The extrema are attained when all (resp., no three) points in C are collinear.

Proof. Obvious for $\sigma = 3$. Further, suppose we have an extremal configuration C and add one point to it. The property of being extremal is not violated if adding a point to C implies removing from the residual design the maximal (resp., minimal) number of lines. □

N	v	k	s	χ_s	Difference set
1	7	3	2	2	0,1,3 mod 7
2	13	4	3	4	0,1,3,9 mod 13
3	21	5	6	6	0,1,6,8,18 mod 21
4	31	7	5	8	0,1,3,8,12,18 mod 31
5	57	8	7	[12..18]	0,1,3,13,32,36,43,52 mod 57
6	73	9	8	[16..25]	0,1,3,7,15,31,36,54,63 mod 73
7	91	10	9	[18..31]	0,1,3,9,27,49,56,61,77,81 mod 91
8	133	12	11	[23..38]	0,1,3,17,29,61,80,86,91,95,113,126 mod 133

Table 1: The main parameters of some key distribution schemes based on finite planes defined by difference sets

Table 1 shows parameters of key distribution schemes defined by projective planes of order 2 to 11 (with the exception of 6 and 10 for which projective planes do not exist). Planes of all these orders can be constructed from difference sets mod v. The corresponding sets are listed in the last column. The first row of the table corresponds to Example 1. For $s \leq 5$, the values of connectivity are given by (11). For the remaining entries, we give estimates from both sides. The lower estimates were found by computer search; the upper ones are obtained straightforwardly by spending one line per each uncovered point, as in Fig. 3. It seems likely that the exact answers are close to the lower estimates.

3   A Comparison with the Fiat-Naor Scheme

Fiat and Naor describe in [1] a way to construct s-resilient schemes that enable the center to communicate a transmission key to any collection of users U outside a coalition of size $\leq s$. In their schemes, the precise coalition does not need to be known to the centre, as long as the members of U are not part of it.

For $s = 1$, in the Fiat-Naor (FN) scheme, the centre selects two secret large primes and makes the product N public. It further selects a secret integer g, $g < N$, of high index. Each user u, $1 \leq u \leq v$, is provided with the secret key $k_u = g^{p_u}$ (mod N) by the centre, where the p_u's are relatively prime and known to all participants.

The common key k for a set U of users is defined by

$$k_U = g^{\prod_{u \in U} p_u} \quad (\text{mod } N).$$

Note that each user can compute this key k_U by raising his own (secret) $k_u = g^{p_u}$ to the power $\prod_{v \in U - \{u\}} p_v$.

Under the assumption that extracting roots modulo composite numbers is hard this scheme is 1-resilient. Note that the centre does not need to send any message (it is a zero-message scheme). This scheme is not 2-resilient, because g can easily be computed from $g = k_1^{a_1} k_2^{a_2}$, where a_1 and a_2 satisfy $a_1 p_1 + a_2 p_2 = 1$ and can be found with Euclid's algorithm.

To create an s-resilient scheme for $s \geq 2$, a family of hash functions f_i, $1 \leq i \leq l$, of $\mathcal{V}$ to $\{1, 2, \ldots, m\}$ with $m < v$ is needed. These functions should have the property that for any s-subset C of $\mathcal{V}$ there exists some index $1 \leq i \leq l$ such that the restriction of f_i to C is a one-to-one function.

Consider a matrix $||R_{ij}||_{i=1,l}^{j=1,m}$, where each R_{ij} represents a 1-resilient scheme. Every user u in $\mathcal{V}$ receives the keys associated with schemes $R_{i,f_i(u)}$ for all $1 \leq i \leq l$. Due to the assumption on the functions f_i, for every coalition C of size s there exists an index i such that any two users in C belong to a different scheme.

In order to send a secret message M, represented as a binary sequence, to a subset U of $\mathcal{V}$, outside some coalition C of size s, the centre writes it as an exclusive-or of messages M_i, $1 \leq i \leq l$. The centre uses the scheme R_{ij} to broadcast the message M_i to all users u with $f_i(u) = j$ for all $1 \leq i \leq l$ and $1 \leq j \leq m$. Each user u in U can receive all messages M_i and compute M. However, for a coalition of at most s users there is at least one M_i that remains unknown.

Of one version of their s-resilient scheme the authors [1] estimate the number of keys that each user has to store as $O(s \log s \log v)$ and the number of messages that the centre needs to transmit as $O(s^2 \log^2 s \log v)$. Let us compare this with our results. Consider schemes based on projective planes. Their maximal resilience, guaranteed by our estimates, is r/c, where $c > 1$ is a real constant. The number of keys that each user has to store is $r = O(\sqrt{v})$. The number of messages transmitted during the key management can be estimated as

$$O(\sqrt{v} \log v),$$

see (10). The constant factor in this estimate is determined by the value $c/(c-1)$ (i.e., c regulates the tradeoff between the resilience and implementation complexity). We conclude that this compares favorably to the s-resilient schemes of [1] with $s \log s$ of order $\sqrt[4]{v}$ or greater[1]. Note also that our schemes are constructive, while the results in [1] involve randomization and are purely existential. Finally, our results do not depend on assumptions of hash or one-way functions. On the other hand, as was already remarked, in the FN scheme the coalition does not need to be known to the centre.

4 A Combination of the Two Approaches

By Proposition 2.3 and (6), the resilience of schemes based of projective planes is not more than the square root of the number of users. This number can be doubled by combining the FN scheme with schemes from 2-designs into what we call *concatenated key distribution schemes*.

Consider an s-resilient scheme based on a 2-design $\mathcal{D} = (\mathcal{V}, \mathcal{B})$. Let C be a coalition of s unauthorized users. A user $u \in \mathcal{V} - C$ can be reached by the centre if there exists a block $B \in \mathcal{B}$ with $u \in B$ and

$$|B \cap C| = 0. \tag{12}$$

We can relax this condition by associating to every block $B \in \mathcal{B}$ a 1-resilient FN scheme with its own parameters $N^{(B)}, g^{(B)}$, which will be used to distribute new master keys K. Then instead of (12) we require that there exist a block B meeting u with

$$|B \cap C| \leq 1.$$

Let us calculate the resilience of concatenated schemes from projective planes. Let $v \in \mathcal{V} - C$, $\mathcal{B}_v = \{B \in \mathcal{B} | v \in B\}$ and suppose B_1 and B_2 are two blocks in $\mathcal{B}_v$ intersecting C at least two points, $\{a, b\}$ and $\{c, d\}$, respectively. Since every pair is contained in exactly one block, we observe that $\{a, b\} \cup \{c, d\} = \emptyset$. Therefore, every pair of points in C forbids at most one block in $\mathcal{B}_u$. Since $|\mathcal{B}_u| = r$, the centre can handle any coalition C of size $\lfloor |C|/2 \rfloor \leq r - 1$. We have proved the following.

Proposition 4.7. *The concatenated scheme with a 1-resilient FN inner scheme and an s-resilient scheme derived from a projective plane of order s as the outer one yields a $2s$-resilient key distribution scheme.*

---

[1] For instance, for a TV-net with 10 million subscribers our estimates are better starting from s of order several dozens.

5 Conclusion

A constructive key distribution scheme for broadcast encryption is described that is suited for a dynamically changing set of compromised users. This scheme makes use of 2-designs. The number of messages to be transmitted by the centre before broadcasting can take place depends on the number σ of compromised users. In particular, for projective planes, if the total number of users v grows and σ remains constant, the number of messages is estimated as $O(\sqrt{v})$, while if σ grows, this number is $O(\sqrt{v} \log v)$.

References

1. A. Fiat and M. Naor, "Broadcast Encryption," *Advances in Cryptology, Proc. CRYPTO'93*, D. R. Stinson, Ed., Lecture Notes in Computer Science 773, Springer Verlag, Berlin, pp. 480–491, 1994.
2. E. F. Assmus and J. D. Key, *Designs and Their Codes*, Cambr. Univ. Press, 1993.
3. L. Lovász, "On the ratio of optimal integral and fractional covers," *Discrete Math.*, **13**, 383–390, 1975.
4. S. K. Stein, "Two combinatorial covering theorems," *J. Comb. Theory*, Ser. A, **16**, 391–397, 1974.
5. Z. Füredi, "Matchings and coverings in hypergraphs," *Graphs and Combinatorics*, **4**, 115–206, 1985.

Minimal Supports in Linear Codes

Alexei Ashikhmin* and Alexander Barg†

*Faculty of Technical Mathematics and Informatics, Delft University of Technology
P.O.Box 5031, 2600 GA Delft, The Netherlands
aea@twi.tudelft.nl
†Faculty of Mathematics and Computer Science, Eidnhoven University of Technology
Den Dolech 2, P.O. Box 513, 5600 MB Eindhoven, The Netherlands
abarg@win.tue.nl

We consider q-ary linear codes of length n.

Definition 1. A nonempty subset $I \subseteq \{0, 1, \ldots, n-1\}$ is called a *minimal support* with respect to a given code C if
(i) there exists a codeword $\mathbf{c} \in C$ with supp $\mathbf{c} = I$,
(ii) there is no nonzero codeword $\mathbf{c}'$ with supp $\mathbf{c}' \subset I$.

We begin with general properties of minimal supports. Then we evaluate the parameters of the distribution of the number of minimal supports in random linear codes, study their asymptotic behaviour, and discuss its consequences for some maximum likelihood decoding algorithms.

In the second part, we construct explicitly sets of minimal supports in Hamming codes and second order Reed-Muller codes.

Next we propose a generalization of minimal supports for codes over finite commutative rings and evaluate them for the quaternary Kerdock code and quaternary Hamming code.

One application of minimal supports is in cryptography, namely, in the so-called linear secret-sharing schemes [1]. We briefly outline this connection and present an analog of the main result of [1] for codes over rings.

Since minimal supports in linear codes correspond to cycles in representable matroids, our results have a clear interpretation in the frame of matroid theory.

Details and proofs can be found in [2, 3].

References

1. G. R. Blakley and G. A. Kabatianskii, "Linear algebra approach to secret sharing schemes," in: *Error Control, Cryptology, and Speech Compression*, A. Chmora and S. Wicker, Eds., Lect. Notes. Comput. Sci., **829**, pp. 33–40 (1994).
2. A. Ashikhmin and A. Barg, "Minimal vectors in linear codes and sharing of secrets," submitted to *IEEE Trans. Inform. Theory*, also Preprint 94-113, SFB 343, Bielefeld University (1994).
3. A. Ashikhmin, A. Barg, G. Cohen, and L. Huguet, "Variations on minimal codewords in linear codes," in: *Applied Algebra, Algebraic Algorithms, and Error-Correcting Codes*, G. Cohen, T. Mora, and M. Giusti, Eds., Lect. Notes Comput. Sci., **948**, pp. 96–105 (1995).

Sequential Decoding for a Subcode of Reed Solomon Codes

Sooyoung Kim Shin and Peter Sweeney

Centre for Satellite Engineering Research
University of Surrey, Guildford
Surrey GU2 5XH U.K

Abstract. This paper describes the performance of sequential decoding as applied to a subcode of Reed Solomon (RS) codes. As an efficient soft decision decoding technique sequential decoding using a modified Fano algorithm has been attempted. An encoding technique for generating subcode is designed to enhance the performance and the computational efficiency of the decoder. The performance has been estimated on an AWGN channel. Sequential decoding performance of the subcode of (15,11) RS code could produce a comparable performance as well as computational efficiency to Viterbi decoding of rate 1/2 convolutional code with constraint length 7.

1 Introduction

RS codes have found wide applicability in situations where impulsive noise is encountered. Conventional assessment of code performance, however, are carried out in terms of BER versus E_b/N_o on the AWGN channel. In such conditions, convolutional codes are found to be superior, at least at moderate values of required BER (around 10^{-5} to 10^{-6}). To a large extent, this superiority derives from a lack of a generally applicable method for soft decision decoding of RS codes.

In many ways, the AWGN channel represents the worst case for RS codes and, once good methods have been devised for such a channel, it may be possible to adapt those methods to other channel types so that the burst error correcting capabilities of the codes may be maintained. Thus the development of soft decision decoding methods for RS codes is an important problem which remains to be solved.

Since trellis decoding methods can accept soft decisions very easily, they can give quite useful improvements in coding gain over hard decision decoding with little increase in complexity over hard decision. Wolf[1] showed that the trellis can be constructed for any q-ary (n, k) block codes with q^{n-k} states, or q^k states if $k < n - k$. Trellis decoding techniques, including reduced search decoding and sequential decoding, have been applied to RS codes [3][6][4].

In this paper, based on the modified Fano sequential decoding technique for RS codes [6], an encoding strategy is introduced to enhance computational efficiency as well as to improve performance at the expense of bandwidth, by use of a subcode. Sequential decoding has been applied here to a subcode of the (15,11) RS code. In section 2, the detailed encoding strategy and its characteristics are discussed. A brief description of Fano sequential decoding for RS codes is represented in section 3. By applying the strategy shown in section 2 and 3, decoding performance on an AWGN channel is simulated and comparison of those of other decoders and encoders are demonstrated in section 4. Conclusion is drawn in section 5.

2   Constructing Subcode

2.1   Impact on Trellis Decoding

The complexity of any trellis decoder for RS codes depends on the *order of the field* where the code is constructed. The order of the field simply represents the number of the elements in the field. Consider the number of branches emanating from a node in the code trellis as the order of the field[3]. This means that the trellis decoder complexity can be reduced if a subcode is constructed with a reduced number of field elements. Use of the subcode is now then able to improve the decoding speed.

In addition to the decoding speed, the subcode may be able to enhance the decoding performance if it is constructed in the way that the weight distribution of the code is improved. This is possible because the number of the codewords will be reduced compared with the original code, and thus the distance between each codeword in the subcode may become large. The performance of a trellis decoder is governed by binary weight distribution of the code[5]. The subcode should be constructed with a better binary weight distribution, that is with a larger minimum distance (d_{min}), in order to improve the decoding performance.

2.2   Encoding Strategy

In our approach, encoding for constructing the subcode is performed by inserting a zero bit in every information symbol, so that the encoder will produce only $2^{(m-1)k}$ possible codewords instead of 2^{mk} codewords, where m is the symbol length and k is the information length of a code. The code rate of the subcode then will be $R = k(m-1)/nm$, where n is the codeword length. Fig. 2.2 shows encoding process of the subcode compared with normal RS encoding process for trellis decoding.

The above encoding process makes the decoder evaluate only a subset of codewords. Because of the reduced number of candidates, the computational efficiency can be increased. Moreover the binary weight distribution of RS subcodes was improved owing to the above encoding method, that is greater minimum distance resulted. This means a subcode of a RS code may produce better performance than the RS code itself.

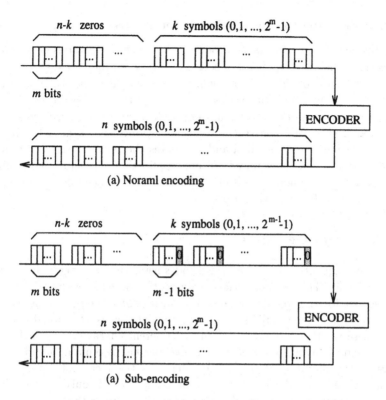

Fig. 1. Encoding process of RS codes for trellis decoding

2.3 Consideration of Systematic Encoding

The performance and computational efficiency improvement resulted from decoding of the subcode can be achieved at the sacrifice of the bandwidth. If systematic encoding is employed, the transmitter does not have to send all m bits per symbol. Since the first k encoded symbols are exactly same as the information symbols, the decoder can fully recover the information by receiving $k(m-1)$bits (information part) plus $(n-k)m$bits (parity part). This way we can increase the code rate up to $R = (k-1)m/(nm-k)$. Higher code rate leads to larger energy per symbol at the transmitting end. This may hopefully result in further improvement in coding gain.

With the above idea, systematic trellis encoding and decoding of the subcode was implemented. However, the performance improvement of the systematic encoding is less than the nonsystematic one. In the systematic encoding method, an encoded codeword is totally dependent on the information word. For this reason, if we decrease the minimum distance of the information words by inserting a zero bit into every symbol, then the d_{min} of the codewords will also be decreased compared to nonsystematic encoding. Accordingly, the decoding performance of systematic encoding cannot be better than that of nonsystematic one.

2.4 Constructing the Best Subcode for Trellis Decoding

The trellis decoding performance produced by the subcode varies depending on the position where a zero bit is inserted in every symbol because the binary weight distribution of each subcode will be changed. Although the performance difference is not large, it will be desirable to find the best subcode which can produce the best weight distribution. In order to find the best subcode a process will be required, which constructs subcodes by inserting a zero bit in every different position in each symbol and calculates the binary weight distribution of each subcode, will be required. In our application, the subcode which was constructed by inserting a zero bit in the first position of each information symbol was used without further evaluation.

3 Sequential Decoding

Soft decision sequential decoding for RS codes using the Fano algorithm was first introduced in [6]. The decoder performs almost the same procedure as the normal Fano decoder for the convolutional codes[2] with some modifications. Here, we will briefly explain the modifications compared with the conventional Fano algorithm. These modifications result from the different characteristics of RS codes compared with convolutional codes. The decoder has two main modifications.

Firstly, the modified sequential decoder uses the fact that Reed Solomon codes have fixed block length. Since the decoder proceeds only to the nth level of the tree, we may expedite the decoder's search by using a threshold, T_n, where n is the block length of a code. T_n determines whether the currently chosen path is correct when the decoder reaches the last level of the code trellis. As is the case for the current threshold value, T_n can be determined empirically. For example, since we know that the metric of the correct path will grow linearly, we may simply set T_n to $k \times \tau$, where τ is the spacing between thresholds and k is the information block length.

Suppose that there is a path whose metric value is greater than that of the correct path at the kth level, then the decoder will select a wrong path. However, the selected path will record a poor metric while the decoder proceeds from the kth to the nth level. If the decoder compares the metric of the chosen path to T_n, then the selected path will fail the threshold test. The decoder will then go backwards in order to try an alternative. Notice that in this instance the decoder must go backwards up to the $(k-1)$th level because there is only one outgoing branch from the kth to the nth level. In very noisy conditions, however, the metric of the correct path may be smaller than T_n. To prevent the loss of the correct path, we should record the path with the best metric among those which have been kept by the decoder at level n.

Secondly, Reed Solomon codes produce a large number of branches at a node, and thus the decoder has many alternatives during a backward search. Since many of the alternatives will have relatively poor correlation, the metrics of the paths extended from those will produce poor metrics. If the decoder lowers the threshold values in order to examine all the alternatives which exist, the threshold value would be too small to determine the correct path, and eventually it may lose its ability. To solve this problem, the decoder uses another empirically determined threshold value, T_a, which is used for rejecting some unlikely alternatives.

Here, in the modified algorithm, we introduced two new thresholds, T_n and T_a. The current threshold at each level, T, is a value which can be varied along the decoder's movement. On the other hand, T_n and T_a are constants with a given system configuration.

Now the algorithm of the modified sequential decoder for (n, k) Reed Solomon codes can be illustrated by the flow chart shown in Fig. 2.

4   Performance Comparison

The subcode of the (15,11) RS code used for sequential decoding has an effective rate of 33/60. An AWGN channel has been modelled to simulate the decoding performances. Maximum likelihood performance was established by simulating the Viterbi algorithm with importance sampling simulation technique (error-event simulation)[5].

The decoding performances of various encoding and decoding schemes are represented in Fig. 3. It can be seen that the sequential decoder for the (15,11) RS subcode produces performance which approximates to that of the Viterbi decoder for a convolutional code (rate = 1/2, constraint length = 7). Compared with decoding performance of the original (15,11) code, that of the subcode offered more than 0.5dB coding gain.

The computational efficiency is compared by estimating the number of path extensions per information bit and is shown in Fig 4. Within the BER range $(10^{-5}$ to $10^{-6})$ of interest, the average computational efficiency of sequential decoder for the RS subcode is superior to the Viterbi decoder for the convolutional code. At BER value of 10^{-5}, for example, the sequential decoder for the subcode needs to extend about 16 paths per information bit on average. On the other hand, the Viterbi decoder for the convolutional code still needs to extend 128 paths per information bit at the same BER range.

5   Conclusion

A new encoding strategy of RS codes for efficient trellis decoding was introduced in this paper. At the expense of bandwidth, the subcode can offer a better performance than the original code. Combined with the sequential decoder, it can also produce more enhanced computational efficiency. With the example of the subcode of (15,11) RS code, we could see some advantages over convolutional codes. It means that there might be a possibility to overcome the disadvantages of RS codes by applying a similar method to those.

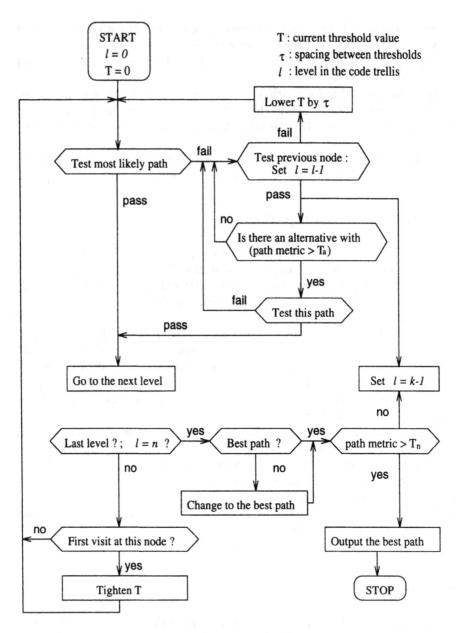

Fig. 2. Modified Fano algorithm for a (n, k) Reed Solomon code

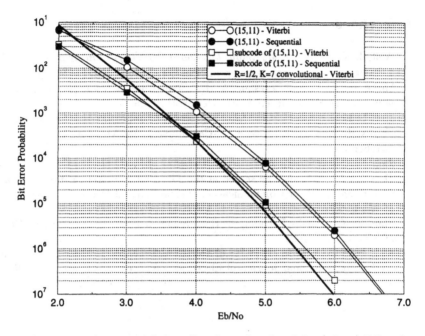

Fig. 3. Performance of sequential decoding for subcode of the (15,11) RS code

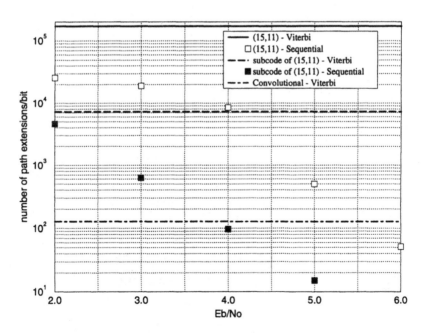

Fig. 4. Computational efficiency of sequential decoding for subcode of the (15,11) RS code

References

1. Wolf, J. K., *"Efficient Maximum Likelihood Decoding of Linear Block Codes Using a Trellis"*, IEEE Transactions on Information Theory, Vol. IT-24, pp. 76-80, 1978
2. Michelson, A. M. and Levesque, A. H., *"Error-Control Techniques for Digital Communications"*, Wiley-Interscience, New York, 1985
3. Shin, S. K. and Sweeney, P., *"Soft Decision Decoding of Reed Solomon Codes Using Trellis Methods"*, IEE Proceedings-Communications, Vol. 141, No. 5, pp. 303-308, October 1994
4. Shin, S. K. and Sweeney, P., *"Trellis Decoding of Reed Solomon Codes by Using Soft Decision Information"*, presented at the Third UK/Australian International Symposium on "DSP for Communication Systems", 12-14 December 1994, University of Warwick, U.K
5. Shin, S. K., "Trellis Decoding of Reed Solomon Codes", Ph.D Thesis, University of Surrey, U.K., 1994
6. Shin, S. K. and Sweeney, P., *"Soft Decision Sequential Decoder for Reed Solomon Codes"*, International Journal of Electronics, Vol. 79, No. 1, pp. 1-6, 1995

Linear Span Analysis of a Set of Periodic Sequence Generators *

P. Caballero-Gil[1] and A. Fúster-Sabater[2]

[1] Department of Statistics, Operations Research and Computation, Faculty of
Maths, University of La Laguna, 38271 La Laguna, Spain
(pcaballero@ull.es).
[2] Laboratory of Cryptography, Department of Information Theory and Coding,
Institute of Electronics of Communications (CSIC), Serrano 144, 28006 Madrid, Spain
(amparo@iec.csic.es).

Abstract. An algorithm for computing lower bounds on the global linear complexity of nonlinearly filtered PN-sequences is presented. Unlike
the existing methods, the algorithm here presented is based on the realization of bit wise logic operations. The numerical results obtained are
valid for any nonlinear function with a unique term of maximum order and for any maximal-length LFSR. To illustrate the power of this
technique, we give some high lower bounds that confirm Rueppel's conclusion about the exponential growth of the linear complexity in filter
generators.

1 Introduction

A fundamental problem in the theory of stream ciphers is the determination
of the global linear complexity of the keystreams. Depending on whether the
keystream involves one or more than one LFSR, the problem of controlling the
global linear complexity can achieve different levels of difficulty. In fact, the
global linear complexity of sequences obtained from a nonlinear combination
of LFSR-sequences is mostly predictable. Such is the case of many well-known
generator proposals (e.g. clock-controlled generators, alternating step generators, cascade generators, etc.) whose global linear complexity is either linear or
exponential in the number of storage cells employed. For a general survey see
[9].

Unlike the combination generators, the global linear complexity of the filter generators depends exclusively on the particular form of the filter and the
LFSR minimal polynomial. Generally speaking, there is no systematic method
to predict the resulting complexity. This is the reason why in the open literature
statements like 'it is extremely difficult to lowerbound (or guarantee) the linear
complexity of the sequences produced by nonlinearly filtering the state of an
LFSR' [8] can be found. Nevertheless, some authors have faced this problem and
several references can be quoted.

* This work was supported by R&D Spanish Programs TIC91-0386 and TIC95-0080.

Apart from the works of Key [2] and Kumar et al. [4], Rueppel [8] established his 'root presence test' for the product of distinct phases of a PN-sequence. This result let Fuster et al. [1] introduce the concept of 'fixed-distance coset' and prove its nondegeneration. Most recent works on this subject, [6] and [7], have focussed on the use of the Discrete Fourier Transform Technique to analyze the linear complexity. The former applies the DFT technique to the case of 2nd-order products while the latter derives a newway of the Rueppel's root presence test which can be applied to the case of 'regular shifts'.

In this work, an algorithm to compute high lower bounds on the global linear complexity of nonlinearly filtered PN-sequences is presented. Unlike the papers mentioned above, the proposed scheme can be applied to any nonlinear function with a unique term of maximum order. The work also provides numerical results of the bounds computed for different values of L (LFSR's length) and k (order of the function). The procedure which is based on the realization of bit wise logic operations seems to be rather adequate to either software simulation or hardware implementation.

The work is organized as follows. Section 2 introduces notation and basic concepts. Section 3 is devoted to the theoretical foundations of the algorithm. In section 4, the algorithm is described first. Next some comments concerning its performance and a table of numerical results are presented. Finally, conclusions in section 5 end the work.

2   Notation and Basic Concepts

Some fundamental concepts and definitions which will be used throughout the paper can be summarized as follows.

S is the output sequence of an LFSR whose minimal polynomial $m_s(x) \in GF(2)[x]$ is primitive.

L is the length of the LFSR.

$\alpha \in GF(2^L)$ is one root of $m_s(x)$.

f denotes the unique maximum order term of a nonlinear kth-order function applied to the LFSR's stages. That is, f is the product of k distinct phases of S, $f = s_{n+t_0} s_{n+t_1} \ldots s_{n+t_{k-1}}$, where the symbols t_j (j=0,1,...,k-1) are integers verifying $0 \le t_0 < t_1 < \ldots < t_{k-1} < 2^L - 1$.

In this work only the contribution of f to the linear complexity of the resulting sequence will be studied.

The root presence test for the product of distinct phases of a PN-sequence can be stated as follows, [8]:

$\alpha^E \in GF(2^L)$ is a root of the minimal polynomial of the generated sequence if and only if

$$A_E = \begin{vmatrix} \alpha^{t_0 2^{e_0}} & \cdots & \alpha^{t_{k-1} 2^{e_0}} \\ \alpha^{t_0 2^{e_1}} & \cdots & \alpha^{t_{k-1} 2^{e_1}} \\ & \cdots & \\ \alpha^{t_0 2^{e_{k-1}}} & \cdots & \alpha^{t_{k-1} 2^{e_{k-1}}} \end{vmatrix} \ne 0$$

Here $\alpha^{t_j} \in GF(2^L)$ (j=0,1,..,k-1) correspond respectively to the k phases (s_{n+t_j}) of the PN-sequence. E, the representative element of the cyclotomic *coset* E, is a positive integer of the form $E = 2^{e_0} + 2^{e_1} + ... + 2^{e_{k-1}}$ with the e_i (i=0,1,..., k-1) all different running in the interval [0,L). Under these conditions, α^E and its conjugate roots contribute to the linear complexity of the nonlinearly filtered sequence. The value of this contribution is equal to the number of elements in such a cyclotomic coset.

Definition 1. The cyclotomic *coset* E is nondegenerate if the corresponding determinant A_E equals zero.

Notice that every cyclotomic *coset* E can be easily associated with the radix-2 form of the integer E. This fact quite naturally suggests the introduction of binary strings of length L and k 1's. Indeed, the cyclotomic *coset* E can be equivalently characterized by:
    (i) the integer E of the form $E = 2^{e_0} + 2^{e_1} + ... + 2^{e_{k-1}}$.
    (ii) an L-bit string whose 1's are placed at the positions $\{e_i\}_{i=0,...,k-1}$.
    (iii) the determinant A_E as defined before.
In the sequel these three characterizations will be used indistinctly. Regarding the use of the strings, some additional notation is introduced.

Let $E = 2^{e_0} + 2^{e_1} + ... + 2^{e_{k-1}}$ and $F = 2^{f_0} + 2^{f_1} + ... + 2^{f_{l-1}}$ be two L-bit strings and k < l. $E \subset F$ means that $\{e_i\}_{i=0,...,k-1} \subset \{f_i\}_{i=0,...,l-1}$. That is, all the 1's in E are also in F.

For a set of L-bit strings $\{E_n\} = \{E_1, ..., E_N\}$, $OR[\{E_n\}]$ denotes the L-bit string resulting from a bit wise OR among the L-bit strings of the set. Obviously, we have that $\forall n \in \{1, ..., N\}$, $E_n \subset OR[\{E_n\}]$.

Definition 2. The cyclotomic *coset* E_d is a fixed-distance coset if its representative element E_d is of the form $E_d = 2^{e_0} + ... + 2^{e_{k-1}}$, $e_i \equiv d \cdot i \pmod{L}$ (i = 0, ..., k-1) where d is a positive integer less than L such that $(d, L) = 1$.

In order to simplify the notation, we will denote by A_d the determinant A_{E_d}.

Let us call jth-1 of E_d to that 1 in the L-bit string associated with the coset E_d which is placed at the position indicated by e_j.

Finally, we quote the following results relating to the linear complexity of a function with a unique term of maximum order, [1].

Theorem 3. *Let f be the maximum-order term defined as before. f is a kth-order function if and only if all the fixed-distance cosets are nondegenerate.*

The proof of this theorem is based on the generation of a normal basis of $GF(2^L)$ over $GF(2)$ by means of the construction of the L-bit strings associated with the fixed-distance cosets.

Corollary 4. *The linear complexity Λ of the resulting sequence is lowerbound by $\Lambda \geq N_L \cdot L$, where $N_L = \frac{\Phi(L)}{2}$ ($\Phi(L)$ being the Euler function). Here N_L represents the number of fixed-distance cosets and L the number of elements in such cosets.*

Remark that both results, which constitute the starting point of the present work, are independent of the LFSR, the order k of the function considered and the particular form of the term f.

3   Theoretical Foundations

Considering a general function with a unique maximum order term f defined as before, the present paper is concerned with a simple idea:

Not many degeneracies can exist simultaneously.

A proof of this general statement can be outlined in three steps. First, a specific set of N cosets are supposed to be simultaneously degenerate. Then, it is proved that only m of such cosets (m < N) can be simultaneously degenerate. Consequently, (N-m) cosets contribute to the linear complexity of the resulting sequence.

This procedure can be expressed in a more formal way as follows. Firstly, a new class of cosets is introduced.

Definition 5. Given a fixed-distance coset $E_d = 2^{e_0} + 2^{e_1} + ... + 2^{e_{k-1}}$ and $j \in \{0, ..., k-1\}$, we will call jth-quasi fixed-distance coset (by short jth-quasi f-d coset) to any cyclotomic coset whose representative element F_d^j is of the form

$$F_d^j = 2^{f_0} + 2^{f_1} + ... + 2^{f_{k-1}} \text{ such that } \{e_i\}_{\substack{i \neq j \\ i=0,...,k-1}} \subset \{f_i\}_{i=0,...,k-1}.$$

That is, a jth-quasi f-d coset F_d^j is any cyclotomic coset whose L-bit string associated contains all the 1's of the L-bit string associated with E_d except for the jth-1.

As we did before, we will denote by A_d^j the determinant $A_{F_d^j}$.

$\{F_{d,n}^j\} = \{F_{d,1}^j, ..., F_{d,N}^j\}$ denotes a set of jth-quasi f-d cosets.

Lemma 6. *Let F_d^j be any jth-quasi f-d coset, then its determinant associated A_d^j has at least a minor of order (k-1) (without the jth-row and a ith-column) that does not equal zero:*

$$\begin{vmatrix} \alpha^{t_0 2^{e_0}} & . & \alpha^{t_{i-1} 2^{e_0}} & \alpha^{t_{i+1} 2^{e_0}} & . & \alpha^{t_{k-1} 2^{e_0}} \\ & . & . & . & . & . \\ \alpha^{t_0 2^{e_{j-1}}} & . & \alpha^{t_{i-1} 2^{e_{j-1}}} & \alpha^{t_{i+1} 2^{e_{j-1}}} & . & \alpha^{t_{k-1} 2^{e_{j-1}}} \\ \alpha^{t_0 2^{e_{j+1}}} & . & \alpha^{t_{i-1} 2^{e_{j+1}}} & \alpha^{t_{i+1} 2^{e_{j+1}}} & . & \alpha^{t_{k-1} 2^{e_{j+1}}} \\ & . & . & . & . & . \\ \alpha^{t_0 2^{e_{k-1}}} & . & \alpha^{t_{i-1} 2^{e_{k-1}}} & \alpha^{t_{i+1} 2^{e_{k-1}}} & . & \alpha^{t_{k-1} 2^{e_{k-1}}} \end{vmatrix} \neq 0.$$

Sketch of proof

The determinants A_d^j and A_d differ exclusively in the jth-row. Expanding both determinants along the jth-row, we can write A_d^j and A_d in terms of the previous (k-1)th-order minors (i=0,1,...k-1). The fact that $A_d \neq 0$ (see Theorem 3) completes the proof.

$\square$

Theorem 7. *Let E_d be any fixed-distance coset and $j \in \{0, ..., k-1\}$. If for some set of jth-quasi f-d cosets $\{F_{d,n}^j\}$ exists at least a fixed-distance coset $E_{d'}$ such that $E_{d'} \subset OR[\{F_{d,n}^j\}]$ then the cosets of $\{F_{d,n}^j\}$ cannot be simultaneously degenerate.*

Sketch of proof

We proceed by contradiction. We assume that $\{F_{d,n}^j\}$ are simultaneously degenerate. This simultaneous degeneration is equivalent to the existence of a set of homogeneous linear systems associated with each determinant $A_{d,n}^j$. All these systems have nontrivial solution and (k-1) equations in common. Furthermore, due to Lemma 6, this solution is proportional to $\alpha^{t_i 2^{e_j}}$. Since there is a joint solution to all the systems, the compatibility of the system composed of all the equations can be easily deduced. Finally, according to the hypothesis of the theorem, the k equations associated to the determinant $A_{d'}$ are among the equations of the complete system. This means that a compatible system has a non-compatible sub-system, which obviously is a contradiction.

$\square$

The algorithm to be presented realizes the previous results by means of the handling of L-bit strings.

4   General Features of the Algorithm

4.1   Notation

The following notation is used throughout the algorithm.

FDC(i) ($i = 1, ..., N_L$) denotes the L-bit string corresponding to the ith-fixed-distance coset.

Δ is the lower bound on the linear complexity.

FCC(i,j) ($j = 1, ..., k - 1$) denotes the L-bit strings obtained from FDC(i) by replacing the jth-1 by a 0. Remark that FCC(i,0) is a shifted version of FCC(i,k-1).

CD(i,j) denotes a group of L-bit strings associated to the jth-quasi f-d cosets $\{F_{d_i,n}^j\}$. Any L-bit string in CD(i,j) previously considered must be eliminated. In order to do that, firstly fixed-distance cosets are discovered by means of AND operations. Secondly, after realizing exclusive-OR (denoted by XOR) operations with every previous FCC, those cosets that produce at least a string with a unique 1 are eliminated from CD(i,j).

m is a decreasing counter whose first value (denoted by mpri) is the number of L-bit strings in CD(i,j) after eliminations.

a(n) $(n = 1, ..., \binom{mpri}{m})$ denotes each mpri-bit string with m 1's.

VOR denotes the string resulting from an OR operation among those m cosets of CD(i,j) indicated by the positions of the 1's in a(n).

VL is a binary variable whose value depends on the AND operation between VOR and each FDC(i).

4.2 The Algorithm

The algorithm INPUTS are L (LFSR's length) and k (order of the function) with $2 < k < L - 2$, and its OUTPUT is the lower bound Δ.

The flow chart of this algorithm is depicted in Fig. 1. Steps 1 and 2 are referred to the following.

Step 1

Compute the N_L values of d.

Generate the FDC(i) $(i = 1, ..., N_L)$.

Initialize the lower bound $\Delta = L \cdot N_L$.

Step 2

Generate FCC(i,j). m=L-k.

Generate the group CD(i,j).

Realize the AND between each FDC(l) $(l = 1, ..., N_L)$ and every coset of CD(i,j).

Each time some logic product coincides with FDC(l), the corresponding coset is eliminated from CD(i,j) and m=m-1.

Realize the XOR between each FCC(o,p) $(o = 1, ..., i-1, p = 1, ..., k-1; o = i, p = 1, ..., j-1)$ and every coset of CD(i,j).

Each time some addition has a unique 1, the corresponding coset is eliminated from CD(i,j) and m=m-1.

4.3 An Illustrative Example

For L=11, k=6, d=1, 2, 3, 4, 5 and N_{11}=5.

(d=1) FDC(1) : 00000111111

(d=2) FDC(2) : 10101010101

(d=3) FDC(3) : 01001011011

(d=4) FDC(4) : 01100110011

(d=5) FDC(5) : 11000111001

* indicates a shifted version

lower bound for the complexity: $\Delta = 11 \cdot 5 = 55$.

For i=1, j=1

$$FCC(1,1) : 00000111101, \quad CD(1,1) : \begin{cases} 00001111101 \\ 00010111101 \\ 00100111101 \\ 01000111101 \\ 10000111101 \end{cases}$$

28

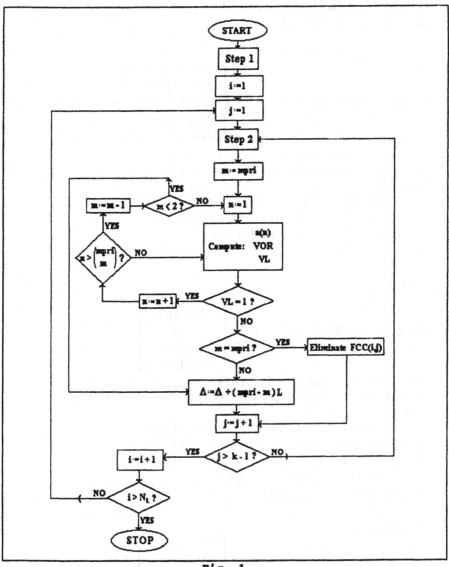

Fig. 1

mpri=5. After analyzing the cases m=5, 4 and 3,
for m=2,
a(1)=00011, VOR : 00011111101, VL=1 by FDC(1)*
a(2)=00101, VOR : 00101111101, VL=1 by FDC(3)*
a(3)=01001, VOR : 01001111101, VL=0
$\Delta = 55 + (5 - 2) \cdot 11 = 88$

For i=1, j=2

$$FCC(1,2) : 00000111011, CD(1,2) : \left\{ \begin{array}{l} 00001111011 \\ 00010111011 \\ 00100111011 \\ 01000111011 \\ 10000111011 \end{array} \right\}$$

00001111011 eliminated by FCC(1,1)
mpri=4. After analyzing the cases m=4 and 3,
for m=2,
a(1)=0011, VOR : 00110111011, VL=1 by FDC(4)*
a(2)=0101, VOR : 01010111011, VL=1 by FDC(2)*
a(3)=1001, VOR : 10010111011, VL=0
$\Delta = 88 + (4 - 2) \cdot 11 = 110$

For i=1, j=3

$$FCC(1,3) : 00000110111, CD(1,3) : \left\{ \begin{array}{l} 00001110111 \\ 00010110111 \\ 00100110111 \\ 01000110111 \\ 10000110111 \end{array} \right\}$$

00001110111 eliminated by FCC(1,2)
mpri=4. After analyzing the cases m=4 and 3,
for m=2,
a(1)=0011, VOR : 00110110111, VL=1 by FDC(4)*
a(2)=1001, VOR : 10010110111, VL=1 by FDC(3)*
a(3)=0110, VOR : 01100110111, VL=1 by FDC(4)
a(4)=1010, VOR : 10100110111, VL=0
$\Delta = 110 + (4 - 2) \cdot 11 = 132$

For i=1, j=4

$$FCC(1,4) : 00000101111, CD(1,4) : \left\{ \begin{array}{l} 00001101111 \\ 00010101111 \\ 00100101111 \\ 01000101111 \\ 10000101111 \end{array} \right\}$$

01000101111 eliminated by FCC(1,1)
00001101111 eliminated by FCC(1,3)
mpri=3. After analyzing the case m=3,
for m=2,

a(1)=011, VOR : 00110101111, VL=0
$$\Delta = 132 + (3 - 2) \cdot 11 = 143$$

For i=1, j=5

$$\text{FCC(1,5)} : 00000011111, \text{CD(1,5)} : \begin{cases} 00001011111 \\ 00010011111 \\ 00100011111 \\ 01000011111 \\ 10000011111 \end{cases}$$

10000011111 eliminated by FDC(1)
01000011111 eliminated by FCC(1,1)
00001011111 eliminated by FCC(1,4)
mpri=2.
For m=2,
a(1)=11, VOR : 00110011111, VL=1 by FDC(4)*
$$\Delta = 143 + (2 - 1) \cdot 11 = 154$$

After similar calculations with i=2, 3, 4, 5 we obtain finally $\Delta = 242$.

This example deserves some remarks:

Since the algorithm is independent of the specific function and polynomial, the lower bound obtained is valid for any nonlinear function with a unique term of maximum order 6 and for any maximal-length LFSR of length 11.

If we had used the root presence test to obtain the same result, we would have had to compute (for each function of order 6 and each maximal-length LFSR of length 11) at least 22 determinants of order 6 in $GF(2^{11})$. This would have implied more than a million arithmetic operations in a finite field, [3].

4.4 Implementation and Computational Complexity

The main facts concerning the performance of the algorithm are summarized in this section.

The algorithm is divided into two stages. The first stage includes the generation and 'debugger' of the cosets to be analyzed. The second stage is concerned with the simultaneous degenerations of the different groups of cosets. In the second stage a 'sweep' of some groups of cosets is carried out, which permits their use later on the algorithm.

Regarding the required memory, note that only the cosets FCC(i,j) but not the cosets CD(i,j) have to be stored. This means keeping one out of (L-k) cosets analyzed.

In order to handle the cosets of CD(i,j), the more suitable dynamic structure of information is a linked linear list. This structure seems also adequate to select, through the codification a(n), the cosets involved in each bit wise OR operation.

But it's beyond doubt that the most remarkable feature of the implementation is the use of backtracking for solving the problem of the generation of the successive strings a(n).

We determine that the algorithm in general has a maximum computational complexity of order $O(2^{L-k})$, where L denotes the length of the LFSR and k is the order of the function. In order to estimate this value, it has been assumed at various points that we are in the 'worst possible case', which involves a number of logic operations given by $N_L(k-1)[\binom{mpri}{mpri} + ... + \binom{mpri}{2})] = N_L(k-1)(2^{mpri} - mpri) \leq N_L(k-1)2^{L-k}$.

From several experimental results it can be deduced that it is not necessary to do all the comparisons. The bound is usually obtained after having analyzed few values of m and, for the minimum value of m, after having analyzed few groups.

Anyway, three considerations have to be taken into account. First, for each L and k the algorithm has to be used only once. Second, it will be used only with relatively small inputs ($L \in [120, 250]$). And third, a high bound obtained for specific values of L and k will encourage the designer of sequence generators to use nonlinear filter with a unique term of maximum order k applied to any LFSR of length L.

4.5 Experimental Results and Discussion

In this section we report the results of computer runs over values of L primes. The effect of this choice is twofold. On the one hand, it simplifies the computation of the N_L values of d. On the other hand, the more fixed-distance cosets there are the higher bounds the algorithm produces.

The following table shows some experimental results obtained respectively in [1] and those obtained with the algorithm presented here. G denotes the number of cosets guaranteed to be nondegenerate.

L k	G	Bound	G	Bound
11 6	15	165	22	242
17 9	32	544	184	3128
23 12	44	1012	363	8349
29 15	70	2030	770	22330
37 19	90	3330	1296	47952
43 22	105	4515	1764	75852
47 24	115	5405	2115	99405
53 27	130	6890	2702	143206

A comparison of both results shows that there is a substantial improvement in the number of cosets guaranteed to be nondegenerate. This implies a high increase in the lower bound. It can be estimated that the rate between both bounds decreases as L increases, what let us estimate a bound above 500,000 for L=89. According to these results, we indeed obtain the required lower bound that allows to confirm Rueppel's conclusion about an exponential growth in the linear complexity, [8]. Furthermore this algorithm solves the problem of the unpredictability of the resulting linear complexity. This may cause many designers of pseudo-random sequence generator use nonlinear state-filter generators with more confidence.

But the most important result is, in fact, that no longer specific classes of functions and LFSR polynomial have to be considered to apply the algorithm. Only the order of the function and the length of the LFSR will be sufficient. This is intuitively pleasing from the realization viewpoint because no special care has to be taken to exclude 'insecure' functions.

On the other hand, this algorithm allows some modifications, e.g. the first loop can be made to the reverse, starting from the maximum value of d, or else taking them randomly. Also the generation of FCC could be done starting from the last, that is moving the 0 from left to right. These changes can produce improvements in the obtained bounds, although we have predicted they will not be quantitative changes. The authors are working towards improving the algorithm by means of changes like these.

5   Conclusions and Open Problems

Our research has highlighted the problem of the global linear complexity of the nonlinear filter generators. In this paper a new algorithm for bounding the linear complexity from below has been presented. This proposal differs from existing schemes in different aspects. Firstly, unlike the well-known Berlekamp-Massey's algorithm [5], we do not analyze the digits of the output sequence, instead we study the characteristics of the nonlinear function applied to the LFSR's stages. Secondly, the proposed algorithm indeed does not require any condition on the shifts, as do [4], [6] and [7], so the obtained bounds are valid for any nonlinear function with a unique term of maximum order. And finally, this work is based on the handling of L-bit strings instead of determinants in a finite field (Rueppel's method, [8]).

The obtained numeric results have shown that the lower bounds for the linear complexity of these sequences are reasonably high without any restriction on function nor polynomial, that confirms Rueppel's conclusion and encourages to the use of nonlinear state-filter generators.

This investigation has left an open problem that is the study of the remaining cosets this algorithm has not analyzed.

References

1. A. Fúster-Sabater and P. Caballero-Gil, 'On the Linear Complexity of Nonlinearly Filtered PN-Sequences', Advances in Cryptology-ASIACRYPT'94, Lecture Notes in Computer Science Vol. 917, Springer-Verlag.
2. E.L. Key, 'An Analysis of the Structure and Complexity of Nonlinear Binary Sequence Generators', IEEE Trans. Inform. Theory, Vol. IT-22, pp. 732-736, Nov. 1976.
3. D.E. Knuth, 'The Art of Computer Programming, Vol. 2: Seminumerical Algorithms', Addison-Wesley, 1981.

4. P.V. Kumar and R.A. Scholtz, 'Bounds on the Linear Span of Bent Sequences', IEEE Transactions on Information Theory, Vol. IT-29, pp. 854-862, Nov. 1983.

5. J.L. Massey, 'Shift-Register Synthesis and BCH Decoding', IEEE Transactions on Information Theory, Vol. IT-15, pp. 122-127, Jan. 1969.

6. J.L. Massey and S. Serconek, 'A Fourier Transform Approach to the Linear Complexity of Nonlinearly Filtered Sequences', Advances in Cryptology-CRYPTO'94, Lecture Notes in Computer Science Vol. 839, pp. 332-340, Springer-Verlag, 1994.

7. K.G. Paterson, 'New Lower Bounds on the Linear Complexity of Nonlinearly Filtered m-Sequences', submitted to IEEE Transactions on Information Theory, 1995.

8. R.A. Rueppel, 'Analysis and Design of Stream Ciphers', Springer-Verlag, New York, 1986.

9. G.J. Simmons (ed.), 'Contemporary Cryptology: The Science of Information Integrity', IEEE Press, 1991.

Minimal Weight k-SR Representations

Yongfei Han, Dieter Gollmann, Chris Mitchell

Information Security Group
Department of Computer Science
Royal Holloway, University of London
Egham
SurreyTW20 0EX
E-mail: {yongfei, dieter, cjm@dcs.rhbnc.ac.uk}

Abstract. An algorithm for a minimal weight string replacement representation for the standard square and multiply exponentiation method is discussed, with a presentation of the design and proof of the algorithm. The performance of this new method is analysed and compared with previously proposed methods. The techniques presented in this paper have applications in speeding up the implementation of public-key cryptographic algorithms such as RSA [3].

1 Introduction

We are investigating the property of a redundant integer representation which can help in reducing the number of multiplications in a square & multiply exponentiation by reducing the weight of exponents [1]. The concept of k-SR representations is based on the idea of replacing 1-runs in the binary representation of an integer by a single digit. We consider runs up to length k, hence the name k string replacement. An obvious conversion algorithm would just substitute 1-runs by k-SR digits. However, it was observed that this does not necessarily lead to optimal results. For example, the number $21 = (10101)_2$ contains three 1-runs but has a 3-SR representation of weight 2, i.e. (77). In this paper, we will examine algorithms for computing (nearly) minimal k-SR representations.

Definition 1. k-**SR representation**: A k-SR (string replacement) representation of an integer e is given by

$$e = \sum_{i=0}^{r} f_i 2^i, \quad \text{with } f_i = 2^{k_i} - 1, \ 0 \le k_i \le k.$$

Note that k-SR representations are not unique, e.g. we have the following 4-SR representations of 21: $10101, 77, 3F$. (F is used to represent 15.) In the following, let $\underline{x}$ be a k-SR number.

Definition 2. The number of 0-runs in $\underline{x}$ is written as $r_0(\underline{x})$. The number of 1-runs in $\underline{x}$ is written as $r_1(\underline{x})$.

Definition 3. Weight: The *weight* of a string $\underline{x}$, written as $w(\underline{x})$, is defined as the number of non-zero symbols in $\underline{x}$. With $w_i(\underline{x})$, we denote the length of the i-th 1-run in $\underline{x}$.

2 Staggered k-SR Representations

In this Section, we will define a standard representation of k-SR numbers and show that we always can find a minimal k-SR representation of this form. These properties can then be the foundation for developing a minimisation algorithm. Let e have a minimal representation

$$e = \sum_{i=0}^{r}(2^{j_i} - 1)2^{l_i}.$$

Note that we have by definition $l_i < l_{i+1}$ for all $i < r$.

Definition 4. A k-SR number $\sum_{i=0}^{r}(2^{j_i} - 1)2^{l_i}$ is called *staggered* if we have $j_i + l_i < j_{i+1} + l_{i+1}$ for all $i < r$.

Lemma 5. *Every integer e has a staggered k-SR representation of minimal weight.*

Proof: Assume we have a minimal k-SR representation with $j_i + l_i \geq j_{i+1} + l_{i+1}$ for some i. We rewrite

$$(2^{j_{i+1}} - 1)2^{l_{i+1}} + (2^{j_i} - 1)2^{l_i} = (2^{j_i+l_i-l_{i+1}} - 1)2^{l_{i+1}} + (2^{j_{i+1}+l_{i+1}-l_i} - 1)2^{l_i}$$

and get a valid k-SR representation as

$$j_i > j_i + l_i - l_{i+1} \geq j_i + l_i - j_{i+1} - l_{i+1} \geq 0, \quad j_i \geq j_{i+1} + l_{i+1} - l_i \geq l_{i+1} - l_i > 0.$$

$\square$

Theorem 6. *In a minimal k-SR representation, $j_0 - l_0$ is the position of a 0 terminating a 1-run or to a 1 in a 1-run of length strictly greater than k.*

Proof: Assume that we can find a minimal staggered k-SR representation where our assumption does not hold. Without loss of generality, assume $l_0 = 0$. First, assume that j_0 points to some other 0 in e, e.g. $j_0 = 4$ in $e = \ldots 00101$.

$$
\begin{array}{r}
1\,1\,1\,1 \\
\ldots 1\,1\,1\,1 \\
\hline
\ldots 0\,\underline{1}\,1\,0\,1
\end{array}
$$

Because we have a staggered representation we will always have a 1 in position $j_0 - 1$ unless we add a correcting digit $2^{j_1} - 1$ in this position. We get an alternative representation of equal weight and reduced j_0 as

$$(2^{j_0} - 1) + (2^{j_1} - 1)2^{j_0-1} = 2^{j_1+j_0-1} + (2^{j_0-1} - 1).$$

Applying this step repeatedly, we can make j_0 point to the terminator of a 1-run. In our second case, we let j_0 point to some 1 in e, e.g. $j_0 = 5$ in $e = \ldots 110101$.

$$
\begin{array}{r}
1\,1\,1\,1\,1 \\
\ldots 1\,1\,1\,1\,1 \\
\hline
\ldots 0\,\underline{1}\,1\,1\,0\,1
\end{array}
$$

Because we have a staggered representation we will always have a 0 in position j_0 unless we add a correcting digit $2^{j_2} - 1$ in this position. Assume we also have to correct a 0 in a position $l_1 < j_0$ by adding $(2^{j_1} - 1)2^{l_1}$. We get an alternative representation of equal weight and reduced j_0 as

$$(2^{j_0} - 1) + (2^{j_2} - 1)2^{j_0} + (2^{j_1} - 1)2^{l_1} = 2^{j_2+j_0} + (2^{j_1-1} - 1)2^{l_1+1} + (2^{l_1} - 1).$$

Applying this step repeatedly, we can make j_0 point to the terminator of a 1-run. If we do not have to correct a 0, we get can increment j_0 as

$$(2^{j_0} - 1) + (2^{j_2} - 1)2^{j_0} = 2^{j_2+j_0} - 1 = (2^{j_0+1} - 1) + (2^{j_2-1} - 1)2^{j_0+1}.$$

□

In the following, we show how the choice of the l_i can be restricted in a minimisation algorithm. We will frequently make use of the fact that we may construct a staggered representation. Adding a new digit $(2^{j_i} - 1)2^{l_i}$ to an intermediate result will have the following effect:

- When the intermediate result has a 1 in position l_i, then $(2^{j_i} - 1)2^{l_i}$ is a *0-corrector*, changing this bit to 0, leaving the remainder of the intermediate result unchanged, and appending a string $10\ldots 0$ at the top.
- When the intermediate result has a 0 in position l_i, then $(2^{j_i} - 1)2^{l_i}$ is a *1-corrector*, changing any string $10\ldots 0$ starting in position l_i to $01\ldots 1$ leaving the remainder of the intermediate result unchanged, and appending a string $10\ldots 0$ at the top.

Adding $(2^{j_i} - 1)2^{l_i}$ to an intermediate result, therefore, will only imply local changes at l_i and the addition of a string at the top.

Lemma 7. *When computing a minimal k-SR representation for a given j_0 and $l_i < j_0 + l_0$, then $j_i + l_i$ cannot be the position of a 0 in the 'first available' 0-run.*

Proof: W.l.o.g. assume that $j_1 + l_1$ is the position of a 0 in the first 0-run above j_0, e.g. $j_1 = 3$ and $l_1 = 1$ in $e = 1001101$.

$$\begin{array}{r} 1\,1\,1\,1 \\ 1\,1\,1\,1 \\ \hline \underline{1}\,0\,1\,1\,0\,1 \end{array}$$

Adding $(2^{j_1} - 1)2^{l_1}$ to the intermediate result will generate a 1 in position $j_1 + l_1$. We do not have to correct any 1 in a position below $j_1 + l_1$. We may have to correct some 0's but because we are constructing a staggered representation, no 0-corrector would change the 1 in position $j_1 + l_1$. Therefore, we have to add a corrector $(2^{j_2} - 1)2^{j_1+l_1}$. We get an alternative representation of equal weight for the digits in position l_0, l_1, l_2 which meets the condition of our Lemma as

$$(2^{j_0} - 1)2^{l_0} + (2^{j_1} - 1)2^{l_1} + (2^{j_2} - 1)2^{j_1+l_1}$$
$$= (2^{l_1-l_0} - 1)2^{l_0} + (2^{j_0+l_0-l_1-1} - 1)2^{l_1+1} + 2^{j_1+j_2+l_1}.$$

□

Lemma 8. *When computing a minimal k-SR representation for a given j_0 and $l_i < j_0 + l_0$, then $j_i + l_i$ can be the position of a 1 only if this 1 is the terminator of a 0-run.*

Proof: W.l.o.g. assume $j_1 + l_1$ is the second position in a 1-run, e.g. $j_1 = 4$ and $l_1 = 1$ in $e = \ldots 110101$.

$$
\begin{array}{r}
1\,1\,1 \\
1\,1\,1\,1 \\
\hline
\ldots 1\underline{0}\,0\,1\,0\,1 \\
\ldots 1\,1 \\
\hline
\ldots \underline{0}\,1\,0\,1\,0\,1
\end{array}
$$

Adding $(2^{j_1} - 1)2^{l_1}$ to the intermediate result will generate a 0 in the position where the 1-run starts. Adding a digit $(2^{j_2} - 1)2^{l_2}$ to correct this 0 will change the 1 in position $j_1 + l_1$ to a 0 and we have to add another digit $(2^{j_3} - 1)2^{j_1 + l_1}$. Note that other digits in a staggered representation will not affect these positions. We get an alternative representation of equal weight for these digits which meets the condition of our Lemma as

$$(2^{j_1} - 1)2^{l_1} + (2^{j_2} - 1)2^{l_2} + (2^{j_3} - 1)2^{j_1 + l_1}$$
$$= (2^{l_2 - l_1} - 1)2^{l_1} + (2^{j_2 - 1} - 1)2^{l_2 + 1} + 2^{j_1 + j_3 + l_1}.$$

$\square$

Lemma 9. *When computing a minimal k-SR representation for a given j_0 and $l_i < j_0 + l_0$, then $j_i + l_i$ can be the position of a 0 only if it this 0 is the terminator of a 1-run.*

Proof: W.l.o.g. assume that l_2 is the start of a 1-run, l_3 is the start of the next 1-run, and $l_2 < l_3 < j_1 + l_1$, e.g. $l_2 = 3, l_3 = 5$, and $j_1 + l_1 = 6$ in $e = \ldots 00110101$.

$$
\begin{array}{r}
1\,1\,1 \\
1\,1\,1\,1\,1\,1 \\
\hline
\ldots 1\,0\,0\,\underline{0}\,0\,1\,0\,1 \\
\ldots 1\,1\,1\,1 \\
\hline
\ldots \underline{0}\,\underline{1}\,1\,1\,0\,1\,0\,1
\end{array}
$$

Adding $(2^{j_1} - 1)2^{l_1}$ to the intermediate result will generate a 0 in position l_2. Adding a digit $(2^{j_2} - 1)2^{l_2}$ to correct this 0 will create 1's in positions l_3 to $j_1 + l_1 - 1$. We have to correct these 1's individually. Consider, for example, position $j_1 + l_1 - 1$. We add a digit $(2^{j_4} - 1)2^{j_1 + l_1 - 1}$ but

$$(2^{j_1} - 1)2^{l_1} + (2^{j_4} - 1)2^{j_1 + l_1 - 1} = (2^{j_1 - 1} - 1)2^{l_1} + 2^{j_1 + j_4 + l_1 - 1}.$$

Applying this argument repeatedly, we get a representation of equal weight and $j_1 + l_1 = l_3$.

$\square$

Lemma 10. *When computing a minimal k-SR representation for a given j_0 and $l_i < j_0 + l_0$, then $j_i + l_i$ cannot be the terminator of a 1-run of length 1.*

Proof: W.l.o.g. assume that $j_1 + l_1$ is the terminator of a 1-run of length 1. We have to add a 1-compensator in position $j_1 + l_1 - 1$ and a 0-compensator in position $j_1 + l_1$, but

$$(2^{j_1} - 1)2^{l_1} + (2^{j_2} - 1)2^{j_1 + l_1 - 1} + (2^{j_3} - 1)2^{j_1 + l_1}$$
$$= (2^{j_1 - 1} - 1)2^{l_1} + (2^{j_2 - 1} - 1)2^{j_1 + l_1} + 2^{j_1 + j_3 + l_1}.$$

$\square$

3   Computing a k-SR representation

We now can formulate strategies for finding minimal k-SR representations of an integer e. We start from the binary representation of e. Let l_0 be the position of the least significant 1 in e.

1. Choose $j_0 + l_0$ to be the terminator of a 1-run or $j_0 = k$ if there is no such terminator;
2. For every l_i with $e_{l_i} = 0$ and $l_0 < l_i < j_0$ add a 0-corrector; we correct from the least significant 0 upwards by matching the 'next available 0-run in e'.
3. Strategy 1: choose j_i so that $j_i + l_i$ is the most significant position in this 0-run;
4. Strategy 2: choose j_i so that $j_i + l_i$ is the most significant position in the 1-run following this 0-run; add the least significant position of this 1-run into a list of positions which have to be corrected;
5. If the next available 0-runs is followed by a 1-run of length 1, then choose Strategy 1.

For a given binary string x with least significant bit 1, we now compute a k-SR representation using Strategy 1 only.

$$\sum_{i=0}^{r} \left(2^{j_i(x)} - 1\right) 2^{l_i(x)}.$$

Let the $l_i(x)$ be the position of the i-th least significant 0 in x. Define $j_0(x) \le k$ to be the terminator of a 1-run and let $N_{j_0}(x)$ be the number of zeroes in $x[j_0(x), 0]$. For $i > 0$ and $l_i(x) < j_0(x)$, define $j_i(x)$ iteratively by the algorithm given in Figure 1. For example, with $k = 15$, $x = 11\ 0111\ 0101\ 0001\ 0010\ 0101\ 0101$, and $j_0(x) = 13$ we get

i	0	1	2	3	4	5	6	7
l_i	0	1	3	5	7	8	10	11
j_i	13	15	15	15	14	14	14	14

and the representation of weight 8,

$$(2^{14} - 1)(2^{14} - 1)0(2^{14} - 1)\ (2^{14} - 1)0(2^{15} - 1)0\ (2^{15} - 1)0(2^{15} - 1)(2^{13} - 1).$$

```
p := j₀(x);       /* p is a pointer */
i := 1;
WHILE lᵢ(x) < j₀(x) DO
        p := p + 1;
        IF xₚ = 1 THEN
            jᵢ(x) := p - lᵢ(x);
            IF jᵢ(x) > k THEN STOP
            i := i + 1;
        END IF
END WHILE
```

Fig. 1. A k-SR replacement algorithm

However, the maximal $j_0(\underline{x})$ is not always the optimal choice, in our example we have also 15-SR representations of weight 7,

$$10\ 0000\ (2^5 - 1)(2^5 - 1)00\ 0000\ 0000\ 00(2^{11} - 1)0\ (2^9 - 1)0(2^8 - 1)(2^7 - 1)$$

or

$$00\ 0000\ (2^6 - 1)0(2^7 - 1)(2^7 - 1)\ 0001\ 0000\ 0000\ (2^6 - 1)0(2^5 - 1)(2^5 - 1).$$

We now prove that the algorithm given in Figure 1 is correct.

Lemma 11. *Assume that $\underline{x}$ represents an odd number. If $j_i(\underline{x})$, $0 \le i \le r$, are successfully computed for some r, then $\underline{x}$ and the binary expansion of*

$$\tilde{\underline{x}}_r = \sum_{i=0}^{r}(2^{j_i(\underline{x})} - 1)2^{l_i(\underline{x})}$$

will coincide in positions 0 to $l_r(\underline{x})$ and positions $j_0(\underline{x}) - 1$ to $l_r(\underline{x}) + j_r(\underline{x})$. Between positions $l_r(\underline{x}) + 1$ and $j_0(\underline{x}) - 1$, the coefficients of the binary expansion of $\tilde{\underline{x}}_r$ are all 1.

Proof (by induction): For $r = 0$, we have $\tilde{\underline{x}}_0 = 2^{j_0(\underline{x})} - 1$ As $\underline{x}$ is odd and $l_0(\underline{x}) = 0$, we have $\underline{x}_0 = 1$, $\underline{x}_{j_0(\underline{x})-1} = 1$, and $\underline{x}_{j_0(\underline{x})} = 0$ so the Lemma holds.

Assume that all the assumptions of the Lemma hold for a given r. The next 0 in $\underline{x}$ will occur in position $l_{r+1}(\underline{x})$. If $l_{r+1}(\underline{x}) \ge j_0(\underline{x})$ we stop and observe that $\underline{x}$ and the binary expansion of $\tilde{\underline{x}}_r$ now coincide in positions 0 to $l_r(\underline{x}) + j_r(\underline{x})$.

If $l_{r+1}(\underline{x}) < j_0(\underline{x})$ and if $j_{r+1}(\underline{x})$ can be computed, then we update $\tilde{\underline{x}}_r$ by adding $(2^{j_{r+1}(\underline{x})} - 1)2^{l_{r+1}(\underline{x})}$. This addition will switch the coefficient of the binary expansion of $\tilde{\underline{x}}_{r+1}$ in position $l_{r+1}(\underline{x})$ to 0 and leave all coefficients in positions $l_{r+1}(\underline{x}) + 1$ to $l_r(\underline{x}) + j_r(\underline{x})$ unchanged. Thus, $\underline{x}$ and the binary expansion of $\tilde{\underline{x}}_{r+1}$ coincide in positions 0 to $l_{r+1}(\underline{x})$ and in positions $j_0(\underline{x})$ to $l_r(\underline{x}) + j_r(\underline{x})$.

Define $\delta := l_{r+1}(\underline{x}) + j_{r+1}(\underline{x}) - (l_r(\underline{x}) + j_r(\underline{x})) - 1$. Note that $\delta = 0$ if $l_{r+1}(\underline{x}) + j_{r+1}(\underline{x}) = l_r(\underline{x}) + j_r(\underline{x}) + 1$ and that δ is the length of a 0-run otherwise. The coefficients of $\tilde{\underline{x}}_{r+1}$ in position $l_r(\underline{x}) + j_r(\underline{x})$ and above are therefore $10^{\delta}1$. We have to consider two cases.

1. $\delta = 0$: then the coefficient of $\underline{x}$ in position $l_r(\underline{x}) + j_r(\underline{x}) + 1 = 1$; hence $\underline{x}$ and the binary expansion of $\tilde{\underline{x}}_{r+1}$ coincide also in position $l_{r+1}(\underline{x}) + j_{r+1}(\underline{x}) = l_r(\underline{x}) + j_r(\underline{x}) + 1$.
2. $\delta \neq 0$: there is a 0-run in $\underline{x}$ which is terminated by a 1 in position $l_{r+1}(\underline{x}) + j_{r+1}(\underline{x})$; again $\underline{x}$ and the binary expansion of $\tilde{\underline{x}}_{r+1}$ coincide in all positions up to $l_{r+1}(\underline{x}) + j_{r+1}(\underline{x})$.

$\square$

Corollary 12. *Assume that $\underline{x}$ represents an odd number and let r be the maximal value with $l_r(\underline{x}) < j_0(\underline{x})$. If the algorithm successfully computes $j_i(\underline{x})$ for all i with $l_i(\underline{x}) < j_0(\underline{x})$, then $\underline{x}[j_r(\underline{x}) + l_r(\underline{x}), 0]$ can be replaced by*

$$\tilde{\underline{x}}_r = \sum_{i=0}^{r}(2^{j_i(\underline{x})} - 1)2^{l_i(\underline{x})}.$$

4 The Algorithm

The algorithm is defined for any n-bit binary digit e and for any $k \geq 2$. We will use following notations.

- $\underline{e}$ is a binary string of length n;
- i is pointer indicating the current position in $\underline{e}$;
- weight$[-1 \ldots n-1]$ is an array of integers initialized to n.

In our algorithm, we first look for the start of 1-runs, i.e. the least significant position of such a run (Fig. 2). Once we have found the start of a 1-run, we examine all possible choices of j_0 (Fig. 3). The subroutine CORE has as input the starting point i of a substring and a position j_0 which terminates a 1-run. It executes the algorithm given in Fig. 1. If it succeeds in computing all j_i's and l_i's, then it signals success = TRUE and returns the weight N_{j_0} and a pointer top to the most significant position which is affected by this computation.
As an example, we compute the weight of a 4-SR representations of the string 1 1111 0101.

i	n	n	n	n	n	n	n	n	n
0	n	n	n	n	2	n	n	n	1
1	n	n	n	n	2	n	n	1	1
2	n	n	n	n	2	n	2	1	1
3	n	n	n	n	2	2	2	1	1
4	n	3	n	n	2	2	2	1	1
5	3	3	n	n	2	2	2	1	1

```
i:= 0;
weight[-1]:= 0;
WHILE i < n do
        IF e_i = 0 THEN
                weight[i]:= MIN(weight[i],weight[i-1]);
                i:= i + 1;
        END THEN
        ELSE REPLACE(i)
        END WHILE
```

Fig. 2. The main algorithm

```
REPLACE(i):
j_0 := position of the next 0 above position i;
IF j_0 > i + k
   THEN weight[i + k]:= MIN(weight[i + k],weight[i-1]+1);
        i:= i + k;
   END THEN
   ELSE weight[j_0]:= MIN(weight[j_0],weight[i-1]+1);
        WHILE j_0 ≤ i + k DO
                j_0:= terminator of the next 1-run;
                CORE(i,j_0,top,N_{j_0},success);
                IF  success=TRUE THEN
                        weight[top]:= MIN(weight[top],weight[i-1]+N_{j_0})
        END WHILE
        i:= i+1
   END ELSE
RETURN
```

Fig. 3. Subroutine REPLACE

Using a backtracking algorithm, we would obtain the 4-SR representation 0 00F0 0077. For our previous example with $k = 15$ and $\underline{x} = 11\ 0111\ 0101\ 0001\ 0010\ 0101\ 0101$ we get

```
 i│1 1 0 1 1 1 0 1 0 1 0 0 0 1 0 0 1 0 0 1 0 1 0 1 0 1
 0│8     6       4         3       2       1
 2│8 8   6 6     4     4   3     3 2   2 1 1
 4│8 8 8 6 6 6   4     4   3     3   2 2 2 1 1
 6│8 8 8 6 6 6   4     4   3     3 2 2 2 2 1 1
 9│8 8 8 6 6 6   4     4     3 3 3 3 2 2 2 2 1 1
12│8 8 8 6 6 6   4   4 3 3 3 3 3 3 2 2 2 2 1 1
16│7 8 8 6 6 6 4 4 4 4 4 3 3 3 3 3 3 2 2 2 2 1 1
18│7 6 8 6 6 5 4 4 4 4 4 4 3 3 3 3 3 3 2 2 2 2 1 1
20│7 6 6 6 6 5 5 4 4 4 4 4 4 3 3 3 3 3 3 2 2 2 2 1 1
```

5   Analysis of the Algorithm

Define the cost of the algorithm to be the number of read operations on sequence elements.

Theorem 13. *The cost of our algorithm is $O(nk^2)$.*

Proof: The main algorithm reads every position exactly once. For each position i, we have to consider less than k choices for j_0. For each choice, we have to correct less than k positions checking less than $2k$ positions overall.   □

Remark: For $k = 6$ and $\underline{x} = 1\,0110\,0101$ our algorithm only finds 6-SR representations of weight 4, despite the existence of a 6-SR representation of weight 3, $70\,00(2^6 - 1)7$. We can change our algorithm so that it chooses Strategy 2 whenever possible. Note that this may force us to check the entire binary representation of e for every choice of j_0. We get

Theorem 14. *The cost of a Strategy 2–algorithm is $O(n^2k^2)$.*

To see the limitations of this approach, consider the number 1101101011. Here, the modified algorithm would not terminate and we have to add further rules defining when to switch from Strategy 2 to Strategy 1.

Open question: Find an indicator to determine which strategy to choose. An 'increase of weights in a 1-run' is a candidate, e.g.

```
1 0 1 1 0 0 1 0 1
4 3 3 2 2 2 2 1 1
```

Finally, we may decide to pursue both strategies for every corrector. This is guaranteed to give a minimal representation but has exponential complexity.

6 Conclusion

We have proposed a standard representation for k-SR numbers and have shown that all integers have a minimal k-SR representation of this form. Furthermore, we have derived properties which will hold for minimal representations and proposed an algorithm based on these properties. The minimal representation can save more modular multiplications in the 'square and multiply' algorithm than signed-digit representations or canonical k-SR representations [1, 2].

References

1. D. Gollmann, Yongfei Han, and C. Mitchell. Redundant integer representations and fast exponentiation. *to appear in International Journal: Designs, Codes and Cryptography.*
2. D.E. Knuth. *The art of computer programming, Volume 2: seminumerical algorithms.* Addison-Wesley, Reading, Mass., 2nd edition, 1981.
3. R.L. Rivest, A. Shamir, and L. Adleman. A method for obtaining digital signatures and public-key cryptosystems. *Communications of the ACM,* **21**:120–126, 1978.

The Main Conjecture for MDS Codes

J. W. P. Hirschfeld

School of Mathematical Sciences, University of Sussex,
Brighton BN1 9QH, U.K.
Email: jwph@sussex.ac.uk

1 Introduction

What is special about the 10×10 matrix M given below?

$$M = \begin{bmatrix} 1 & 1 & 1 & 1 & 1 & 1 & 1 & 1 & 1 & 1 \\ 1 & 10 & 13 & 5 & 4 & 16 & 11 & 12 & 17 & 2 \\ 10 & 13 & 5 & 4 & 16 & 11 & 12 & 17 & 2 & 7 \\ 13 & 5 & 4 & 16 & 11 & 12 & 17 & 2 & 7 & 8 \\ 5 & 4 & 16 & 11 & 12 & 17 & 2 & 7 & 8 & 3 \\ 4 & 16 & 11 & 12 & 17 & 2 & 7 & 8 & 3 & 15 \\ 16 & 11 & 12 & 17 & 2 & 7 & 8 & 3 & 15 & 14 \\ 11 & 12 & 17 & 2 & 7 & 8 & 3 & 15 & 14 & 6 \\ 12 & 17 & 2 & 7 & 8 & 3 & 15 & 14 & 6 & 9 \\ 17 & 2 & 7 & 8 & 3 & 15 & 14 & 6 & 9 & 18 \end{bmatrix}$$

The answer is that it has the property that *every* minor is non-zero modulo 19; further, there is no matrix of larger size with this property.

The aim of this paper is to express this property also in terms of coding theory, vector space, and projective space for an arbitrary finite field $\mathbf{F}_q$ and to give the most recent results.

The following four notions are equivalent for $n \geq k$:

1. (CODING THEORY) a *maximum distance separable* (MDS) linear code C of length n, dimension k and hence minimum distance $d = n - k + 1$, that is, an $[n, k, n - k + 1]$ code over $\mathbf{F}_q$;
2. (MATRIX THEORY) a $k \times n - k$ matrix A with entries in $\mathbf{F}_q$ such that every minor is non-zero;
3. (VECTOR SPACE) a set K' of n vectors in $V(k, q)$, the vector space of k dimensions over $\mathbf{F}_q$, with any k linearly independent;
4. (PROJECTIVE SPACE) an n–arc in $PG(k - 1, q)$, that is, a set K of n points with at most $k - 1$ in any hyperplane of the projective space of $k - 1$ dimensions over $\mathbf{F}_q$.

To show the equivalence of these four concepts, consider a generator matrix G for such a code C in canonical form:

$$n$$

$$k \quad \begin{bmatrix} 1\,0\ldots0\ a_{11}\ \ldots\ a_{1,n-k} \\ 0\,1\ldots0\ a_{21}\ \ldots\ a_{2,n-k} \\ \vdots\ \vdots \quad \vdots\ \vdots \qquad \vdots \\ 0\,0\ldots1\ a_{k1}\ \ldots\ a_{k,n-k} \end{bmatrix} = G.$$

Since C has minimum distance $n - k + 1$, any linear combination of the rows of G has at most $k - 1$ zeros; that is, considering the columns of G as a set K' of n vectors in $V(k,q)$, any k are linearly independent. Regarding the columns of G as a set K of points of $PG(k-1,q)$ means that no k lie in a hyperplane; equivalently, any k points of K are linearly independent. This, in turn, implies that every minor of A is non-zero.

For given k and q, let $M(k,q)$ be the maximum value of n for such a code. Then

$$M(k,q) = k + 1 \text{ for } q \le k.$$

A suitable set of vectors in $V(k,q)$ is

$$(1,0,\ldots,0),\ (0,1,0,\ldots,0),\ \ldots,(0,\ldots,0,1),(1,1,\ldots,1);$$

that is, for $q \le k$, every element of $V(k,q)$ is a linear combination of at most $k - 1$ of these $k + 1$ vectors.

The Main Conjecture MC_k for MDS Codes, always taking $q > k$, is the following:

$$M(k,q) = \begin{cases} q + 2 \text{ for } k = 3 \text{ and } k = q - 1 \text{ both with } q \text{ even,} \\ q + 1 \text{ in all other cases.} \end{cases}$$

It will be convenient to have the notation $m(k-1,q) = M(k,q)$.

Since the problem involves linear dependence, the projective space setting is more economical than the vector space setting, since the redundant scalars are factored out.

Segre [19] enunciated three problems:

I. For given k and q, what is the maximum value of n such that an n–arc exists in $PG(k-1,q)$? What are the n–arcs corresponding to this value of n?

II. Are there values of k and q with $q > k$ such that every $(q+1)$–arc of $PG(k-1,q)$ is a normal rational curve?

III. For given k and q with $q > k$, what are the values of $n(\le q)$ such that each n–arc is contained in a normal rational curve of $PG(k-1,q)$? In how many such curves is the n–arc contained?

An n–arc is *complete* if it is maximal with respect to inclusion; that is, it is not contained in an $(n+1)$–arc. Implicit in Problem III is Problem IV, which may be enunciated as follows:

IV. What are the values of n for which a complete n–arc exists in $PG(k-1,q)$? In particular, what is the size of the second largest complete arc in $PG(k-1,q)$?

From above, $m(r,q)$ is the maximum size of an arc in $PG(r,q)$; also, let $m'(r,q)$ denote the size of the second largest complete arc in $PG(r,q)$. Then an

n–arc in $PG(r,q)$ with $n > m'(r,q)$ is contained in an $m(r,q)$–arc. This is an important inductive tool.

For other recent surveys, see [28], [12]; for a survey of other similar constants, see [14].

2   Preliminary Results

A *normal rational curve* in $PG(r,q)$ is any subset of $PG(r,q)$ which is projectively equivalent to

$$\{(t^r, t^{r-1}, \ldots, t, 1) \in PG(r,q) \mid t \in GF(q) \cup \{\infty\}\}.$$

For $r = 2$, it is a *conic*; for $r = 3$, it is a *twisted cubic*. There exists a unique normal rational curve through an $(r+3)$–arc in $PG(r,q)$ provided $q \geq r+2$ [11, p.229]. An arc is called *rational* or *classical* if it is a subset of a normal rational curve.

First, in Table 1, the number of all, projectively distinct, complete arcs in $PG(2,q)$ for $q \leq 13$ is given.

For $q \leq 9$, see [9]; for $q = 11$, see [20]; for $q = 13$, see [1], [8], [21], [2]. Then in Table 2, the value of $m'(2,q)$ is given for $7 \leq q \leq 25$. For $q = 17, 19$, see [4]; for $q = 23, 25, 27$, see [5].

The next is a useful result which gives simultaneous information on pairs of dimensions.

Theorem 1. (i) *The dual code of an MDS code is also MDS.*
(ii) *An n–arc exists in $PG(k-1,q)$ if and only if an n–arc exists in $PG(n-k-1,q)$.*

Proof If $G = [I_k \mid A]$ is a generator matrix for the code C, then $H = [-A^* \mid I_{n-k}]$ is a generator matix for the dual code $C^\perp$, where A^* is the transpose of A.

Corollary 2. (i) *A $(q+1)$–arc exists in $PG(k-1,q)$ if and only if a $(q+1)$–arc exists in $PG(q-k,q)$.*
(ii) *A $(q+2)$–arc exists in $PG(k-1,q)$ if and only if a $(q+2)$–arc exists in $PG(q-k+1,q)$.*
(iii) *A $(q+3)$–arc exists in $PG(k-1,q)$ if and only if a $(q+3)$–arc exists in $PG(q-k+2,q)$.*

Theorem 3. ([26], [15, §27.5])
Let $a_{n,r}$ be the number of n–arcs in $PG(r,q)$ and let ν_r be the number of normal rational curves in $PG(r,q)$. Then

$$a_{n,n-2-r}/a_{n,r} = \nu_{n-2-r}/\nu_r.$$

The crucial result for approaching the Main Conjecture is the fact that, in a certain dimension, a $(q+1)$–arc is a normal rational curve.

Theorem 4. (Segre [9, Theorem 8.2.4])
In $PG(2,q)$, q odd, a $(q+1)$–arc is a conic.

Table 1. Complete n–arcs in $PG(2,q)$ for $q \leq 13$

q	n	Number of $PGL(3,q)$–orbits
2	4	1
3	4	1
4	6	1
5	6	1
7	6	2
	8	1
8	6	3
	10	1
9	6	1
	7	1
	8	1
	10	1
11	7	1
	8	9
	9	3
	10	1
	12	1
13	8	2
	9	30
	10	21
	12	1
	14	1

Table 2. The size of the second largest arc in $PG(2,q)$

q	7	8	9	11	13	16	17	19	23	25	27
$m'(2,q)$	6	6	8	10	12	13	14	14	17	21	22

Theorem 5. (Casse and Glynn [15, Theorem 27.6.10])
In $PG(4,q)$, q even, a $(q+1)$-arc is a normal rational curve.

It is elementary that $m(2,q) = M(3,q) = \begin{cases} q+1 \text{ for } q \text{ odd,} \\ q+2 \text{ for } q \text{ even.} \end{cases}$ Theorem 4 characterizes the corresponding sets for odd q. For even q, although a *regular* $(q+2)$-arc is a conic plus its nucleus, the meet of its tangents, the converse is true if and only if $q \leq 8$. A classification of $(q+2)$-arcs has been done for $q \leq 32$; for a survey and other references, see [14].

These two theorems imply a dependency of $m(r,q)$ and $m'(2,q)$ for q odd and of $m(r,q)$ and $m'(4,q)$ for q even. Define the integer functions F and G by the purely arithmetic conditions:

$$q + 1 > m'(2, q) + r - 2 \iff q > F(r);$$
$$q + 1 > m'(4, q) + r - 4 \iff q > G(r).$$

There are two results that produce an inductive argument on dimension.

Theorem 6. ([16]) *Let K be an n-arc in $PG(r, q)$ with $q + 1 \geq n \geq r + 3 \geq 6$ and suppose there exist $P_0, P_1 \in K$ and a hyperplane π containing neither P_0 nor P_1 such that, for $i = 0, 1$, the projection K_i of K onto π is rational in π. Then the arc K is contained in one and only one normal rational curve in $PG(r, q)$.*

Theorem 7. ([16]) *Let K be a $(q + 2)$-arc in $PG(r, q)$ with $q + 1 \geq r + 3 \geq 6$. If a hyperplane π of $PG(r, q)$ contains neither of the points P_0, P_1 of K, then it cannot happen that both projections K_i of K from P_i, $i = 0, 1$, onto π are rational in π. In particular, if every $(q + 1)$-arc in $PG(r - 1, q)$ is rational, then $m(r, q) = q + 1$.*

These two results have as corollaries the following way of approaching solutions to problems I, II, III.

Theorem 8. *In $PG(r, q)$, q odd, $r \geq 3$,*
(i) if K is an n-arc with $n > m'(2, q) + r - 2$, then K lies on a unique normal rational curve;
(ii) if $q > F(r)$, then every $(q + 1)$-arc is a normal rational curve;
(iii) if $q > F(r - 1)$, then $m(r, q) = q + 1$.

Theorem 9. *In $PG(r, q)$, q even, $q > 2$, $r \geq 4$,*
(i) if K is an n-arc with $n > m'(4, q) + r - 4$, then K lies on a unique normal rational curve;
(ii) if $q > G(r)$, then every $(q + 1)$-arc is a normal rational curve;
(iii) if $q > G(r - 1)$, then $m(r, q) = q + 1$.

Now, bounds for $m'(2, q)$ and $m'(4, q)$ are considered. Better upper bounds for $m'(2, q)$ all start from Segre's result connecting an n-arc with an algebraic envelope.

Theorem 10. *The tangents to an n-arc K in $PG(2, q)$ belong to an algebraic envelope Γ of class t or $2t$ according as q is even or odd, where $t = q + 2 - n$.*

Bounds for $m'(2, q)$ depend on estimating the number of $\mathbf{F}_q$-rational elements of Γ. The best upper bounds known for $m'(2, q)$ and $m'(4, q)$ are given in the next two results. These bounds give similar bounds for $m'(r, q)$ in higher dimensions. For a historical survey, see [10].

Theorem 11. (i) $m'(2, q) = q - \sqrt{q} + 1$ for $q = 2^{2e}$, $e > 1$.
(ii) $m'(2, q) \leq q - \frac{1}{4}\sqrt{q} + \frac{25}{16}$ for q odd.
(iii) $m'(2, q) \leq \frac{44}{45}q + \frac{8}{9}$ for q prime, $q > 5$.
(iv) $m'(2, q) \leq q - \frac{1}{2}\sqrt{q} + 5$ for $q = p^h$ with $p \geq 5$.
(v) $m'(2, q) \leq q - \frac{1}{4}\sqrt{pq} + \frac{29}{16}p + 1$ for $q = p^{2e+1}$, p odd, $e \geq 1$.
(vi) $m'(2, q) \leq q - \sqrt{2q} + 2$ for $q = 2^{2e+1}$, $e \geq 1$.

Proof (i) See [9, Theorem 10.3.3] and [6].
(ii) See [27].
(iii) See [29].
(iv) See [13].
(v), (vi) See [30].

Theorem 12. (Storme and Thas [25], [15, Theorem 27.7.22])
If q is even and $q > 2$, then $m'(4, q) \leq q - \frac{1}{2}\sqrt{q} + \frac{13}{4}$.

When Theorems 11 and 12 are applied to Theorems 8 and 9, the following are obtained.

Theorem 13. (Thas [15, Theorem 27.6.4], Voloch [29])
For q odd, $r \geq 3$,
(i) $F(r) \leq (4r - \frac{23}{4})^2$;
(ii) *for q prime, $F(r) \leq 5(9r - 19)$.*

Theorem 14. (Storme and Thas [25], [15, Theorem 27.7.23])
For q even, $q > 2$, $r \geq 4$,
$$G(r) \leq (2r - \tfrac{7}{2})^2.$$

However, it should be noted that the classification of $(q+1)-$arcs for $r \geq 4$ will not produce a simple result.

Theorem 15. (Glynn [7]) *In $PG(4, 9)$, there are precisely two projectively distinct $10-$arcs, a normal rational curve and a non-classical arc.*

3   Results on the Main Conjecture

The summary of results concerning the Main Conjecture are given in the next two theorems. For the convenience of different readers, the results are given both in terms of the vector space and coding dimension k and the projective space dimension $r = k - 1$.

Theorem 16. *For $q = p^h$, p odd, the Main Conjecture holds for the values of k and $r = k - 1$ given in Table 3.*

Corollary 17. *For q odd, the Main Conjecture holds as follows:*
(i) *when $r = 5$ and $k = 6$ for $q \geq 7$ except possibly for*
    $q \in \{29, 31, 37, 41, 43, 47, 49, 53, 59, 61, 67, 71, 73, 79, 81, 83\}$;
(ii) *when $r = 6$ and $k = 7$ for $q \geq 9$ except possibly for $q \in \{29, \ldots, 127\} \cup \{169\}$.*

Corollary 18. *For q odd and $q \leq 27$, the largest size of square matrix over $\mathbf{F}_q$ with every minor non-zero is $\frac{1}{2}(q + 1)$.*

Theorem 19. *For $q = 2^h$, the Main Conjecture holds for the values of k and $r = k - 1$ given in Table 4.*

Table 3. The Main Conjecture for fields of odd order

k	q	r	q
$\leq q$	$q \leq 27$	$\leq q-1$	$q \leq 27$
3	all q	2	all q
4	$q > 3$	3	$q > 3$
5	$q > 5$	4	$q > 5$
$q-1$	$q > 3$	$q-2$	$q > 3$
$q-2$	$q > 3$	$q-3$	$q > 3$
$q-3$	$q > 5$	$q-4$	$q > 5$
$q > (4k - 55/4)^2$		$q > (4r - 39/4)^2$	
$q > 5(9k - 37)$, q prime		$q > 5(9r - 28)$, q prime	
$4\sqrt{pq} - 29p > 16(k-4)$, h odd		$4\sqrt{pq} - 29p > 16(r-3)$, h odd	
$q > 4k^2$, $p \geq 5$		$q > 4(r+1)^2$, $p \geq 5$	
$k > q - (1/4)\sqrt{q} - 23/16$		$r > q - (1/4)\sqrt{q} - 39/16$	
$k > (44/45)q - 19/9$, q prime		$r > (44/45)q - 28/9$, q prime	
$k > q - (1/4)\sqrt{q} + (29/16)p - 2$, h odd		$r > q - (1/4)\sqrt{q} + (29/16)p - 3$, h odd	
$k > q - (1/2)\sqrt{q} + 2$, $p \geq 5$		$r > q - (1/2)\sqrt{q} + 1$, $p \geq 5$	

Table 4. The Main Conjecture for fields of even order

k	q	r	q
$\leq q-2$	$q \leq 16$	$\leq q-3$	$q \leq 16$
3	all q	2	all q
4	$q > 2$	3	$q > 2$
5	$q > 4$	4	$q > 4$
6	$q > 4$	5	$q > 4$
7	$q > 4$	6	$q > 4$
$q-1$	$q > 2$	$q-2$	$q > 2$
$q-2$	$q > 4$	$q-3$	$q > 4$
$q-3$	$q > 4$	$q-4$	$q > 4$
$q-4$	$q > 4$	$q-5$	$q > 4$
$q-5$	$q > 4$	$q-6$	$q > 4$
$q > (2k - 15/2)^2$		$q > (2r - 11/2)^2$	
$k > q - (1/2)\sqrt{q} - 7/4$		$r > q - (1/2)\sqrt{q} - 11/4$	

References

1. A.H. Ali, Classification of arcs in the plane of order 13, Ph.D. thesis, University of Sussex, 1993.
2. A.H. Ali, J.W.P. Hirschfeld and H. Kaneta, The automorphism group of a complete $(q-1)$-arc in $PG(2,q)$, *J. Combin. Designs* **2** (1994), 131-145.
3. A.H. Ali, J.W.P. Hirschfeld and H. Kaneta, On the size of arcs in projective spaces, *IEEE Trans. Inform. Theory*, to appear.
4. J.M. Chao and H. Kaneta, Rational arcs in $PG(r,q)$ for $11 \leq q \leq 19$, preprint.
5. J.M. Chao and H. Kaneta, Rational arcs in $PG(r,q)$ for $23 \leq q \leq 27$, preprint.
6. J.C. Fisher. J.W.P. Hirschfeld and J.A. Thas, Complete arcs in planes of square order, *Combinatorics '84*, Ann. Discrete Math. **30**, North-Holland Mathematics Studies **123**, North-Holland, Amsterdam, 1986, 243-250.
7. D.G. Glynn, The non-classical 10-arcs of $PG(4,9)$, *Discrete Math.* **59** (1986), 43-51.
8. C.E. Gordon, Orbits of arcs in projective spaces, *Finite Geometry and Combinatorics*, London Math. Soc. Lecture Notes **191**, Cambridge University Press, Cambridge, 1993, pp. 161-171.
9. J.W.P. Hirschfeld, *Projective Geometries over Finite Fields*, Oxford University Press, Oxford, 1979.
10. J.W.P. Hirschfeld, Maximum sets in finite projective spaces, *Surveys in Combinatorics*, London Math. Soc. Lecture Note Series **82**, Cambridge University Press, Cambridge, 1983, pp. 55-76.
11. J.W.P. Hirschfeld, *Finite Projective Spaces of Three Dimensions*, Oxford University Press, Oxford, 1985.
12. J.W.P. Hirschfeld, Complete arcs, *Combinatorics '94*, to appear.
13. J.W.P. Hirschfeld and G. Korchmáros, On the embedding of an arc into a conic in a finite plane, *Finite Fields Appl.*, to appear.
14. J.W.P. Hirschfeld and L. Storme, The packing problem in statistics, coding theory and finite projective spaces, *J. Statist. Plann. Inference*, submitted.
15. J.W.P. Hirschfeld and J.A. Thas, *General Galois Geometries*, Oxford University Press, Oxford, 1991.
16. H. Kaneta and T. Maruta, An elementary proof and an extension of Thas' theorem on k-arcs, *Math. Proc. Cambridge. Philos. Soc.* **105** (1989), 459-462.
17. F.J. MacWilliams and N.J.A. Sloane, *The Theory of Error-Correcting Codes*, North-Holland, Amsterdam, 1977.
18. T. Maruta and H. Kaneta, On the uniqueness of $(q+1)$-arcs of $PG(5,q)$, $q = 2^h$, $h \geq 4$, *Math. Proc. Cambridge Philos. Soc.* **110** (1991), 91-94.
19. B. Segre, Curve razionali normali e k-archi negli spazi finiti, *Ann. Mat. Pura Appl.* **39** (1955), 357-379.
20. A.R. Sadeh, The classification of k-arcs and cubic surfaces with twenty-seven lines over the field of eleven elments, Ph.D. thesis, University of Sussex, 1984.
21. M. Scipioni, Sugli archi completi nei piani desarguesiani, Tesi di laurea, University of Rome "La Sapienza", 1990.
22. L. Storme, Completeness of normal rational curves, *J. Algebraic Combin.* **1** (1992), 197-202.
23. L. Storme and J.A. Thas, Generalized Reed-Solomon codes and normal rational curves: an improvement of results by Seroussi and Roth, *Advances in Finite Geometries and Designs*, Oxford University Press, 1991, 369-389.
24. L. Storme and J.A. Thas, Complete k-arcs in $PG(n,q)$, q even, *Discrete Math.* **106/107** (1992), 455-469.

25. L. Storme and J.A. Thas, M.D.S. codes and arcs in $PG(n, q)$ with q even: an improvement of the bounds of Bruen, Thas, and Blokhuis, *J. Combin. Theory Ser. A* **62** (1993), 139-154.

26. J.A. Thas, Connection between the Grassmannian $G_{k-1;n}$ and the set of k–arcs of the Galois space $S_{n,q}$, *Rend. Mat.* **2** (1969), 121-134.

27. J.A. Thas, Complete arcs and algebraic curves in $PG(2, q)$, *J. Algebra* **106** (1987), 451-464.

28. J.A. Thas, M.D.S. codes and arcs in projective spaces, *Le Matematiche* **47** (1992), 315-328.

29. J.F. Voloch, Arcs in projective planes over prime fields, *J. Geom.* **38** (1990), 198-200.

30. J.F. Voloch, Complete arcs in Galois planes of non-square order, *Advances in Finite Geometries and Designs*, Oxford University Press, Oxford, 1991, pp. 401-406.

Some Decoding Applications of Minimal Realization*

Graham Norton

Centre for Communications Research, University of Bristol, England.

Abstract. We show that minimal realization (MR) of a finite sequence and the associated MR algorithm [10] provide new solutions to a number of decoding problems: BCH and Reed–Solomon codes, errors and erasures, classical Goppa codes and negacyclic codes. We concentrate on the MR of the DFT of an error polynomial, thus avoiding the "key equation" and Forney's procedure. We also discuss simplification of the theory in characteristic two and an extension of the MR theory to several sequences, obtaining a new *simultaneous MR* algorithm.

1 Introduction

In [10] we gave a conceptually simple theory of *minimal realization* (MR) of a finite sequence over a domain R; the corresponding Algorithm MR ([10, Algorithm 4.6]) is useful in Mathematical Systems Theory (solving e.g. the Partial Realization Problem over R and linear systems with polynomial entries) and Algebraic Combinatorics (finding e.g. a complete path weight enumerator).

In this paper we give a number of decoding applications of MR, further strengthening connections between Coding and Mathematical Systems Theory. Our method is to reduce decoding to solving an MR problem in $R((X^{-1}))$, the domain of Laurent series in X^{-1} over R. This is the natural home for *negatively indexed* sequences, $R[X]$ and its action on these sequences. This action is closely related to $R[X]$ as standard submodule of $R((X^{-1}))$ (see Proposition 3). We use the exponential valuation on $R((X^{-1}))$, which extends the degree function on $R[X]$. This underlying coherence (which does not obtain in $R[[X]]$) completely avoids using the reciprocal of a polynomial, the order of a series etc. and simplifies many definitions, theorems and their proofs. See also [9].

An abbreviated version ([10, Algorithm 3.19]) of Algorithm MR computes a minimal polynomial (MP) of a sequence; we call this Algorithm MP and assume that over a field, both algorithms produce a *monic* MP. It was shown in [10, Propositions 3.23, 4.8] that Algorithms MP and MR require at most $L(3L+1)/2$ and $L(5L + 1)/2$ R-multiplications respectively for a sequence of length L, so that we have given up very little computational efficiency for conceptual clarity.

The MR algorithm differs from the Berlekamp–Massey algorithm [1, 6] in a number of ways: (i) it is division–free and valid over any domain R (ii) it computes the error locator *and* evaluator polynomials (iii) it computes an MP

---

* Research supported by U.K. Science and Engineering Research Grant GR/H15141.

of a sequence rather than the length and connection polynomial of a shortest shift–register generating the sequence (iv) it does not require that the initial term be non–zero (v) the initialization is canonical (vi) it requires $L \geq 1$ terms.

Note that in addition to the applications mentioned in the Abstract, we may also decode beyond the designed distance by computing with Algorithm MR in $R = F_q[S_u, \ldots, S_{u'}]$, where $S_u, \ldots, S_{u'}$ denote unknown syndromes.

An outline of our approach to decoding a Reed–Solomon code and to finding a minimal polynomial of several sequences was presented in [8].

2 Minimal Realization

We let A denote a commutative ring with 1, R a domain with 1 and F a field. The letters f, g denote $f, g \in A[X]$ and δf is the degree of f, with $\delta 0 = -\infty$. For $G \in A((X^{-1})) \setminus \{0\}$, the *support* of G is $\text{Supp}(G) = \{a \in \mathbb{Z} : G_a \neq 0\}$, where G_a denotes a coefficient of G; $\text{Supp}(0) = \emptyset$. We also write $[G]_a$ for a coefficient of G. Thus for $G \in A((X^{-1})) \setminus \{0\}$, $\delta G = \max \text{Supp}(G)$. If d is an integer, $[d]$ denotes the residue class $d \bmod 2$ and for $G \in A((X^{-1}))$, $G^{[0]}$ denotes the even part of G. The odd part of G is $G^{[1]} = G - G^{[0]}$.

We denote by $S^1(A)$ the set of A–sequences indexed by $\{0, -1, \ldots\}$; the letter s always denotes $s \in S^1(A)$. The *generating function* of s is $\Gamma(s) = \sum_{i \leq 0} s_i X^i \in A[[X^{-1}]]$.

Definition 1. For $\delta f \geq 1$, we call $\beta(f, s)(X) = \sum_{1 \leq i \leq \delta f} (f\Gamma(s))_i X^i$ the *border polynomial* of f and s.

Some elementary properties of the border polynomial are in [10, Prop. 2.3].

Proposition 2. If $\delta f \geq 1$ then $\beta(f, s) = \sum_{i=0}^{\delta f - 1} s_{-i} \, \nu^{(i+1)}(f)$ where ν denotes Newton's divided difference operator.

We make $S^1(A)$ into an $A[X]$–module as follows: if $f = \sum_{i=0}^{\delta f} f_i X^i$ then $(f \circ s)_j = \sum_{i=0}^{\delta f} f_i s_{j-i}$ where $j \leq 0$. Recall that $Ann(s) = \{f : f \circ s = 0\}$ is the *annihilator ideal* of s, and s is a a *linear recurring sequence (lrs)* if there is an $f \in Ann(s) \setminus \{0\}$. If s is an lrs over $\dot{F}$, then $Ann(s)$ has a unique monic generator.

The following obvious but important identity is the *raison d'être* for $\circ$:

Proposition 3. In $A((X^{-1}))$, $f\Gamma(s) = \beta(f, s) + \Gamma(f \circ s)$.

We use $(s|L)$ to denote a sequence of L elements of A, with $L \geq 1$, indexed by $0, -1, \ldots, -L+1$ and $\Gamma(s|L) = \sum_{i=-L+1}^{0} s_i X^i$.

Definition 4. We say that (f, g) *is a realization of* $(s|L)$ or that (f, g) *realizes* $(s|L)$ if $f, g \in A[X], f \neq 0, \text{Supp}(g) \subseteq [1, \delta f]$ and $\delta(f\Gamma(s|L) - g) \leq -L + \delta f$.

For $f \in F[X] \setminus \{0\}$, (f, g) realizes $(s|L)$ iff there is an $h \in F[X]$ such that $\delta h \leq \delta f - 1$ and the order of $f^* \Gamma(s)(X) - h$ is at least L, where $f^* = X^{\delta f} f(X^{-1})$.

If $\delta f \geq L$, any $(f, \beta(f, s))$ realizes $(s|L)$, so we concentrate on realizations (f, g) with $\delta f \leq L - 1$. It also follows from Proposition 3 that if s is an lrs and $f \in Ann(s)$, then $(f, \beta(f, s))$ realizes $(s|L)$ for any $L \geq 1$.

We set $Ann(s|L) = \{f : (f \circ s)_i = 0 \text{ for } -L+1+\delta f \leq i \leq 0\}$, which vacuously contains all polynomials of degree at least L, and $\Delta(s|L)(f) = (f \circ s)_{-L+\delta f}$ is the *discrepancy* of f.

Definition 5. A realization (f, g) of $(s|L)$ is called *minimal* if $\delta f = \min\{\delta f' : (f', g') \text{ realizes } (s|L)\}$ and the minimal value δf is called the *complexity* $\kappa(s|L)$. We let $Min(s|L)$ denote the set of $MP's$ of $(s|L)$.

See [10, Definitions 3.12, 3.16] for the *antecedent* $\alpha = \alpha(s|L)$ of $\kappa(s|L)$.

3 Decoding BCH and Reed–Solomon Codes

We use a variant of the usual approach, avoiding reciprocal polynomials and roots. Let α be an nth root of unity generating F^*. We identify $(e_0, \ldots, e_{n-1}) \in F^n$ and $\sum_{i=0}^{n-1} e_i X^i \in F[X]$. We let ϕ denote a discrete Fourier transform.

Definition 6. Let $b \geq 0$ and $\phi_b : F^n \to F[[X^{-1}]]$ be the F–linear map given by $\phi_b(v) = \sum_{j \leq 0} v(\alpha^{b-j}) X^j$.

Proposition 7. *If* $\omega(X) = \sum_{i \in Supp(v)} v_i \, \alpha^{bi} \prod_{j \in Supp(v), \, j \neq i}(X - \alpha^j)$ *and* $\sigma(X) = \prod_{i \in Supp(v)}(X - \alpha^i)$, *then* $\phi_b(v) = X\omega(X)/\sigma(X)$ *where* $\gcd(\sigma, \omega) = 1$ *and* σ *is the monic* MP *of* $\phi_b(v)$.

Recall that the *formal derivative* of σ is $\sigma' = \sum_{i \in Supp(v)} \prod_{j \in Supp(v), \, j \neq i}(X - \alpha^j)$. Note that if $b = 0$ and each $v_i = 1$, then $\omega = \sigma'$.

Proposition 8. *We can recover v from $\phi_b(v)$ via (i) $Supp(v) = \{i \mid \sigma(\alpha^i) = 0\}$ and (ii) for $i \in Supp(v)$, $v_i = \omega(\alpha^i)/(\alpha^{ib}\sigma'(\alpha^i))$.*

Unique factorization and [10, Corollary 3.24] imply

Proposition 9. *If* $\Gamma(s) = \beta/\mu = \eta/\nu$ *and* μ, ν *are* $MP's$, *then* $\mu = c\nu$ *and* $\beta = c\eta$ *for some* $c \in F^*$.

Suppose now that $\gcd(n, q) = 1$ and that C is an (n, k) cyclic code over $GF(q)$ generated by $g \in GF(q)[X]$. We suppose in addition that the roots of g are α^b, $\alpha^{b+1}, \ldots, \alpha^{b+2t-1}$ for some $b \geq 0$ and $t \geq 1$. Thus C is a t-error correcting code. Let a transmitted codeword $c \in C$ be received as $r = c + e$, where $e \neq 0$. Proposition 3.4 and [10, Corollary 3.27 or Theorem 4.16] yield:

Proposition 10. *If there are at most t errors, then* $(\sigma, X\omega) = (\mu(\phi_b(e)|2t), \beta)$.

Algorithm 11 *(Decoding a BCH code with zeroes $\alpha^b, \ldots, \alpha^{b+2t-1}$, $b \geq 0$.)*
    Input: Received r.
    Output: If at most t errors occur, the nearest codeword c.
    1. Compute the syndromes $S_i = r(\alpha^{b-i})$, $-2t + 1 \leq i \leq 0$.
    2. If some $S_i \neq 0$, compute $\mu(S|2t)$ using Algorithm MP.
    3. Find e using Proposition 8(i).
    4. Return $r - e$.

Example 1. For the 3-error correcting $(15,3)$ BCH code of [7, Example 8.4, p.181], $r =$ (111 000 110 011 110) and $(S|6) = \alpha^7, \alpha^{14}, \alpha^{11}, \alpha^{13}, 1, \alpha^7$.
    Algorithm MP yields $\mu(S|6) = X^3 + \alpha^7 X^2 + \alpha^4 X + \alpha^6$, with roots α^3, α^6 and α^{12}. Thus $e_i = 1$ iff $i = 3, 6, 12$ and the decoder returns (111 100 010 011 010).

Algorithm 12 *(Decoding an RS code with zeroes $\alpha^b, \ldots, \alpha^{b+2t-1}$, $b \geq 0$.)*
    Input: Received vector r.
    Output: If at most t errors occur, the nearest codeword c.
    1. Compute the syndromes $S_i = r(\alpha^{b-i})$, $-2t + 1 \leq i \leq 0$.
    2. If some $S_i \neq 0$, compute $(\mu(S|2t), \beta)$ using Algorithm MR.
    3. Find e using Proposition 8.
    4. Return $r - e$.

Example 2. For the 2-error correcting $(7,3)$ RS code over $GF(8)$ of [7, Example 8.5, p182], $(\alpha^3, \alpha, 1, \alpha^2, 0, \alpha^3, 1)$ is received, and $(S|4) = \alpha^3, \alpha^4, \alpha^4, 0$.
    Algorithm MR yields $(\sigma, X\omega_1) = (X^2 + \alpha^5 X + \alpha^5, \ \alpha^3 X^2 + \alpha^2 X)$, and σ has roots α^2, α^3. Thus $e_2 = \alpha^3$, $e_3 = \alpha^6$ and $e_i = 0$ otherwise, so that the decoder returns $(\alpha^3, \alpha, \alpha, 1, 0, \alpha^3, 1)$.

4   Decoding Errors and Erasures

It is worth emphasizing that Proposition 7 does not require that there be at most t errors, which has implications for erasures–only decoding.
    Suppose that there are U errors and K erasures. Let σ_U be the *(Unknown)* error locator polynomial and let σ_K be the *(Known)* erasure locator polynomial, where if $K = 0$, we set $\sigma_K = 1$. Thus $U = \delta\sigma_U$, $K = \delta\sigma_K$ and $\sigma = \sigma_U \sigma_K$.
    In the simple erasures–only case, we need only compute the $2t$ syndromes $r(\alpha^b), \ldots, r(\alpha^{b+2t-1})$, and not the MP. Thus $1 \leq K \leq 2t$ is permissible.

Proposition 13. *Let $U = 0$ and $K \leq 2t$ so that $\sigma = \sigma_K$. If $i \in Supp(e)$ and $a = \alpha^i$ then $\sigma'_K(a) \neq 0$ and $e_i = \omega(a)/(a^b \sigma'_K(a))$ where $\omega = \sum_{i=0}^{K-1} [\sigma_K S]_{i+1} X^i$.*

Of course ω may also be computed as in Proposition 2.

Theorem 14. *If $(S_U)_i = [\sigma_K S]_i$ for $i \leq 0$, then $Ann(S_U) = (\sigma_U)$. If there are U errors, K erasures and $2 \leq 2U \leq 2t - K$, then $\sigma_U = \mu(S_U|2t - K)$ and $X\omega = \beta(\sigma_K \sigma_U, S)$.*

It follows that, once we know the roots of σ_U, we can compute the erasure and error *values* using Proposition 8. We recover errors–only decoding as the special case $\sigma_K = 1$ and $K = 0$.

Example 3. Consider the (15,9) Reed–Solomon code over GF(16) of Example 1, p 322 of [12], which has $g(X) = \prod_{i=1}^{6}(X - \alpha^i)$, so that $b = 1$ and $t = 3$. The error e has $e_3 = \alpha^7$, $e_7 = \alpha^2$ (the erasure symbol) and $e_{10} = \alpha^{11}$. Then $(S|6) = 1, \alpha^{13}, \alpha^{14}, \alpha^{11}, \alpha, 0$ and $\sigma_K = X + \alpha^7$, so that $(S_U|4) = \alpha^5, \alpha^{12}, \alpha, \alpha^9$.

Algorithm MP yields $\sigma_U = X^2 + \alpha^{12}X + \alpha^{13}$, which has roots α^3 and α^{10}. Consequently $\sigma = (x + \alpha^7)\sigma_U = X^3 + \alpha^2 X^2 + \alpha^{11}X + \alpha^5$, and $\omega_1 = X^2 + \alpha^{14}X + \alpha^5$. This gives non–zero error values $e_3 = \alpha^7, e_7 = \alpha^2$ and $e_{10} = \alpha^{11}$.

Finally, for $1 \leq i \leq 2t - K$, $\Delta(S_U|i)(f) = \Delta(S|i + K)(\sigma_K f)$ and we may also compute $\sigma = \sigma_K \sigma_U$ by initializing μ to σ_K in Algorithm MP, and iterating from $K + 1$ to $2t$. Let M_1 be the naive upper bound for the number of F–multiplications using Theorem 14 and M_2 using the second method. We obtain $M_1 - M_2 < 2t(2 - K) - K$, which suggests that the second method may not always be more efficient, *cf.* [12].

5 Decoding Classical Goppa Codes

We present a simplification of [11] using the key equation derived in [2].

Proposition 15. *Let* $g, S \in A[X]$, g *monic and* $0 \leq \delta S \leq \delta g - 1$. *Put* $T = XS/g$. *Then (i)* $\sigma S \equiv \omega \bmod g$, $\delta\omega \leq \delta\sigma - 1 \iff (\sigma, \beta) \in MR(T|\delta g)$ *and* $\omega = \sigma S - g\beta/X$ *(ii)* $T_0 = S_{\delta g - 1}$ *and for* $i \leq -1$, $T_i = [(X^{-i}S) \bmod g]_{\delta g - 1}$.

Definition 16. *For* $-1 \leq i \leq \delta g$, *let* $\omega_i = S\mu_i - g\beta_i/X$, *where* μ_i *and* β_i *are computed using Algorithm* MR *applied to* $(XS/g \,|\delta g)$.

An analogue of [10, Theorem 4.5] for the ω_i now yields

Algorithm 17 *(Solve classical key equation)*
  Input: $g, S \in R[X]$, $0 \leq \delta S \leq \delta g - 1$, g *monic, and* $\kappa(XS/g \,|\delta g) \leq \delta g/2$.
  Output: σ, ω *with* $\omega = S\sigma \bmod g$, $\delta\omega \leq \delta\sigma - 1 \leq (\delta g - 1)/2$ *and* $\delta\sigma$ *minimal.*

Compute $T_j = [(X^{-j}S) \bmod g]_{\delta g - 1}$ *for* $-\delta g + 1 \leq j \leq 0$;

$\sigma_\alpha := 0; \omega_\alpha := g; \Delta_\alpha := 1$;
$\sigma := 1; \ \omega := S; \ m := -1$;

for $j := 1$ *to* δg *do begin*
    $\Delta := \Delta(T|j - 1)(\sigma)$;
    if $\Delta \neq 0$ *then if* $m \geq 0$ *then begin*
$$\sigma := \Delta_\alpha \, \sigma - \Delta X^m \, \sigma_\alpha;$$
$$\omega := \Delta_\alpha \, \omega - \Delta X^m \, \omega_\alpha; \ end;$$

$$else \ \ begin$$
$$temp := \sigma;$$
$$\sigma := \Delta_\alpha X^{-m}\sigma \ - \ \Delta \ \sigma_\alpha;$$
$$\sigma_\alpha := temp;$$

$$temp := \omega;$$
$$\omega := \Delta_\alpha X^{-m}\omega \ - \ \Delta \ \omega_\alpha;$$
$$\omega_\alpha := temp;$$

$$\Delta_\alpha := \Delta; m := -m; \ end;$$

$$m := m - 1;$$
$$end;$$
$$return \ (\sigma, \omega).$$

The upper bound of $7\delta g(\delta g + 3)/2$ on the number of multiplications in the 2nd part of the preceding algorithm can probably be lowered. This algorithm also applies to decoding irreducible binary Goppa codes as in [11, Section V].

6 Simplifications in Characteristic Two

It is well–known that Berlekamp's algorithm can be simplified for decoding binary BCH codes. Our analogue is Theorem 21 below, which we derive using several preliminary results. Throughout this section, $char(F) = 2$.

Proposition 18. If $m \geq 1$ is odd, then $X^m \ f^{[d]} = (X^m \ f)^{[1+d]}$.

Lemma 19. (i) (cf. [1, Lemma 7.61]) If $\theta(1 + X^{-1}\Gamma) = 1$, then $\theta\Gamma = X\theta^{[1]}$ and $\theta = 1 + \theta^{[1]}$
(ii) If $f \in Ann(s|L)$, then $\Delta_L(f) = (\theta^{[1]}\{X \ f + \beta(f, s)\})_{-L+\delta f}$
(iii) If L is odd and $(X\mu_L + \beta_L)^{[\delta\mu_L]} = 0$, then $\Delta_L(f) = 0$ i.e. $\mu_{L+1} = \mu_L$.

It only remains to determine inductively when $(X\mu_L + \beta_L)^{[\delta\mu_L]} = 0$:

Lemma 20. Let L be odd, $\Delta_{L-1} \neq 0$ and $\alpha = \alpha(L-1)$. Then $(\beta_L + X\mu_L)^{[d]} = 0$ if either (i) $2\delta_{L-1} < L$ and $(\beta_\alpha + X\mu_\alpha)^{[d]} = (\beta_{L-1} + X\mu_{L-1})^{[1+d]} = 0$ or
(ii) $2\delta_{L-1} > L$ and $(\beta_\alpha + X\mu_\alpha)^{[1+d]} = (\beta_{L-1} + X\mu_{L-1})^{[d]} = 0$.

Combining these results, we obtain:

Theorem 21. Let L be odd and let $(s|L)$ be a finite sequence satisfying $s_0 \neq 0$ and $s_{2k-1} = s_k^2$ for $(-L+1)/2 \leq k \leq 0$. Then for odd i, $3 \leq i \leq L$, $\mu(s|i+1) = \mu(s|i)$ and $\beta(\mu(s|i+1), s) = \beta(\mu(s|i), s)$.

We note that Theorem 21 fails if $s_0 = 0$, as the following example shows:

Example 4. Let $(s|6) = 0, 0, 1, 0, 1, 1$. We obtain $\mu_1 = \mu_2 = 1$, $\mu_3 = \mu_4 = X^3$, but $\mu_5 = X^3 + 1$ and $\mu_6 = X^3 + X + 1$. The reader may verify that the connection polynomials of [6] also differ for $(s|6)$ at $L = 5$ and $L = 6$.

One possible way to circumvent the dependance of Theorem 21 on s_0 is to work with $(1, s|2t + 1)$, as is done in [1]. We have verified the following for $1 \le t \le 5$:

Conjecture 22 *If $s_{2k-1} = s_k^2$ for $k \le 0$ and $\mu = \mu(s|2t)$, $\mu^+ = \mu(1, s|2t + 1)$ are computed using Algorithm MR, then either (i) $\delta\mu^+ = \delta\mu$ and $\mu^+ = \mu$ or (ii) $\delta\mu^+ > \delta\mu$ and $\mu^+ = X \, \mu$.*

We could then combine Theorem 21 and Algorithm 11 to decode a BCH code, working with $(1, S|2t + 1)$, skipping even iterations after the second one and adding the test *"If $\mu_0^+ \ne 0$ then $\sigma = \mu^+$ else $\sigma = \mu^+/X$"*, since $\sigma_0 \ne 0$.

7 Decoding Negacyclic Codes

Algorithm MR gives an alternative approach to establishing a designed distance for negacyclic codes, as well as a decoding method. This section is based on [1, Chapter 10].

Definition 23. Let p be an odd prime and $n \not\equiv 0 \bmod p$. A negacyclic code of length n is an ideal of $GF(p)[X]/(X^n + 1)$.

If α denotes a primitive $2n^{th}$ root of 1, then roots of $X^n + 1$ are odd powers of α. We only know the *odd* syndromes and each $e_i = 1$, so that $\phi_0(e) = X\sigma'/\sigma$.

Lemma 24. *Let $S^{[1]}$ include the syndromes corresponding to $\alpha, \alpha^3, \ldots, \alpha^{2t-1}$, where $2t-1 < p$ and $T(X^{-2}) = \int \left(-X^{-1}T(1 + T) + (T^2 - X^{-2}(T + 1)^2)S^{[1]} \right) dX$. If $MR(T|t) = (\mu, \beta)$, then $\sigma^{[1]} = -\beta(X^2)/X$ and $\sigma^{[0]} = \mu(X^2) + \beta(X^2)/X^2$.*

(For example, the first two terms are S_{-1} and $(3S_{-1}^2 - S_{-1}^3 + S_{-3})/3$.) The lemma now implies:

Theorem 25. *([1, Theorem 9.4]) If the roots of the generator polynomial of a negacyclic code over $GF(p)$ include $\alpha, \alpha^3, \ldots, \alpha^{2t-1}$, where $2t - 1 < p$, then that negacyclic code has designed Lee–distance t.*

Example 5. (cf. [1, Problem 9.2]) 3 is a primitive 30th root of unity and so by Theorem 25, the negacyclic code of length 15 over GF(31) with generator polynomial $(X - 3)(X - 3^3)(X - 3^5)(X - 3^7)$ is 4–error–correcting. If $r = (20, 20, 11, 5, 16, 9, 14, 22, 19, 23, 5, 17, 1, 6, 15)$ and at most 4 errors occur, we can correct them as follows: $(S^{[1]}|4) = 14, 30, 6, 11$ and $(T|4) = 14, 25, 1, 3$.
We obtain $MR(T|4) = (X^2 + 5X + 22, X(14X + 2))$ so that $\sigma^{[1]} = 17X^3 + 29X$, $\sigma^{[0]} = X^4 + 19X^2 + 24$, and $\sigma = X^4 + 17X^3 + 19X^2 + 29X + 24 = (X - 3^3)(X + 3^7)(X + 3^9)^2$, yielding $e = (0,0,0,1,0,0,0,-1,0,-2,0,0,0,0,0)$, as expected.

8   Simultaneous Minimal Realization

We simplify and extend the "Fundamental Iterative Algorithm" [4, 5] and "Massey's Conjectured Algorithm" [3] using [3, Appendix] and [10]. See also [10, Sect. 3.3].

Suppose we have M sequences $s^{(i)}$, each of length L, where $M, L \geq 1$. Let $[s|L]$ be the $M \times L$ matrix with $[s|L]_{i,j} = s^{(i)}_{-j+1}$ where $1 \leq i \leq M$ and $1 \leq j \leq L$. Thus we may think of $[s|L]$ as being the array of M sequences. We set $Ann[s|L] = \bigcap_{i=1}^{M} Ann(s^{(i)}|L)$ and $Min[s|L] = \bigcap_{i=1}^{M} Min(s^{(i)}|L)$.

Definition 26. Let $f_0 = 1$, $f_j \in F[X] \setminus \{0\}, \delta f_j \leq j$ for $1 \leq j \leq L$. For $c \in F^L$, we define

$$[f_L, f_0, \dots, f_{L-1}]_c = X^{\delta_c - \delta f_L} f_L - \sum_{j=0}^{L-1} c_j X^{\delta_c - L + j - \delta f_j} f_j$$

where $\delta_c = \max\{\delta f_L, \max\{L - j + \delta f_j : 0 \leq j \leq L - 1, c_j \neq 0\}\}$.

One checks that $\delta[f_L, f_0, \dots, f_{L-1}]_c = \delta_c \leq L$. For $0 \leq j \leq L$, let the jth *discrepancy column vector* Δ_j be given by $(\Delta_j)_i = (f_j \circ s^{(i)})_{-j+\delta f_j}$ for $1 \leq i \leq M$.

A straightforward verification yields

Proposition 27. *Let* $f_0 = 1$ *and* $f_j \in Ann[s|j] \setminus Ann[s|j + 1]$, $\delta f_j \leq j$ *for* $1 \leq j \leq L$. *If* $\Delta_L = (\Delta_0 \Delta_1 \dots \Delta_{L-1})c^T$ *for some* $c \in F^L$, *then*
(i) $f_{L+1} = [f_L, f_0, \dots, f_{L-1}]_c \in Ann[s|L]$
(ii) $\beta(f_{L+1}, s^{(i)}) = X^{\delta_c - \delta f_L} \beta(f_L, s^{(i)}) - \sum_{j=0}^{L-1} c_j X^{\delta_c - L + j - \delta f_j} \beta(f_j, s^{(i)})$.

The next two results are our analogues of [3, Theorems A.1–A.6]:

Theorem 28. *Let* $f_0 = 1$, $f_j \in F[X] \setminus \{0\}$, $\delta f_j \leq j$ *for* $1 \leq j \leq L$, *with* f_L *monic. If* $f \in Ann[s|L + 1]$ *is monic and* $\delta f \leq L$, *then for some* $c \in F^L$
(i) $\Delta_L = (\Delta_0, \Delta_1, \dots, \Delta_{L-1})c^T$
(ii) $f = X^{\delta f - \delta_c}[f_L, f_0, \dots, f_{L-1}]_c$. *Also, if* $f_L \in Min[s|L]$, *then* $\delta f \geq \delta_c$.

Corollary 29. *Let* $f_0 = 1$, $f_j \in Ann[s|j] \setminus Ann[s|j + 1]$, $\delta f_j \leq j$ *for* $1 \leq j \leq L$, *and* $f_L \in Min[s|L]$. *Then*
(i) $\kappa[s|L + 1] = L + 1$ *iff the linear system* $\Delta_L = (\Delta_0, \Delta_1, \dots, \Delta_{L-1})c^T$ *has no solution* $c \in F^L$
(ii) if $\kappa[s|L+1] \leq L$, *then* $\kappa[s|L+1] = \min\{\delta_{c'} : \Delta_L = (\Delta_0, \Delta_1, \dots, \Delta_{L-1})c'^T\}$.
Thus if further $\delta_c = \kappa[s|L + 1]$, *then* $[f_L, f_0, \dots, f_{L-1}]_c \in Min[s|L + 1]$.

Example 6. (cf. [5, Example, p.1276]): $F = GF(2)$, $s^{(1)} = (0, 0, 1, 1, 0, 0)$, $s^{(2)} = (1, 0, 0, 0, 0, 0)$, $s^{(3)} = (0, 0, 0, 0, 0, 0)$, $s^{(4)} = (0, 1, 0, 1, 0, 1)$, $s^{(5)} = (0, 0, 0, 1, 1, 0)$, $s^{(6)} = (0, 0, 1, 0, 1, 0)$.

We obtain $\Delta_0^T = (0, 1, 0, 0, 0, 0)$ and $\mu_1 = X$, $\Delta_1^T = (0, 0, 0, 1, 0, 0)$ and $\mu_2 = X^2$, $\Delta_2^T = (1, 0, 0, 0, 0, 1)$ and $\mu_3 = X^3$, $\Delta_3^T = (1, 0, 0, 1, 1, 0)$ and $\mu_4 = X^4$, $\Delta_4^T = (0, 0, 0, 0, 1, 1) = \Delta_1^T + \Delta_2^T + \Delta_3^T$, $\kappa[s|5] = 4$ and $\mu_5 = X^4 + (X + X^2 + X^3)$. Finally, $\Delta_5 = 0$ and so $\mu_6 = \mu_5$. (We note that in [5], only the first five terms of the sequence are used i.e. $L = 5$ and the shift–register obtained has *connection polynomial* $1 + X + X^2 + X^3$ and length 4, as expected.)

Combining the results of this section, we obtain:

Algorithm 30 *(SMR — Simultaneous minimal realization)*
   Input: Sequences $s^{(i)}$, each of length L, where $L \geq 1$ and $1 \leq i \leq M$.
   Output: $(\mu_L, \beta_L^{(i)})$, where $\mu_L \in Min[s|L]$, $\beta_L^{(i)} = \beta(\mu_L, s^{(i)})$ and $1 \leq i \leq M$.

$\mu_0 := 1; \beta_0^{(i)} := 0;$
for $j := 0$ to $L - 1$ do begin
   Compute Δ_j;
   If $\Delta_j = 0$ then begin $\mu_{j+1} := \mu_j$;
         $\beta(\mu_{j+1}, s^{(i)}) := \beta(\mu_j, s^{(i)})$; *end;*
      else begin Solve $(\Delta_0 \ldots \Delta_{j-1})c^T = \Delta_j$;
      if no solution then begin $\mu_{j+1} = X^{j+1}$;
                  $\beta_{j+1}^{(i)} = \sum_{k=0}^{j} s_{-k}^{(i)} X^{j+1-k}$;*end;*
      else begin Find $\delta_c = \min\{\delta_{c'} : c'$ is a solution$\}$;
      $\mu_{j+1} := [\mu_j, \mu_0, \ldots, \mu_{j-1}]c$;
      $\beta_{j+1}^{(i)} := X^{\delta_c - \delta\mu_j} \beta_j^{(i)} - \sum_{k=0}^{j-1} c_k X^{\delta_c - j + k - \delta\mu_k} \beta_k^{(i)}$;
      end;

            end;

   end;
return $(\mu_L, \beta_L^{(i)})$.

Acknowledgements. The author gratefully acknowledges financial support from the UK Science and Engineering Research Council under grants GR/H15141, GR/K27728. Some of the above algorithms were implemented in MAPLE by A.Au. Thanks to members of Projet Codes, INRIA for some useful references and to Tim Blackmore for a careful reading.

References

1. Berlekamp, E.R. (1968). *Algebraic Coding Theory*. Mc–Graw Hill, New York.
2. Berlekamp, E.R. (1973) Goppa Codes. *IEEE Trans. Information Theory* **19**, 590–592.
3. Ding, C., Xiao, G., Shan, W. (1991). The stability theory of stream ciphers. *Springer Lecture Notes in Computer Science* **561**.
4. Feng, G.L., Rao, T.R.N. (1993) Decoding algebraic–geometric codes up to designed minimum distance. *IEEE Trans. Information Theory* **39**, 37–45.
5. Feng, G.L. and Tzeng, K.K. (1991). A generalization of the Berlekamp–Massey algorithm for multisequence shift register sequence synthesis with applications to decoding cyclic codes. *IEEE Trans. Information Theory* **37**, 1274 – 1287.
6. Massey, J.L. (1969). Shift register synthesis and BCH decoding. *IEEE Trans. Information Theory* **15**, 122–127.
7. McEliece, R. (1977). *The Theory of Information and Coding (Encyclopedia of Mathematics and its Applications 3)*. Addison–Wesley, Reading, Mass.
8. Norton, G.H. (1994). On the minimal realizations of a finite sequence. (Extended Abstract). Eurocodes '94 (Côte d'Or, October 1994), 203–208.

9. Norton, G.H. (1995). On n–dimensional sequences. I. *J. Symbolic Computation.* In press.

10. Norton, G.H. (1995). On the minimal realizations of a finite sequence. *J. Symbolic Computation.* In press.

11. Patterson, N.J.(1975). The algebraic decoding of Goppa codes. *IEEE Trans. IT* **21**, 203–207.

12. Truong, T.K., Eastman, W.L., Reed, I.S., Hsu, I.S. (1988). Simplified procedure for correcting both errors and erasures of Reed–Solomon code using Euclidean algorithm. *IEE Proceedings* **135**, 318–324.

The Synthesis of Perfect Sequences

P. Z. Fan and M. Darnell

Dept of Electronic & Electrical Engineering
Leeds University, Leeds, LS2 9JT, UK, Fax: 0113 233 2032
Email: p.fan@ieee.org, miked@elec-eng.leeds.ac.uk

Abstract. Perfect sequences find application in many areas including synchronisation techniques, channel estimation, fast start-up equalization, pulse compression radars and CDMA systems. This paper will first discuss the necessary and sufficient condition for, and some useful properties of, perfect sequences. Then, a comprehensive description of various perfect sequences is given. The emphasis will be on the synthesis of different perfect sequences, including two-valued perfect sequences, ternary perfect sequences, polyphase perfect sequences and modulatable perfect sequences. The perfect array and other related topics are also discussed briefly.

1 Introduction

The ultimate goal in periodic sequence design is the sequence set $\{a_n^{(r)}\}$ of period N satisfying the ideal periodic correlation requirements

$$\text{(a) ACF: } R_r(\tau) = \sum_{n=0}^{N-1} a_n^{(r)} a_{n+\tau}^{(r)*} = \begin{cases} E, & \tau = 0 \\ 0, & \tau \neq 0 \end{cases} ; \tag{1}$$

$$\text{(b) CCF: } R_{r,s}(\tau) = \sum_{n=0}^{N-1} a_n^{(r)} a_{n+\tau}^{(s)*} = 0, \quad \text{for all } \tau, \ r \neq s. \tag{2}$$

There are several cases that should be identified:

1. For binary sequences with elements ± 1, it is almost certain that only one sequence with the ideal ACF exists, i.e. $\{a_n\} = (+1, +1, +1, -1)$. For longer periods, the best ACF that can be achieved for binary sequences, as is the case with m-sequences or Lengendre sequences, is $|R_r(\tau \neq 0)| = 1$.
2. For the non-binary case, it is possible to find sequences with ideal ACFs. These sequences are called *perfect sequences* and their synthesis will be considered in detail in this paper.
3. For both binary and non-binary cases, it is possible to synthesize sequence sets with ideal CCFs, although it is not possible for each of the sequences to have perfect ACF at the same time.
4. It is impossible to find set of sequences with both ideal periodic ACF and CCF properties. The best sequence set that can be achieved is the set of perfect sequences with the following crosscorrelation properties: $R_{rs}(\tau) = \sqrt{N}, \ \forall \tau, \ r \neq s$.

Perfect sequences find applications in many areas, including synchronisation techniques, channel estimation, fast start-up equalization, pulse compression radars and CDMA systems [1, 2, 3, 4, 5, 6]. In this paper, we will first consider the necessary and sufficient condition for a perfect sequence, and give some useful properties of perfect sequences. Then, a comprehensive description of various perfect sequences is given.

2 Perfect Sequences and Their Properties

Let us first consider the necessary and sufficient condition for a sequence $\{a_n\}$ to be perfect. For any sequence $\{a_n\}$ of period N, its discrete Fourier transform will also generate a discrete periodic spectrum $\{F_k\}$ of period N, where $F_k = \sum_{n=0}^{N-1} a_n e^{-i\frac{2\pi nk}{N}}$, $0 \le k < N$. Note the relationship between an ACF and its Fourier transform: $R(\tau) \longleftrightarrow F_k^* F_k = |F_k|^2$. If the sequence is perfect, that is $R(\tau) = E$ for $\tau = 0$ and $R(\tau) = 0$ otherwise, then

$$R(\tau) \longleftrightarrow \sum_{\tau=0}^{N-1} R(\tau)e^{i\frac{2\pi \tau k}{N}} = R(0)e^{i\frac{2\pi k 0}{N}} = E \qquad (3)$$

or $|F_k| = \sqrt{R(0)} = \sqrt{E}$; thus a sequence $\{a_n\}$ is a perfect sequence if and only if all components of a sequence $\{F_k\}$ have the same magnitude $|F_k| = \sqrt{E}$.

According to Sarwate's inequality [9], for any sequence set of period N and size M,

$$\frac{R_{cm}^2}{N} + \frac{N-1}{N(M-1)}\frac{R_{am}^2}{N} \ge 1 \qquad (4)$$

where R_{am} and R_{cm} are the maximum magnitudes of the periodic ACF and CCF respectively. When the sequence set is perfect, i.e. $R_{am} = 0$, it follows that the lower bound for the maximum magnitude of the periodic CCF is equal to $\sqrt{N}$.

For most applications, the perfect sequences whose element amplitudes are not constant should possess a good energy efficiency η defined by

$$\eta = \frac{\sum_{n=0}^{N-1} |a_n|^2}{N \max\{|a_n|^2\}} = \frac{E}{N \max\{|a_n|^2\}} \qquad (5)$$

If sequence $a = \{a_n\}$ of period N_1 and sequence $b = \{b_n\}$ of period N_2 are perfect and $gcd(N_1, N_2) = 1$, then by repeating the sequence $\{a_n\}$ N_1 times and sequence $\{b_n\}$ N_2 times, and multiplying them together digit-by-digit, we obtain a product sequence $c = a \cdot b = \{a_n \cdot b_n\}$ of period $N = N_1 N_2$ which is also a perfect sequence.

If $\{a_n\}$ is a polyphase perfect sequence, i.e., $a_n = e^{i\frac{2\pi}{N}f(n)} = \alpha^{f(n)}$, where α is a primitive Nth root of unity, then so are

1. $\{a_{m\pm n}\}$, where m is any integer and the subscript is expressed modulo N;
2. $\{ca_n\}$, where c is any complex constant;
3. $\{a_n\beta^{nm}\}$, where m is any integer and β is an qth root of 1;
4. $\{a_n^*\}$, where a_n^* denotes complex conjugation;
5. $\{F(k)\}$, the discrete Fourier transform of a_n.

3   Two-Valued Perfect Sequences

Although binary perfect sequences would be particularly useful in practice, as stated previously only one such binary sequence exists, i.e. $(+1+1+1-1)$. If the sequence elements are non-binary real or complex, then there are many perfect sequences that can be synthesized, as is shown by Golomb and Lüke [10, 2]. Fortunately, it is possible to transmit a non-perfect binary sequence and still achieve essentially the same correlation properties as obtained by transmitting a perfect binary sequence by using a so-called mismatched filter [1] and a perfect real- valued sequence at the receiver.

It can be shown [10] that any integer difference set (v, k, λ) corresponds to a binary pseudo-random sequence $\{a_n\}$ of period $N = v$ whose ACF is two-valued, i.e.

$$R(\tau) = \begin{cases} v, & \tau = 0 \\ v - 4(k - \lambda), & \tau \neq 0 \end{cases} \tag{6}$$

In general, the out-of-phase value of $R(\tau) = v - 4(k - \lambda)$, $\tau \neq 0$, will not be zero for sequences longer than $N = v = 4$. However, if we replace $a_n = -1$ by a suitable real or complex number β, it is possible to obtain a perfect sequence with $R(\tau) = 0$, $\tau \neq 0$.

Let us consider a general two-valued complex sequence of period $N = v$ by setting $a_n = \alpha$ for $n \in D$ and $a_n = \beta$ for $n \notin D$, where $D = (d_1, d_2, \cdots, d_k)$ is any integer difference set (v, k, λ). In each period of the sequence, it can be proved that

$$R(\tau) = \begin{cases} k|\alpha|^2 + (v - k)|\beta|^2, & \tau = 0 \\ \lambda|\alpha|^2 + (v - 2k + \lambda)|\beta|^2 + (k - \lambda)(\alpha\beta^* + \alpha^*\beta), & \tau \neq 0 \end{cases} \tag{7}$$

In order to qualify as a perfect sequence, the out-of-phase values must be zero, i.e. $R(\tau) = 0$, $\tau \neq 0$. Two cases of special interest have been discussed by Golomb [10]: α and β are real, and α and β lie on the unit circle in the complex plane. For example, given an integer difference set $(v, k, \lambda) = (13, 4, 1)$ and $(v, k, \lambda) = (7, 4, 2)$ the following two-valued perfect sequences can be obtained $\{a_n\} = (1, 1, -0.79, 1, -0.79, -0.79, -0.79, -0.79, -0.79, 1, -0.79, -0.79, -0.79)$, $\{a_n\} = (1, e^{i\,2.41886}, e^{i\,2.41886}, 1, 1, 1, e^{i\,2.41886})$.

4   Ternary Perfect Sequences

For any ternary perfect sequence $\{a_n\}$ with elements in $\{0, +1, -1\}$, if there are m_+ +1s, m_- −1s and m_0 0s, giving a total number of digits $N = m_+ + m_- + m_0$, then the following relationship can be shown to hold: $(m_+ - m_-)^2 = m_+ + m_-$. Perfect ternary sequences have been studied by Godfrey [5], Chang [11], Moharir [12], Ipatov [13, 14], Shed and Sarwate [15], Hoholdt and Justesen [16].

Based on non-binary m-sequences or non-binary vector m-sequences of period $q^m - 1$, Ipatov [13, 14] derived a large class of ternary perfect sequences of period $N = \frac{q^m - 1}{q - 1}$, where m is an odd number and $q = p^s$, p is an odd prime and s is an integer. Let $\{b_n\}$ be an m-sequence generated by the primitive polynomial $h(x)$. Because any nonzero elements of $GF(q)$ can be represented as powers of a primitive element α; then $b_n = \alpha^u$ if $b_n \neq 0$ because the sequence elements b_n are also elements of $GF(q)$, u is an integer. If we define

$$c_n = \begin{cases} 0, & \text{if } b_n = 0 \\ (-1)^u, & \text{if } b_n = \alpha^u \end{cases} \tag{8}$$

then it can be shown that the ternary sequence $a_n = (-1)^n c_n$ is perfect and has period $N = \frac{q^m - 1}{q - 1}$. Its ACF and energy efficiency are given respectively by $R_a(\tau \neq) = q^{m-1}$ and $\eta = \frac{q^m - q^{m-1}}{q^m - 1}$.

Let $\{b_n\}$ and $\{c_n\}$ be two sequences of the same period N with two-valued ACFs, i.e.

$$R_b(\tau) = \begin{cases} A_b, & \tau = 0 \\ B_b, & \tau \neq 0 \end{cases} \qquad R_c(\tau) = \begin{cases} A_c, & \tau = 0 \\ B_c, & \tau \neq 0 \end{cases} \tag{9}$$

where A_b, B_b, A_c and B_c are real. Let $a_n = R_{b,c}(n)$, according to Sarwate identity [15], we have

$$R_a(\tau) = \begin{cases} A_b A_c + (N-1) B_b B_c, & \tau = 0 \\ A_b B_c + A_c B_b + (N-2) B_b B_c, & \tau \neq 0 \end{cases} \tag{10}$$

It is clear that if $\{b_n\}$ and $\{c_n\}$ are perfect sequences, i.e., $B_b = B_c = 0$, then $\{a_n\}$ is also a perfect sequence. Further, it can be shown that, given any sequence with two-valued ACF, it is easy to find a number δ such that adding δ to each element of the sequence produces a perfect sequence. Based on this idea, Hoholdt and Justesen [16] derived a class of ternary perfect sequences of period $N = \frac{q^{2r+1} - 1}{q - 1}$ by using two difference sets with parameters $(v, k, \lambda) = \left(\frac{q^{2r+1} - 1}{q - 1}, \frac{q^{2r} - 1}{q - 1}, \frac{q^{2r-1} - 1}{q - 1} \right)$, $q = 2^s$. For example, a ternary perfect sequence $(1, 1, 1, 1, 1, -1, 1, 0, 1, 0, -1, 1, 1, -1, 0, 0, 1, -1, 0, -1, -1)$ can be derived in this way.

The ternary sequences described by Godfrey [5] and Chang [11] are special cases of Ipatov sequences [13] for $q = 3$. The sequences discussed by Shed and Sarwate [15] are special cases of Hoholdt and Justesen sequences.

Instead of ternary perfect sequences with elements 0 and ± 1, it is also useful to be able to synthesize perfect sequences with three-valued real or complex elements. Lüke [2], Bömer and Antweiler [17] investigated general three-valued perfect sequences. Lüke gives a list of multi-valued perfect sequences with best energy efficiencies by using various methods, including computer search [2]. Fan and Darnell also investigated perfect Gaussian integer sequences [18, 19].

5 Polyphase Perfect Sequences

Frank sequences $F = \{a^{(1)}, \cdots, a^{(r)}, \cdots, a^{(q-1)}\}$ are a class of polyphase sequences of length $N = q^2$, in which the qth roots of unity are the elements of the sequence $a^{(r)} = (a_0^{(r)}, a_1^{(r)}, \cdots, a_{N-1}^{(r)})$, i.e.

$$a_n^{(r)} = a_{jq+k}^{(r)} = e^{\frac{i2\pi}{q}rkj} = \alpha^{kj}, \quad 0 \le k, j < q; \tag{11}$$

where $\alpha = e^{\frac{i2\pi}{q}}$, $gcd(r,q) = 1$, $0 \le n \le q^2 - 1$ and q is any integer. It can be shown that Frank sequences are perfect sequences, i.e. $R_{a^{(r)}}(\tau) = 0, \tau \ne 0 \ (mod \ N)$. If N is odd and $gcd(r-s, N) = 1$, where $gcd(r, N) = 1, gcd(s, N) = 1$, then its CCFs are optimal, $R_{a^{(r)}, a^{(s)}}(\tau) = \sqrt{N}, \forall \tau, r \ne s$.

Frank sequences were first published in 1961 by Heimiller [20]. In his paper, Heimiller described polyphase sequences of length $N = q^2$, where p is a prime number. Later, Frank showed that he had obtained the same sequences more than 9 years earlier, but without the restriction that q be prime [21, 22, 23].

Besides the well-known Frank sequences, Chu sequences [24] also have ideal periodic ACFs and optimal CCFs. For *Chu sequences* of length N, $C = \{a^{(1)}, \cdots, a^{(r)}, \cdots, a^{(N-1)}\}$, the elements of the sequence $a^{(r)} = (a_0^{(r)}, a_1^{(r)}, \cdots, a_{N-1}^{(r)})$ are given by

$$a_n^{(r)} = \begin{cases} e^{\frac{i\pi}{N}r(n+1)n+mn}, & N \text{ odd}, \\ e^{\frac{i\pi}{N}rn^2+mn}, & N \text{ even}, \end{cases} \quad 0 \le n < N; \quad (r, N) = 1. \tag{12}$$

where m is any integer.

Note that Chu sequences exist for every integer $N > 1$, rather than solely at lengths which are perfect squares, as is the case with Frank sequences. One year after Chu published his perfect sequences, Frank remarked that the same sequences had been discovered by Zadoff many years previously [25]; so Chu sequences are also termed as Zadoff-Chu sequences in some references [26].

In 1980, Alltop proposed a class of perfect quadric phase sequences [27] for odd N

$$a_n^{(r)} = \alpha^{rn^2}, \quad \alpha = e^{i\frac{2\pi}{N}}, \ 0 \le n < N \tag{13}$$

He noted that the quadric phase sequences are similar to Chu sequences. In fact the quadric phase sequences are equivalent to Chu sequences of odd length N, i.e.

$$a_n^{(r)} = \left(\alpha^2\right)^{(n^2+n-n)/2} = \left(\alpha^2\right)^{n(n+1)/2} \alpha^{-n}, \quad \alpha = e^{i\frac{2\pi}{N}} \tag{14}$$

Note that when N is odd, 2 is relatively prime to N; in other words, α^2 is a primitive Nth root of unity. Hence the left-hand side of Eqn 13 corresponds to Chu sequences of odd N.

Ipatov sequences [28] are defined by

$$a_n^{(r)} = \alpha^{rn^2+mk}, \quad \alpha = e^{i\frac{2\pi}{N}}, \ 0 \leq n < N \tag{15}$$

which are obviously the same as quadric phase sequences. The quadric phase sequences can also be derived from generalized bent function sequences [29, 30].

In 1982, Lewis and Kretschmer proposed two classes of polyphase sequences, P3 and P4 sequences [31, 7]. For any positive integer N, P3 and P4 sequences are defined as

$$\text{P3: } a_n = e^{i\frac{\pi}{N}n^2}, \quad \text{P4: } a_n = e^{i\frac{\pi}{N}n^2+\pi n}, \quad 0 \leq n < N \tag{16}$$

It is simple to show that Lewis-Kretschmer sequences are also equivalent to Chu sequences. In fact, P3 codes can be obtained from the definition of Chu sequences for even N. For N even, the P4 sequences can be obtained by $m = N/2$ in Eqn 12; for N odd, the P4 sequences can be obtained by substituting $m = (N-1)/2$ in Eqn 12 [32].

By introducing a new "window autocorrelation", Golomb defined another similar class of perfect polyphase sequences [33] for arbitrary integer N

$$a_n^{(r)} = \alpha^{rk(k-1)/2}, \quad \alpha = e^{i\frac{\pi}{N}}, \ 1 \leq n < N, \ (r, N) = 1. \tag{17}$$

When N is odd, Golomb sequences are the same as Chu sequences; when N is even, the period of Golomb sequences is defined as $2N$ and a window size of N is used in the correlation calculation, which gives the same perfect ACF properties.

Although the correlation properties of Frank and Chu sequences are very good, it is noted that the number of Frank sequences and Chu sequences available for a given length L is relatively small if large families of sequences with good ACFs and small CCF values are required. To meet this requirement, sets of combined Frank/Chu sequences, which contain a larger number of sequences than either of the two constituent sets, are considered [34]. It is shown analytically that the CCFs are similar to those of the original sets, with one exception, whilst the ACFs remain perfectly impulsive.

In connection with his study of channel estimation and fast start-up equalization, Milewski proposed a class of new polyphase sequences with a greater number of phases than Frank sequences but fewer phases than Chu sequences [4]. For any positive integer $q > 1$ and $m \geq 1$, a q^{m+1}-phase Milewski sequence of period q^{2m+1} can be constructed. Let $\{b_n^{(r)}\}$ be a Chu sequence of period q, then we can form a *Milewski sequence* $\{a_n\} = \{a_{jq^m+k}\}$ of period $N = q^{2m+1}$ by

$$c_n^{(r)} = c_{jq^m+k}^{(r)} = b_{j(mod\ q)}^{(r)}\alpha^{rjk}, \quad \alpha = e^{2\pi i/q^{m+1}} \tag{18}$$

where $j = 0, 1, \cdots, q^{m+1} - 1; k = 0, 1, \cdots, q^m - 1$, and $gcd(r, q) = 1$.

Recently, Gabidulin proposed another class of perfect polyphase sequences of period $N = p^{2m+1}$, p prime [35],

$$a_n = b_v \beta^{uc} \alpha^{vc}, \quad 0 \le n < N, \quad \alpha = e^{i\frac{2\pi}{p^{m+1}}}, \quad \beta = e^{i\frac{2\pi}{p^m}} \tag{19}$$

where $n = up^{m+1} + vp^m + c$, $0 \le u < p^m$, $0 \le v < p$, $0 \le c < p^m$, and b_v is a perfect sequence of length p. However it can be shown that Gabidulin sequences are equivalent to Milewski sequences if we let b_v be as for Chu sequences of length p; note, however, that there is no restriction on p for Milewski sequences. Let $v' = (pu + v)$; then the Gabidulin sequences become

$$a_n = b_v \alpha^{(pu+v)c} = b_{v'(\text{mod } p)} \alpha^{v'c}, \quad 0 \le n < N, \quad \alpha = e^{i\frac{2\pi}{p^{m+1}}} \tag{20}$$

which has the form of Milewski sequences, where $n = v'p^m + c$, $0 \le v' < p^{m+1}$, $0 \le c < p^m$.

In the study of generalized bent function sequences, Kumar et al pointed out that their out-of-phase ACF is identically zero [30, 29]. A generalized bent function can be described as a function $f(Z)$ which maps the vector space V_s of q-ary s-tuples into V_1 of integers modulo q, where the Fourier transform coefficients $F(\Lambda)$ of function $f(Z)$ are all of unit magnitude. When $s = 1$, the generalized bent function $f(n)$ is a mapping from integer $n \in Z_q$ to integer $f(n) \in Z_q$. If q is an integer which is neither the product of distinct primes nor equal to 2 mod 4, then it can be proved that the function $f(n)$ defined by

$$f(n+1) = f(n) + b_n, \quad b_n \in Z_q, \quad f(0) \text{ is arbitrary modulo } q \tag{21}$$

is bent if the integers b_n satisfy the dual conditions $\sum_{n=0}^{r-1} b_n = 0$ and $b_{n+mr} = b_n + cmr \pmod{q}$, where $gcd(c, q) = 1$ and r is any integer greater than one that has the same parity as q and whose square divides q (i.e. $q = kr^2$). In other words, the polyphase sequence $\{a_n\}$ defined by $a_n = \alpha^{f(n)}$ is a class of generalized bent function sequence. It is interesting to note that this class of perfect sequences includes Frank sequences and a subset of Chu sequences as special cases.

Recently Mow [36] proposed a unified construction of perfect polyphase sequences which includes all the above classes of perfect polyphase sequences as special cases. At the same time, Gabidulin [37] obtained a full description of one-dimensional bent function sequences.

6 Modulatable Perfect Sequences

In 1988, Suehiro and Hatori presented a generalized class of Frank sequences with the same ideal periodic ACFs and optimum periodic CCFs [8]. The generalized Frank sequences can be considered as modulated sequences obtained by modulating one of the corresponding original Frank sequences with complex numbers of absolute value 1. Because of the fact that the ACF/CCF properties are not changed by this modulation process, these generalized Frank sequences are called modulatable sequences. The modulating string of complex numbers within the modulatable sequences can be used for information transmission in direct sequence spread-spectrum communications, ensuring at the same time the optimum conditions for the code synchronisation during information transmission.

Let $\{a_n\}$ be a Frank sequence of period $N = q^2$, q is any positive integer; then the modulatable Frank sequences are defined as

$$s_n^{(r)} = b_k a_{jq+k}^{(r)} = b_k e^{\frac{i2\pi}{q}rkj}, \quad 0 \le k, \, j < q; \quad (r, q) = 1 \tag{22}$$

where b_k, $0 \le k < q$, are arbitrary complex numbers with absolute values of 1. Let $\{d_k\} = (d_0, d_1, \cdots, d_{q-1})$ be the information to be carried, where $d_k = 0, 1, \cdots, Q - 1$; then a valid choice of b_k is given by $b_k = e^{i\frac{2\pi}{Q}d_k}$.

The sequences modulated from the same original Frank sequence are defined as forming a class of modulated Frank sequences which have the following properties: (1) each class includes an infinite number of modulated Frank sequences of period $N = q^2$, all of them being perfect sequences; (2) the absolute value of the CCFs between any two sequences in different classes is constant at $\sqrt{N}$; (3) the CCFs between any two sequences in the same class is 0 for every time shift that is not a multiple of q.

Later, Gabidulin proposed independently a similar generalization of Frank sequences which is a special case of modulatable Frank sequences for $q = p^k$, p being an odd prime [35].

Chu sequences can be generalized to obtain a quite general class of perfect sequences, called modulatable Chu sequences or *generalized chirp-like sequences* [26]. Let $\{a_n\}$ be a Chu sequence of period $N = sm^2$, where m and s are any positive integers; let $\{b_n\}$ be any sequence of m complex numbers having absolute values equal to 1. The modulatable Chu sequences $\{s_n\}$ are defined as

$$s_n^{(r)} = b_{n(\bmod\ m)}a_n^{(r)} = \begin{cases} b_{n(\bmod\ m)}e^{\frac{i\pi}{N}r(n+1)n}, & N \text{ odd}, \\ b_{n(\bmod\ m)}e^{\frac{i\pi}{N}rn^2}, & N \text{ even}, \end{cases} \quad 0 \le n < N; \ (r, N) = 1.$$
$$\tag{23}$$

Let $\{d_n\} = (d_0, d_1, \cdots, d_{m-1})$ be the information to be carried, where $d_k = 0, 1, \cdots, Q - 1$; then one can also choose $b_n = e^{i\frac{2\pi}{Q}d_k}$. It is interested to note that the modulatable Chu sequences include modulatable Frank sequences and Milewski sequences as subclasses [38, 32].

7 Perfect Arrays

The concept of perfect sequences can readily be extended to perfect arrays which have found applications in coded aperture imaging, time-frequency- coding, built-in test of VLSI-circuits, etc. [2]. Let $s(n, m)$ be a periodic rectangular array of complex numbers with dimensions $N \times M$. Its periodic repetition is with the periods N in n-direction and M in m- direction. An array is denoted perfect if one period of its periodic ACF is given by

$$R(\tau,\zeta) = \sum_{n=0}^{N-1} \sum_{m=0}^{M-1} s(n,m)s^*(n+\tau,m+\zeta) = \begin{cases} E, & (\tau,\zeta) = (0,0) \\ 0, & (\tau,\zeta) \neq (0,0) \end{cases} \qquad (24)$$

where $E = \sum_{n=0}^{N-1} \sum_{m=0}^{M-1} s^2(n,m)$, $0 \leq \tau < N$, $0 \leq \zeta M$.

It can be shown that a perfect binary array with element $s(m,n) \in \{1,-1\}$ can only exist for an area $N \times M$ which is an even square number. Perfect binary arrays of size $2 \times 2, 4 \times 4, 2 \times 8, 6 \times 6, 8 \times 8, 4 \times 16, 3 \times 12, 6 \times 24, 12 \times 12$ and so on, have been constructed by Calabro, Wolf, Lüke, Bömer, Antweiler and other workers [39, 2, 40]. Methods of constructing perfect ternary arrays and perfect polyphase arrays have been considered by Lüke, Bömer and Antweiler [41, 42].

8 Concluding Remarks

Over the past 30 years, numerous constructions of various perfect sequences and arrays have been proposed. It is anticipated that many more perfect sequences and arrays will be constructed, especially for generalised real and complex perfect sequences and arrays. In the above, we have attempted to unify the different sequence types where possible; it has been shown that some of the perfect sequences can be derived from the others due to their equivalent properties. An important unsolved problem in the theory of perfect sequences is finding non-equivalent sequences or finding the dimension of a set of perfect sequences. For perfect phase-shift keyed (PSK) sequences which include polyphase sequences as a subset, Gabidulin [43, 37] proved that there are only finitely many non-equivalent PSK sequences of prime length and obtained the full classification of perfect PSK sequences for lengths which are powers of prime. The general classification of perfect PSK sequences of arbitrary length still remains unknown.

It should be noted that some of the perfect sequences also possess very favourable non-periodic ACFs, as is the case with Frank sequences and Chu sequences [44, 45]. Compared with the synthesis of perfect sequences, the derivation of non-periodic ACFs/CCFs of perfect sequences is much more difficult.

References

1. H. Rohling and W. Plagge, "Mismatched-filter design for periodical binary phased signals", *IEEE Trans. on Aerospace and Electronic Systems*, vol. AES-25, no. 6, pp. 890–897, November 1989.

2. H. D. Lüke, "Sequences and arrays with perfect periodic correlation", *IEEE Trans. on Aerospace and Electron. Systems*, vol. AES-24, no. 3, pp. 287–294, May 1988.

3. S. U. H. Qureshi, "Fast start-up equalization with periodic training sequences", *IEEE Trans. on Information Theory*, vol. IT - 23, pp. 553–563, 1977.

4. A. Milewski, "Periodic sequences with optimal properties for channel estimation and fast start-up equalization", *IBM J. RES. DEVELOP.*, vol. 27, no. 5, pp. 426–431, Sept. 1983.

5. K. R. Godfrey, "Three-level m sequences", *Electron. Lett.*, vol. 2, no. 7, pp. 241–243, July 1966.
6. N. Levanon and A. Freedman, "Periodic ambiguity function of CW signals with perfect periodic autocorrelation", *IEEE Trans. on Aerospace and Electronic Systems*, vol. AES-28, no. 2, pp. 387–395, April 1992.
7. F. F. Kretschmer Jr. and K. Gerlach, "Low sidelobe radar waveforms derived from orthogonal matrices", *IEEE Trans. on AES*, vol. AES-27, no. 1, pp. 92–101, Jan. 1991.
8. N. Suehiro and M. Hatori, "Modulatable orthogonal sequences and their application to SSMA systems", *IEEE Trans. on Information Theory*, vol. IT - 34, pp. 93–100, Jan. 1988.
9. D. V. Sarwate, "Bounds on crosscorrelation and autocorrelation of sequences", *IEEE Transactions on Information Theory*, vol. 25, pp. 720–724, 1979.
10. S. W. Golomb, "Two-valued sequences with perfect periodic autocorrelation", *IEEE Trans. on Aerospace and Electronic Systems*, vol. AES-28, no. 2, pp. 383–386, April 1992.
11. J. A. Chang, "Ternary sequence with zero correlation", *Proceedings of the IEEE*, vol. 55, no. 7, pp. 1211–1213, July 1967.
12. P. S. Moharir, "Generalized PN sequences", *IEEE Trans. on Inform. Theory*, vol. IT-23, no. 6, pp. 782–784, Nov. 1977.
13. V. P. Ipatov, "Ternary sequences with ideal autocorrelation properties", *Radio Eng. Electron. Phys.*, vol. 24, pp. 75–79, October 1979.
14. V. P. Ipatov, "Contribution to the theory of sequences with perfect periodic autocorrelation properties", *Radio Eng. Electron. Phys.*, vol. 25, pp. 31–34, April 1980.
15. D. A. Shedd and D. V. Sarwate, "Construction of sequences with good correlation properties", *IEEE Transactions on Information Theory*, vol. 25, no. 1, pp. 94–97, January 1979.
16. T. Hoholdt and J. Justesen, "Ternary sequences with perfect periodic autocorrelation", *IEEE Trans. on IT*, vol. 29, no. 4, pp. 597–600, July 1983.
17. L. Bomer and M. Antweiler, "New perfect threelevel and threephase sequences", in *IEEE Int. Symp. Inform. Theory, Budapest, Hungary*, June 24-28 1991, p. 280.
18. P. Z. Fan and M. Darnell, "Maximal length sequences over Gaussian integers", *Electron. Lett.*, vol. 30, no. 16, pp. 1286–1287, August 1994.
19. M. Darnell, P. Z. Fan, and F. Jin, "New classes of perfect sequences derived from m- sequences", *1995 IEEE International Symposium on Information Theory, Whistler, Canada*, Sept 17-22 1994.
20. R. C. Heimiller, "Phase shift pulse codes with good periodic correlation properties", *IRE Trans. on IT*, vol. IT-7, pp. 254–257, 1961.
21. R. L. Frank and S. A. Zadoff, "Phase shift pulse codes with good periodic correlation properties", *IRE Trans. on Information Theory*, vol. IT - 8, pp. 381–382, October 1962.
22. R. L. Frank, "Polyphase codes with good nonperiodic correlation properties", *IEEE Trans. on Information Theory*, vol. IT - 9, pp. 43–45, January 1963.
23. R. C. Heimiller, "Author's comment", *IRE Trans. on IT*, vol. IT-8, pp. 382, Oct. 1962.
24. D. C. Chu, "Polyphase codes with good periodic correlation properties", *IEEE Trans. on Information Theory*, vol. IT - 18, pp. 531–533, July 1972.
25. R. L. Frank, "Comments on Polyphase codes with good correlation properties", *IEEE Trans. on Information Theory*, vol. IT - 19, pp. 244, March 1973.

26. B. M. Popović, "Generalized chirp-like polyphase sequences with optimum correlation properties", *IEEE Trans. on Information Theory*, vol. IT - 38, no. 4, pp. 1406–1409, July 1992.

27. W. O. Alltop, "Complex sequences with low periodic correlations", *IEEE Trans. Inform. Theory*, vol. IT-26, no. 3, pp. 350–354, May 1980.

28. V. P. Ipatov, "Multiphase sequences spectrums", *Izvestiya VUZ. Radioelektronika (Radioelectronics and Communications Systems)*, vol. 22, no. 9, pp. 80–82, 1979.

29. H. Chung and P. V. Kumar, "A new general construction for generalized bent functions", *IEEE Trans. on Information Theory*, vol. IT - 35, no. 1, pp. 206–209, Jan. 1989.

30. P.V. Kumar, R. A. Scholtz, and L. R. Welch, "Generalized bent functions and their properties", *J. Combinat. Theory, series A*, vol. 40, no. 1, pp. 90- -107, 1985.

31. B. L. Lewis and F. F. Kretschmer, "Linear frequency modulation derived polyphase pulse compression", *IEEE Trans. on AES*, vol. AES-18, no. 5, pp. 637–641, Sept. 1982.

32. B. M. Popović, "Efficient matched filter for the generalized chirp-like polyphase sequences", *IEEE Trans. on Aerospace and Electronic Systems*, vol. 30, no. 3, pp. 769–777, July 1994.

33. N. Zhang and S. W. Golomb, "Polyphase sequence with low autocorrelations", *IEEE Trans. on Information Theory*, vol. IT - 39, pp. 1085–1089, May 1993.

34. P. Z. Fan, M. Darnell, and B. Honary, "Crosscorrelations of frank sequences and chu sequences", *Electron. Lett.*, vol. 30, no. 6, pp. 477–478, March 1994.

35. E. M. Gabidulin, "Non-binary sequences with the perfect periodic auto- correlation and with optimal periodic cross-correlation", in *Proc. IEEE Int. Symp. Inform. Theory*, San Antonio, USA, January 1993, p. 412.

36. W. H. Mow, "A unified construction of perfect polyphase sequences", in *Proc. IEEE Int. Symp. Inform. Theory*, Whistler, B.C. Canada, Sept 17-22 1995.

37. E. M. Gabidulin, "Partial classification of sequences with perfect auto- correlation and bent functions", in *Proc. IEEE Int. Symp. Inform. Theory*, Whistler, B.C. Canada, Sept 17-22 1995.

38. B. M. Popović, "GCL polyphase sequences with minimum alphabets", *Electron. Lett.*, vol. 30, no. 2, pp. 106–107, January 1994.

39. D. Calabro and J.K. Wolf, "On the synthesis of two-dimensional arrays with desirable correlation properties", *Information and Control*, vol. 11, pp. 537–560, 1968.

40. L. Börner and M. Antweiler, "Two-dimensional perfect binary arrays with 64 elements", *IEEE Trans. on IT*, vol. IT-36, pp. 411–414, 1990.

41. H. D. Lüke, "Perfect ternary arrays", *IEEE Trans. on Information Theory*, vol. IT-36, no. 3, pp. 696–705, May 1990.

42. L. Börner and M. Antweiler, "Perfect N-phase sequences and arrays", *IEEE Journal on Selected Areas in Communications*, vol. 10, pp. 782–789, 1992.

43. E. M. Gabidulin, "There are only finitely many perfect auto-correlation polyphase sequences of prime length", in *Proc. IEEE Int. Symp. Inform. Theory*, Trondheim, Norway, June 27-July 1 1994, p. 282.

44. P. Z. Fan and M. Darnell, "Aperiodic autocorrelation of Frank sequences", *IEE Proceedings on Communications*, vol. 142, no. 4, pp. 210–215, 1995.

45. E. M. Gabidulin, P. Z. Fan, and M. Darnell, "On the autocorrelation of Golomb sequences", *IEE Proceedings on Communications*, vol. 143, no. 1, 1996.

Computation of Low-Weight Parity Checks for Correlation Attacks on Stream Ciphers

W T Penzhorn and G J Kühn

Department of Electrical and Electronic Engineering
University of Pretoria, 0002 PRETORIA, South Africa

Abstract. The fast correlation attack described by Meier and Staffelbach [1] on certain stream ciphers requires that the number of taps of the characteristic polynomial must be small, typically less than 10. The attack can be extended to characteristic polynomials with an arbitrary number of taps if an efficient algorithm is available to compute low-weight polynomial multiples of the feed-back polynomial. Given an arbitrary polynomial of degree k over $GF(2)$, a method based on the "birthday paradox" was suggested in [1] to find weight–4 polynomial multiples of the polynomial having degree $\leq 2^{k/4}$. The computational complexity is $O(2^{k/2})$, and a table size of $O(2^{k/2})$ is used. In this paper a technique based on Zech's logarithm is described, which uses a significantly smaller table size but at the cost of increased computational complexity. It is shown that weight–4 polynomials of degree k can be found requiring a table size of $O(2^{k/3})$ and computational complexity $O(2^{k/3}) + DL[O(2^{k/3})]$, where the second term denotes the complexity to compute discrete logarithms in $GF(2^k)$ of $O(2^{k/3})$ field elements. If weight–4 polynomials of degree $\leq O(2^{k/4})$ are required, the table size is $O(2^{3k/8})$ and the computational complexity $O(2^{3k/8}) + DL[O(2^{3k/8})]$. To compute the discrete logarithm of field elements the use of Coppersmith's algorithm [2] is suggested.

1 Introduction

In secret-key cryptosystems, pseudonoise generators based on binary linear feedback shift-registers (LFSRs) are used as running key generators. The required keystream $z = (z_n)$ is obtained by combining a fixed number of say, R, LFSRs by means of a combining function f, which is chosen to be non-linear in order to avoid cryptanalytic attacks using the Berlekamp-Massey algorithm [3].

For encryption the plaintext sequence $m = (m_n)$ is added modulo 2 to the keystream sequence $z = (z_n)$ on a bit-by-bit basis to give the ciphertext sequence $c = (c_n)$. This is illustrated in Figure 1.

The characteristic polynomial of each LFSR of length $k_i, i = 1, 2, \ldots, R$ is chosen to be *primitive*, and is assumed to be known to the analyst. Furthermore, it is assumed that the secret key of the cryptosystem specifies the initial states of each LFSR. The total number of keybits required to specify the initial states of the stream cipher generator is $\sum_{i=1}^{R} k_i$. In a brute force attack the $\prod_{i=1}^{R} 2^{k_i}$ possible states of the LFSRs will have to be examined, which is not feasible in

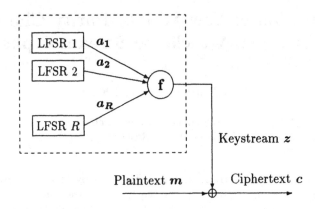

Fig. 1. Stream cipher using non-linearly combined LFSRs

practical systems. However, Siegenthaler [4] has shown that if there exists a measure of correlation between the keystream sequence and the outputs of individual LFSRs, it is possible to determine the initial state of each LFSR independently, thereby reducing the cryptanalytic attack to a divide-and-conquer attack, with approximate complexity $\sum_{i=1}^{R} 2^{k_i}$. Siegenthaler has successfully demonstrated the correlation attack for a number of combining functions proposed in the literature, viz. Brüer [5]Geffe [6]Pless [7]. Siegenthaler's attack amounts to an exhaustive search through the state space of each individual LFSR. In spite of its effectiveness, this attack is only feasible for values of k up to approximately 50.

Recently, it was shown by Meier and Staffelbach [1] that if the number of taps t of the characteristic polynomial is small it is possible to determine the initial LFSR states by means of an iterative algorithm, having complexity much less than an exhaustive search. The algorithm exploits the fact that the bits of a sequence generated by a LFSR satisfy a number of linear relationships, referred to as *parity check equations*, or simply *parity checks*.

The algorithm proposed by Meier and Staffelbach [1] operates roughly as follows: The bits of a sequence generated by LFSR satisfy a number of linear equations. If the bits of the observed ciphertext c_n are substituted in these relations instead of the (unknown) bits of the LFSR sequence a_n, some of the relations will not be satisfied. The algorithm consists of two phases: a *computational phase*, during which conditional probabilities are computed iteratively, and a *complementation phase* where suspect bits are complemented. These two phases are repeated until all the bits in the observed ciphertext sequence are changed so as to satisfy the linear relations. Then the changed ciphertext will be identical to the original unknown LFSR sequence, and hence the initial states of the LFSRs are known.

The algorithm has asymptotic complexity $O(k)$, when the number of taps t is fixed. However, if t grows linearly with LFSR length k, the algorithm has complexity exponential in k. Consequently, the algorithm is only suitable for LFSRs with relatively few taps, i.e. parity checks having low weight $t < 10$. This is one of the most important factors which limits the effectiveness of the algorithm. Several researchers have proposed extensions of the original algorithm to overcome this limitation (see for example [8, 9, 10]).

The object of this paper is to present a systematic method for the generation of low-weight parity checks, making use of Zech's logarithm and the discrete logarithm in $GF(2^k)$.

2 Review of the Statistical Model

Assume that a segment of N keystream digits is being observed by the cryptanalyst. From a practical viewpoint it is desirable that the value of N should be as small as possible; i.e. $N \ll L = 2^{k_i} - 1$. The LFSR sequence a_n is correlated with probability $q > 0.5$ to the keystream sequence, i.e.

$$P(z_n = a_n) = q = 1 - p > 0.5 \quad n = 1, 2, \ldots \tag{1}$$

If it is assumed that the value of z_n depends only on the input a_n and the observation at time index n, the effect of the stream cipher on the shift register sequence a_n may be modelled as a memoryless binary symmetric channel (BSC) with cross-over probability p. This is illustrated in Figure 2.

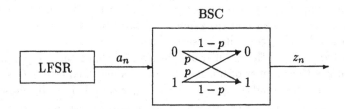

Fig. 2. Binary symmetric channel model for correlation attacks

The corruption of the LFSR sequence a_n due to the other LFSRs in the stream cipher as well as the effect of the plaintext can be modelled as "channel errors". In the case of binary-valued digits, the channel model may be simplified, by letting $z_n = a_n + e_n$, $n = 1, 2, \ldots$. For this channel model holds

$$P(a_n = z_n) = P(e_n = 0) = q = 1 - p \quad n = 0, 1, 2, \ldots \tag{2}$$

which is assumed to be the same for all indices n. Furthermore, we assume that the "error digits" e_n are i.i.d. random variables, which can be expressed in terms of the *correlation coefficient* ξ,

$$P(e_n = 0) = q = 1 - p = \tfrac{1}{2}(1 + \xi) \tag{3}$$

The problem of the cryptanalyst is to restore the unknown LFSR sequence $a_n = (a_n)$ from the observed keystream sequence $z_n = (z_n)$. Meier and Staffelbach [1] first noted that this can be achieved efficiently by exploiting the linear relationships known to exist in a linear recurring sequence. The sequence a_n is produced by a LFSR with primitive characteristic polynomial $f(x)$ of degree k with t taps

$$f(x) = c_0 + c_1 + c_2 x^2 + \cdots + c_k x^k \tag{4}$$

with $c_0 = 1$ and $c_1, c_2, \ldots c_k \in \{0, 1\}$. The output sequence a_n is given by the linear recursion relation

$$a_n = c_1 a_{n-1} + c_2 a_{n-2} + \cdots + c_k a_{n-k} = \sum_{i=1}^{k} c_i a_{n-i} \qquad n = k, k+1, k+2, \ldots \tag{5}$$

The number of taps t of the LFSR is equal to the number of non-zero coefficients $\{c_1, c_2, \ldots, c_k\}$ of the characteristic polynomial $f(x)$. In the case of linear recurring sequences over $GF(2)$ the non-zero coefficients have values $c_i = 1$. Therefore, the linear relation described by (5) can be written as a parity check equation consisting of the $t + 1$ nonzero terms of the LFSR sequence a_n:

$$L = a_0 + a_1 + a_2 + \cdots + a_t = 0 \tag{6}$$

where the a_i denote those sequence digits multiplied by the nonzero coefficients c_i. Following [1], we can replace the bits of the LFSR sequence with the bits of the keystream sequence at the same index positions. Obviously, the parity check equation will not necessarily be satisfied.

As was noted in [1, 8], the number of taps t of the characteristic polynomial, which determines the number of nonzero terms in a check equation L, strongly influences the computational complexity of the algorithm. It is easily shown that the probability of a parity check equation consisting of $t + 1$ non-zero terms is satisfied is given by:

$$P(a_0 + a_1 + a_2 + \cdots + a_t = 0) = \tfrac{1}{2}(1 + \xi^{t+1}) \tag{7}$$

To appreciate the import of this result consider, for example, a stream cipher generator which comprises a LFSR of length $k = 31$. Assume that the number of taps is $t \approx k/2 = 15$. Suppose that for this generator a value of $q = P(a_n = z_n) = 0.75$ is observed. This corresponds to a correlation coefficient of $\xi = 2q - 1 = 0.5$. For a parity check consisting of $t = 15$ terms, it follows that the resulting correlation is:

$$P(a_0 + a_1 + \cdots + a_t = 0) = \tfrac{1}{2}(1 + \xi^{t+1})$$
$$= \tfrac{1}{2}\left(1 + (0.5)^{16}\right) = 0.500007629 \ldots$$

The observed correlation of the parity check has been reduced to a value close to 0.5 which for all practical purposes cannot be exploited in a correlation attack. However, if each parity check consists of, say, three non-zero terms, the observed correlation is

$$P(a_1 + a_2 + a_3 = 0) = \tfrac{1}{2}(1 + \xi^3)$$
$$= \tfrac{1}{2}\left(1 + (0.5)^3\right) = 0.5625$$

which can still be utilised. Therefore, in order to fully exploit the observed correlation between the LFSR sequence a_n and the keystream sequence z_n, it is of crucial importance to find low-weight parity checks, having a small number of non-zero terms. The purpose of this paper is to introduce a systematic procedure for obtaining low-weight parity checks.

3   Maximum a-priori Probability Detection

Suppose a_0 is a fixed digit being observed, where for notational convenience the index has been set to $n = 0$. From (6) it follows that it is possible to obtain a parity check equation of the form

$$L = \sum c_i z_i = \sum c_i(a_i + e_i) = \sum c_i e_i = e_0 + e_1 + \cdots e_t \qquad (8)$$

It follows that

$$P(L = 1) = P(e_0 + e_1 + \cdots + e_t = 1) = \tfrac{1}{2}(1 - \xi^{t+1}) \qquad (9)$$

The following conditional probabilities are easily determined:

$$P(L = 0|e_0 = \tfrac{1}{2}(1 + \xi^t)) \qquad P(L = 0|e_0 = \tfrac{1}{2}(1 - \xi^t))$$
$$P(L = 1|e_0 = \tfrac{1}{2}(1 - \xi^t)) \qquad P(L = 1|e_0 = \tfrac{1}{2}(1 + \xi^t)) \qquad (10)$$

Now, assume that m parity checks have been determined, with the LFSR digits a_n substituted by keystream digits z_n:

$$L_1 = z_n + z_{11} + z_{12} + \cdots + z_{1t}$$
$$L_1 = z_n + z_{21} + z_{22} + \cdots + z_{2t}$$
$$\vdots$$
$$L_m = z_n + z_{m1} + z_{m2} + \cdots + z_{mt} \qquad (11)$$

Suppose h of these values are observed to be zero. We are interested in deriving a threshold value b on the number of nonzero parity checks, which is to be used as follows: Choose $a = z$ if $h \geq b$, otherwise choose $a = \bar{z}$ if $h < b$. For notational convenience assume that the first h equations are equal to 0, and the remaining $m - h$ equations are equal to 1. Denote the condition on the L-values by

$$\mathcal{L} : L_1 = L_2 = \cdots = L_h = 0, L_{h+1} = \cdots = L_m = 1 \tag{12}$$

Clearly

$$P(z = a) = P(e_0 = 0) = \tfrac{1}{2}(1 + \xi) \tag{13}$$

We want to choose $a = z$ or $a = \bar{z}$. Choose $a = z$ if

$$P(e_0 = 0|\mathcal{L}) \geq P(e_0 = 1|\mathcal{L}) \tag{14}$$

Using Bayes' theorem, choose $a = z$ if

$$P(\mathcal{L}|e_0 = 0)P(e_0 = 0) \geq P(\mathcal{L}|e_0 = 1)P(e_0 = 1) \tag{15}$$

To compute $P(\mathcal{L}|e_0)$ assume that $L_1, L_2, \cdots, L_m$ are statistically independent. Then

$$P(\mathcal{L}|e_0) = \prod_{i=1}^{m} P(L_i|e_0) \tag{16}$$

Substituting, we find the condition for choosing $a = z$ is given by

$$P(L = 0|e_0 = 0)^h P(L = 1|e_0 = 0)^{m-h} P(e_0 = 0) \geq$$
$$P(L = 0|e_0 = 1)^h P(L = 1|e_0 = 1)^{m-h} P(e_0 = 1) \tag{17}$$

or,

$$\left(\frac{1 + \xi^t}{1 - \xi^t}\right)^{2h-m} \geq \frac{1 - \xi}{1 + \xi} \tag{18}$$

Therefore, the *a posteriori* value is maximised if b is chosen, satisfying

$$\left(\frac{1 + \xi^t}{1 - \xi^t}\right)^{m-2b} = \frac{1 + \xi}{1 - \xi} \tag{19}$$

Solving for b yields

$$b = \tfrac{1}{2}\left[m - \ln\left(\frac{1 + \xi}{1 - \xi}\right) \Big/ \ln\left(\frac{1 + \xi^t}{1 - \xi^t}\right)\right] \tag{20}$$

Since $b > 0$ (otherwise the best choice for z is to always put $z = a$), it follows that

$$m > \ln\left(\frac{1 + \xi}{1 - \xi}\right) \Big/ \ln\left(\frac{1 + \xi^t}{1 - \xi^t}\right) \tag{21}$$

If $|\xi| \ll 1$, using the approximation $\ln(1 + \xi) \approx \xi$ gives $m > 1/\xi^{t-1}$. In practice a value of m will be chosen such that

$$m = \frac{K}{\xi^{t-1}} \tag{22}$$

for some suitable value of K. Substituting this value for m into (20) gives the required threshold b. It remains to specify a suitable value for K, which in turn determines the value of m, according to (22). Note that if m does not satisfy inequality (21), i.e there are too few equations, it will not be possible to correct the keystream sequence. It is therefore necessary to consider the probability of an incorrect decision, P_e, i.e. a bit z_n was erroneously inverted:

$$
\begin{aligned}
P_e &= P(h < b|z \neq a)P(z \neq a) + P(h \geq b|z = a)P(z = a) \\
&= (\tfrac{1}{2})^{m+1}(1+\xi) \sum_{i=0}^{b-1} \binom{m}{i}(1+\xi^t)^i(1-\xi^t)^{m-i} \\
&\quad + (\tfrac{1}{2})^{m+1}(1-\xi) \sum_{i=b}^{m} \binom{m}{i}(1-\xi^t)(1+\xi^t)^{m-i}
\end{aligned}
\tag{23}
$$

Based on this analysis we introduce an algorithm for correcting the digits z_n in the keystream sequence. The condition for convergence is $P_e < q = (1-p)$.

1. Set the iteration counter $j = 0$, and $q_{(j)} = 1 - p$.
2. Set $j = j + 1$.
3. Calculate $\xi_{(j)} = 1 - 2q_{(j)}$.
4. Determine $b_{(j)}$ according to (20). Make sure that m satisfies (21).
5. Do for $n = 0, 1, 2, \ldots N$: Count the number $h_{(j)}$ of parity checks which are equal to 0. If $h_{(j)} < b_{(j)}$, invert z_n.
6. Determine the probability of an incorrect decision $q_{(j)} = P_e(j)$ according to (23).
7. If $q_{(j)} > q_{(j)-1}$ Exit: The algorithm will not converge.
8. If $q_{(j)} \approx 0$ Stop.
9. Else, return to Step 2 and repeat.

4 Approach Based on Zech's Logarithm

In their paper Meier and Staffelbach [1] propose the use of a coincidence method to find low-weight polynomial multiples of the LFSR characteristic polynomial $f(x)$, which may be an arbitrary polynomial of degree k with coefficients in $GF(2)$. A table of polynomials $x^a + x^b \bmod f(x)$, where $0 \leq a, b \leq 2^{k/4}$, is generated. Each reduced polynomial $\bmod f(x)$ has degree $k - 1$ or less. The "birthday paradox" [11] states that there is probability better than $1 - e^{-1}$ of at least one coincidence in the list. If the coincidence is $x^a + x^b \equiv x^c + x^d$, then $x^a + x^b + x^c + x^d = 0 \bmod f(x)$ is a polynomial multiple of $f(x)$ of degree $\leq 2^{k/4}$. The size of the table, as well as the computational complexity, is $O(2^{k/2})$.

In this paper we present a method based on Zech's logarithm which requires a significantly smaller table size, but at the cost of increased computational complexity.

Let x be a primitive element in $GF(2^k)$. Then Zech's logarithm is defined by $x^{z(i)} = 1 + x^i$, where $i, z(i) \in \{0, 1, 2, \ldots, 2^k - 2\} \cup \{-\infty\}$. Thus $z(i)$ is the logarithm of $1 + x^i$ to the base x. By convention $z(0) = -\infty$. The following properties of $z(i)$ are easily proved [12]:

1. $z(z(i)) = i$ (self-inverse).
2. $z(2i) = 2z(i) \bmod 2^k - 1$.
3. $z(-i) = z(i) - i \bmod 2^k - 1$.

Define $z_i = z(i)$ and let $Z_T = \{z_i, 1 \le i \le T\}$. We assume that $z_i \in Z_T$ is randomly distributed over the interval $(0, L - 1)$, where $L = 2^k$, and that the intervals between adjacent values have a negative exponential distribution, with rate parameter T/L. The assumptions are motivated by the observed random distribution of $z(i)$ in numerical calculations over finite fields $GF(2^k)$ for small values of k, viz. $k = 8, \ldots, 24$.

Finding weight–3 polynomials

The polynomial $x^{z(i)} + x^i + 1$ is a multiple of $f(x)$ having weight 3. Let ν_{max} be the maximum degree of the weight–3 polynomial which is useful in the cryptanalysis algorithm. The probability that $z(i) \le \nu_{max}$ under the negative exponential distribution assumption is $P[z(i) \le \nu_{max}] = 1 - e^{-\nu_{max}T/L}$.

If $\nu_{max} = O(2^{k/4})$, then the required table size for a probability greater than $1 - e^{-1}$ of finding a weight-3 polynomial is $T = O(k^{3/4})$. This large table size would limit the technique to values of k less than approximately 60.

Finding weight–4 polynomials

Suppose two values i, j are found such that $z_i - z_j = d > 0$. Then

$$x^{z_i} + x^i + 1 + x^d(x^{z_j} + x^j + 1) = x^i + x^{j+d} + x^d + 1 \tag{24}$$

is a weight–4 polynomial having degree $\nu = \max(i, j + d)$. If $0 < i, j \le T$, then $\nu_{max} \le T + d$. The probability that two adjacent Zech logarithm values are within a distance d of each other is $P(z_i - z_j = d; T, L) = 1 - e^{-dT/T}$, $d > 0$. The expected number of pairs spaced a distance d is

$$m = T(1 - e^{-dT/L}) \approx dT^2/L \tag{25}$$

Choosing $d = T = L^{1/3}$ gives a table size of $O(2^{k/3})$ and an expected weight-4 polynomial of degree $\nu_{max} \le 2 \cdot 2^{k/3}$. The computational complexity is $O(2^{k/3}) + DL[O(2^{k/3})]$, where $DL[O(2^{k/3})]$ is the complexity of computing the discrete logarithms of $O(2^{k/3})$ field elements. The complexity to compute discrete logarithms will be discussed in the next section.

If the maximum degree of the weight-4 polynomial must be $\nu_{max} \le O(L^{1/4})$, the required parameters are obtained by setting $d = L^{1/4}$ in (25), giving the values $\nu_{max} \le O(2^{k/4})$, $T = O(2^{3k/8})$ and computational complexity $O(2^{3k/8}) + DL[O(2^{3k/8})]$.

5 Coppersmith's Algorithm for Discrete Logarithms

Let $f(x)$ be a primitive polynomial of degree k over $GF(2)$. The elements of $GF(2^k)$ are all polynomials over $GF(2)$ of degree at most $k-1$. Any field element $a(x) \in GF(2^k)$ can be represented as $a(x) \equiv x^s \bmod f(x)$. The *discrete logarithm problem* may be formulated as follows: given any $a(x) \in GF(2^k)$, find the unique integer s, $0 \leq s \leq 2^k - 2$, such that $a(x) \equiv x^s \bmod f(x)$. Coppersmith [2] has developed an efficient algorithm for the calculation of the discrete logarithm for any arbitrary field element in $GF(2^k)$. The algorithm consists of two stages.

Stage I
In the first stage, the discrete logarithms of all irreducible polynomials of degree less than or equal to a suitable bound b are calculated and placed in a database. A polynomial $a(x)$ is said to be *smooth with respect to b* (or simply *smooth*) when $a(x)$ is the product of irreducible factors having degree at most b. The asymptotic running time for this stage is $O(\exp((1 + O(1))k^{1/3} \ln^{2/3} k))$.

Stage II
Next, the discrete logarithm of a given field element $a(x)$ is calculated. The given $a(x)$ is processed until it is smooth, i.e. decomposable into irreducible factors of degree at most b. The corresponding discrete logarithms for each factor are then obtained from the database calculated in Stage I. The asymptotic running time for Stage II to compute one logarithm is given approximately as $O(\exp(\sqrt{k/b} \ln k/b + ln^2 k))$.

Using estimates from [13] for $k = 80$ and $b = 20$, the number of operations required to set up the database is approximately 6×10^9. This amount of computing can be done by a supercomputer in less than a day.

For our purposes it is more important to consider the time needed for the Stage II calculations, as they are repeated for each element in the table of Zech's logarithms. A substantial improvement in the speed of the second stage is proposed in [14]. Assuming $k = 80$ and a bound $b = 20$, it is found from [13] that approximately 10 extended Euclidean operations over $GF(2)$ per table entry is required. Taking a table size of $2^{80/3} \approx 10^8$ entries of 80-bit words, the total number of operations is approximately 10^9. This number is expected to be computed within a day using a supercomputer.

6 Conclusion

In this paper we have presented a method based on Zech's logarithms for finding low-weight parity checks for linear recurring sequences generated by linear feedback shift registers. Finding weight–3 checks is only be feasible for shift register lengths of approximately 60 bits. Weight–4 checks can be obtained by combining two weight–3 checks, which can be found by means of Zech's logarithm and calculation of the discrete logarithm in $GF(2^k)$ with Coppersmith's algorithm.

This method is independent of the number of feedback taps of the LFSR, and its applicability is limited to a large extent by the computational complexity of finding discrete logarithms of finite field elements. The proposed method was successfully applied to the case of $GF(2^{31})$, and it is expected to be feasible for shift register lengths of up to approximately 100 bits.

References

1. W. Meier and O. Staffelbach, "Fast correlation attacks on certain stream ciphers", *Journal of Cryptology*, vol. 1, no. 3, pp. 159–176, 1989.
2. D. Coppersmith, "Fast evaluation of logarithms in fields of characteristic two", *IEEE Trans. on Information Theory*, vol. IT-30, no. 4, pp. 587–594, July 1984.
3. J. L. Massey, "Shift–register synthesis and BCH decoding", *IEEE Trans. Information Theory*, vol. IT-15, pp. 122–127, 1969.
4. T. Siegenthaler, "Decrypting a class of stream ciphers using ciphertext only", *IEEE Trans. Computers*, vol. C-34, pp. 81–85, 1985.
5. J. O. Brüer, "On nonlinear combinations of linear shift register sequences", in *Proc. IEEE ISIT*, les Arcs, France, June 21-25 1982.
6. P. R. Geffe, "How to protect data with ciphers that are really hard to break", *Electronics*, pp. 99–101, January 1973.
7. V. S. Pless, "Encryption schemes for computer confidentiality", *IEEE Trans. Computers*, vol. C-26, pp. 1133–1136, November 1977.
8. C. Chepyzhov and B. Smeets, "On a fast correlation attack on stream ciphers", in *Advances in Cryptology – EUROCRYPT '91.* 1991, pp. 176–185, Springer-Verlag.
9. M. J. Mihaljevic and J. Golic, "A comparison of cryptanalytic principles based on iterative error-correction", in *Advances in Cryptology – EUROCRYPT '91.* 1991, pp. 527–531, Springer-Verlag.
10. K. Zeng and M. Huang, "On the linear syndrome method in cryptanalysis", in *Advances in Cryptology – CRYPTO '88.* 1990, pp. 469–478, Springer-Verlag.
11. K. Nishimura and M. Sibuya, "Probability to meet in the middle", *J. of Cryptology*, vol. 2, no. 1, pp. 13–22, 1990.
12. K. Huber, "Some comments on Zech's logarithm", *IEEE Trans. Information Theory*, vol. IT-36, no. 4, pp. 946–950, July 1990.
13. A. M. Odlyzko, "Discrete logarithms and their cryptographic significance", in *Advances in Cryptology – EUROCRYPT '84.* 1985, pp. 224–314, Springer-Verlag.
14. I. F. Blake, R. Fuji-Hara, R. C. Mullin, and S. A. Vanstone, "Computing logarithms in finite fields of characteristic two", *SIAM Journal on Algebraic Discrete Methods*, vol. 5, pp. 276–285, 1985.

A storage complexity based analogue of Maurer key establishment using public channels

C.J. Mitchell[1]

Royal Holloway, University of London, Egham, Surrey TW20 0EX, U.K.

Abstract. We describe a key agreement system based on the assumption that there exists a public broadcast channel transmitting data at such a rate that an eavesdropper cannot economically store all the data sent over a certain time period. The two legitimate parties select bits randomly from this channel, and use as key bits those which they have selected in common. The work is inspired by recent work of Maurer, [3].

1   Introduction

In a recent paper, [3], Maurer has described a number of related methods for providing secret key agreement between two parties using only publicly available information. These methods are based on a development of Wyner's ideas, [4].

The particular method which has inspired the work described here relies on the two parties wishing to agree a secret key, and any eavesdropper, all receiving a signal from some channel, for which each party only receives a noisy version of the originally transmitted signal. For the system to operate, the common information derived from the channel by the two legitimate parties, A and B say, must not be a subset of the information derived from the channel by any eavesdropper. Note that it is not necessary for A or B to receive a 'better' version of the signal than the eavesdropper.

This requirement is very simply met in any situation where the channel errors are statistically independent for each of the three parties. However, in practice this requirement may be rather difficult to guarantee. As an example consider a radio channel: the eavesdropper might be able to position his antenna close to one of the legitimate parties, and hence obtain a strictly better version of the signal than one of the legitimate parties.

It is also necessary for the parties A and B to share an error-free 'authenticated channel', although this may also be intercepted by the eavesdropper. By an authenticated channel we mean one in which B is able to verify that all the data received on the channel originated from A in exactly the same form as was received, and vice versa. This may also be non-trivial to provide in practice.

In this paper we consider a slightly different key agreement system, which avoids the first requirement above. This system is based on the assumption that there exists a public broadcast channel transmitting data at such a rate that an eavesdropper cannot economically store all the data sent over a certain time period. The two legitimate parties randomly choose which bits to store from this broadcast channel, and then compare notes (using an authenticated channel which need not provide privacy) after the chosen time period has passed as to which bits they have both selected, which then constitute the shared secret key. Unless the eavesdropper has stored a high proportion of all the bits transmitted on the public broadcast channel, it will almost certainly have no more than a small proportion of the secret key bits. This idea is analogous to the encryption system described by Maurer in [2], where the idea of a noise source producing data in quantities which cannot economically be stored is also exploited. However, the idea in [2] is rather different, in that the legitimate users do not divulge which of the bits they have used.

In the scheme described here the provable guarantees of security which can be obtained for Maurer's schemes, [3], are thus exchanged for arguments regarding the cost (in providing large amounts of data storage) to a third party of obtaining the key agreed by A and B. It is important to note that this scheme, like all the schemes in [3], still requires the error-free authenticated channel.

Before proceeding observe that, for clarity and brevity of presentation, we only provide informal arguments to support certain of our main results. Formal proofs can be constructed using information theoretic arguments.

2 The basic scheme

In order to describe our key agreement system we first need to describe our model of the communicating parties. We suppose that A and B, the two legitimate parties wishing to establish a shared secret key, both have access to an (error free) public broadcast channel sending random binary data at a high rate, say R bits/sec. We suppose also that A and B

- share an authenticated error-free channel,
- have the means to store n_A and n_B bits respectively, and
- are 'synchronised' with respect to the broadcast channel, i.e. they have the means to refer to a single bit sent on this channel.

By an authenticated channel shared by A and B we mean a channel for which A can be sure that any bits received claiming to be from B are genuinely from B, and have not been manipulated in transit (and vice versa).

The eavesdropper, who we call C, also has error-free access to both the public broadcast channel and the authenticated channel between A and B.

The basic key agreement system works as follows. Note that in a subsequent section we describe some improvements on this basic scheme.

Algorithm K

1. A and B both monitor the broadcast channel for an agreed interval of time of duration T. The start and end points of this interval can be agreed using the authenticated channel. We assume that the exact details of the time interval are also known to the eavesdropper C. Note that this means that $N = TR$ bits are transmitted during the agreed interval.

2. During this interval A selects n_A of the bits sent over the broadcast channel at random and stores them. Similarly, and independently, B selects n_B bits at random and stores them.
3. At the end of the interval (and not before) A sends B the 'indices' of the bits that it has stored, where the bits sent over the public channel during the time interval are successively given the indices $0, 1, \ldots$, etc. Having received these indices, B examines them and compares them with the indices of the bits it has stored and makes special note of any coincidences.
4. B now sends back to A a list of all coincidences, and the bits corresponding to these coincident indices become the key bits (which both A and B have).

As we now show, given that A and B make genuinely random selections, and T, R, n_A and n_B are chosen appropriately (with $N = TR$), the eavesdropper will be obliged to store many more bits than either A or B to have a good probability of knowing more than a very few of the key bits. To demonstrate this we first establish the following simple result.

Theorem 1. *Suppose A and B follow Algorithm K, and an eavesdropper stores n_C bits randomly selected from the broadcast channel during the agreed time interval. Suppose also that n_A and n_B are both very much smaller than N (the number of bits sent during the agreed time interval). Then the following will hold.*

(i) *The expected number of bits of key shared by A and B at the end of the process will be approximately $n_A n_B / N$.*
(ii) *The expected number of key bits available to C is approximately $n_A n_B n_C / N^2$.*

Proof. Both results follow from elementary probability considerations.

(i) For any given bit of the n_A selected by A, the probability that it is also selected by B is n_B / N. Given that n_A is very small with respect to N, we may ignore the fact that the probabilities are not independent and hence say that the expected size of the set of bits selected by both A and B will be *approximately n_A* times the above probability, and *(i)* follows.
(ii) follows by a precisely analogous argument.

Before proceeding note that A will actually need $(L + 1)n_A$ bits of storage, where $L = \lceil \log_2 N \rceil$, i.e. L is the number of bits in the index values. Similarly B will need $(L + 1)n_B$ bits of storage. In addition A will need to send Ln_A bits over the authenticated channel as part of the key agreement process.

We have thus presented a system which provides a cost difference between legitimate key agreement and unauthorised interception of key material.

3 A simple example of the system

To illustrate how such a system might operate we consider a simple example.

Suppose T is 10^5 seconds (i.e. approximately one day) and R is 10^{10} bits per second, i.e. 10 Gbits/sec, and hence $N = TR = 10^{15}$ and $L = 50$. Now suppose that $n_A = n_B = 3 \times 10^8$ and hence A and B will need to have $51 \times 3 \times 10^8$ bits of storage, i.e. a little under 2 Gigabytes. At current prices, high speed magnetic disk storage of this capacity will cost approximately £500, and prices are likely to continue to fall. At the same time, A will need to send a little under 2 Gigabytes of information to B over the shared authenticated channel. By Theorem 1, at the end of the process A and B will expect to share a key of approximately $(3 \times 10^8)^2 / 10^{15} = 90$ bits.

We next consider what strategy the eavesdropper might adopt to try and learn significant amounts of key information. In order to obtain, say, 10% of the key bits, by Theorem 1 the eavesdropper will need to store 10% of the bits sent over the public broadcast channel. This will require 10^{14} bits of storage, i.e. approximately 12,000 Gbytes. At today's prices, low cost storage (e.g. magnetic tape) still costs significantly more than £10 per Gbyte, and hence such storage will cost the eavesdropper well in excess of £120,000, and, by similar arguments, to obtain 50% or 100% of the key bits would cost in excess of £600,000 or £1,200,000 respectively.

If A and B were concerned about the possibility that an eavesdropper could make use of a small number of the key bits, then security could be increased by using a one-way hash function to produce, say, a 64-bit key from the 90 bits derived from the key exchange process.

4 Extensions of the basic scheme

There are a number of ways in which the basic system can be modified to increase the cost differential between the legitimate parties and the eavesdropper. We consider two such possibilities. Both offer methods by which the storage requirements for A and B can be reduced from $(L+1)n_A$ and $(L+1)n_B$ to around n_A and n_B respectively.

4.1 Pseudo-random selection of bits by A and B

Suppose A and B have agreed in advance the choice of a cryptographically secure pseudo-random number generator. By this we mean a generator which, given a secret key as input, produces a sequence of pseudo-random numbers as output and for which, given knowledge of some of the output sequence, it is computationally infeasible to compute any more information regarding the output sequence (in particular it will be computationally infeasible to deduce the key used to generate this sequence).

We now describe a modified version of the basic system described in Algorithm K, which makes use of such a pseudo-random number generator.

Algorithm L

1. Before starting the process A and B select random keys for their chosen pseudo-random number generator, which we call R_A and R_B respectively.

2. A and B both monitor the broadcast channel for an agreed interval of time of duration T (and $N = TR$ as previously). The start and end points of this interval can be agreed using the authenticated channel. We assume that the exact details of the time interval are also known to the eavesdropper C. For the purposes of this discussion we suppose that the bits sent over the broadcast channel during the selected time interval are labelled

$$b_0, b_1, \ldots, b_{N-1}.$$

3. Before starting the monitoring A and B also choose *step values* s_A and s_B respectively, where $s_A = \lfloor N/n_A \rfloor$ and $s_B = \lfloor N/n_B \rfloor$. If necessary, at some point (either before, during or after the monitoring period) A and B exchange these step values.

4. During the time interval, A uses the chosen pseudo-random number generator and its secret key R_A to produce a sequence $t_0, t_1, \ldots, t_{n_A-1}$ of pseudo-random numbers, where each pseudo-random number t_i is chosen from the range $0, 1, \ldots, s_A - 1$ (with uniform probabilities). A then selects the following n_A bits sent over the broadcast channel and stores them:

$$b_{t_0}, b_{s_A+t_1}, b_{2s_A+t_2}, \ldots, b_{(n_A-1)s_A+t_{n_A-1}}.$$

Similarly, and independently, B uses the pseudo-random number generator and its secret key R_B to produce a sequence $u_0, u_1, \ldots, u_{n_B-1}$ of pseudo-random numbers, where each pseudo-random number u_i is chosen from the range $0, 1, \ldots, s_B - 1$ (with uniform probabilities). B then selects the following n_B bits sent over the broadcast channel and stores them:

$$b_{u_0}, b_{s_B+u_1}, b_{2s_B+u_2}, \ldots, b_{(n_B-1)s_B+u_{n_B-1}}.$$

5. At the end of the interval (and not before) A sends B its secret key R_A, used to help select which bits from the public channel it has stored during the time interval. Similarly B sends A its secret key R_B.

6. A can then use R_A and R_B to find those values of i and j ($0 \le i < n_A$, $0 \le j < n_B$) for which $is_A + t_i = js_B + u_j$. This can be done with the minimum of storage by at any point retaining the values of i, t_i, j and u_j, and then

   - replacing the pair (i, t_i) with the pair $(i+1, t_{i+1})$ if $is_A + t_i < js_B + u_j$,
   - replacing the pair (j, u_j) with the pair $(j+1, u_{j+1})$ if $is_A + t_i > js_B + u_j$, and
   - storing the values $is_A + t_i$ whenever $is_A + t_i = js_B + u_j$.

   B can do precisely the same calculations, leaving A and B with a known set of mutually held bits (which can be used to create a key).

Before attempting to describe the performance of this modified scheme, we first consider the best strategy for an eavesdropper wishing to find as many key bits as possible using the minimum amount of storage. There would appear to be three obvious strategies for the eavesdropper, C. Firstly C could choose to store a random selection of bits from the channel (without regard to the 'step values'). Secondly C could store a fixed number of bits from each range, $b_t = (b_{ts_A}, b_{ts_A+1}, \ldots, b_{(t+1)s_A-1})$, for $t = 0, 1, \ldots, n_A - 1$. Thirdly, C could select a number of ranges b_t for various values of t, $(0 \leq t < n_A)$, and store all the bits for the selected ranges. Note that, alternative versions of the second and third strategies would involve replacing s_A and n_A with s_B and n_B.

Now, since there is at most one key bit within any range b_t, $(0 \leq t < n_A)$, the second strategy would seem to be the best (although all strategies yield very similar results when n_A and n_B are small relative to N). In the following simple result, in which we derive the performance of this revised scheme, we therefore assume that the eavesdropper is using the second of the above strategies.

Theorem 2. *Suppose A and B follow Algorithm L, and an eavesdropper C stores n_C bits selected from the public broadcast channel during the agreed time interval. Suppose also that $n_C = dn_A$ for some integer d, and that C stores d bits from each range b_t, $(0 \leq t < n_A)$. As previously we assume that n_A and n_B are both very much smaller than N. Then the following will hold.*

(i) *The expected number of bits of key shared by A and B at the end of the process will be approximately $n_A n_B / N$.*

(ii) *The expected number of key bits available to C is approximately $n_A n_B n_C / N^2$.*

Proof. (i) Consider any range: b_t (for some value of t satisfying $0 \leq t < n_A$). At the end of the agreed time interval, A will store exactly one bit from this range. The probability that B will also store this bit is equal to n_B / N. Given that there are n_A such ranges, and assuming that these probabilities are independent (which is a reasonable approximation given n_A and n_B are small with respect to N), we see that the expected number of bits held by both A and B at the end of the agreed time interval is approximately equal to $n_A n_B / N$, as required.

(ii) As previously, consider any range: b_t (for some value of t satisfying $0 \leq t < n_A$). At the end of the agreed time interval, A will store exactly one bit from this range. The probability that B and C will also store this bit is equal to $(n_B / N)(d / s_A)$. Given that there are n_A such ranges, and assuming that these probabilities are independent (which is a reasonable approximation given n_A and n_B are small with respect to N), we see that the expected number of bits held by all of A, B and C at the end of the agreed time interval is approximately equal to $n_A n_B d / N s_A = n_A n_B n_C / N^2$, as required.

Before proceeding note that A will only need n_A bits of storage. Similarly B will only need n_B bits of storage. In addition A and B will only need to send their respective secret keys R_A and R_B over the authenticated channel as part of the key agreement process.

Hence this amended procedure reduces the storage for A and B to n_A and n_B respectively, minimises use of the authenticated channel, and also makes the match-finding process a simple one. This is at the cost of making the security dependent on the computational security of the pseudo-random number generator employed by A and B. To show how effective this improvement is we consider a modified version of our previous example; we use the same cost assumptions as in Section 3.

Suppose T is 10^5 seconds (i.e. approximately one day) and R is 10^{12} bits per second, i.e. 1000 Gbits/sec, and hence $N = TR = 10^{17}$. Now suppose that $n_A = n_B = 3 \times 10^9$ and hence A and B will need to have 3×10^9 bits of storage, i.e. a little under 400 Megabytes, costing no more than £100. At the same time, A will need to send only one key (of say 128 bits) to B over the shared authenticated channel (and vice versa). By Theorem 2, at the end of the process A and B will expect to share a key of approximately $(3 \times 10^9)^2/10^{17} = 90$ bits.

We next consider the position of the eavesdropper. In order to obtain, say, 10% of the key bits, the eavesdropper will need to store 10% of the bits sent over the public broadcast channel. This will require 10^{16} bits of storage, i.e. approximately 1,200,000 Gbytes. Such storage will cost the eavesdropper well in excess of £12,000,000, and, by similar arguments, to obtain 50% or 100% of the key bits would cost in excess of £60,000,000 or £120,000,000 respectively.

4.2   Block-wise selection of bits

In the previous section we described a system which minimises both the storage requirements for A and B and the use of the authenticated channel for A and B. This was at the cost of making the security depend on the cryptographic properties of a pseudo-random number generator. We now consider a slightly different modification of the basic scheme which retains many of the advantages of the scheme described in Section 4.1, but which does not rely on any computational security assumptions. The procedure is as follows.

Algorithm M

1. A and B both monitor the broadcast channel for an agreed interval of time of duration T. The start and end points of this interval can be agreed using the authenticated channel. We assume that the exact details of the time interval are also known to the eavesdropper C. Suppose that the bits sent over the broadcast channel during the selected time interval are labelled

$$b_0, b_1, \ldots, b_{N-1}.$$

   To make our discussions simpler we also assume that $n_A | N$ and $n_B | N$, and hence define s_A and s_B by $s_A n_A = s_B n_B = N$. Suppose moreover that $s_A s_B | N$, and define w by $s_A s_B w = N$. We assume that all these parameters are known to A and B before the start of the agreed time interval.

2. During the time interval A randomly chooses w values $p_0, p_1, \ldots, p_{w-1}$, where each value p_i satisfies $0 \le p_i < s_A$. A then stores the following w sets of s_B bits during the agreed time interval (i.e. a total of n_A bits):

$$b_{i s_A s_B + p_i s_B}, b_{i s_A s_B + p_i s_B + 1}, \ldots, b_{i s_A s_B + p_i s_B + s_B - 1}$$

   for $i = 0, 1, \ldots, w - 1$.

Similarly, and independently, B randomly chooses w values $q_0, q_1, \ldots, q_{w-1}$, where each value q_i satisfies $0 \leq q_i < s_B$. B then stores the following w sets of s_A bits during the agreed time interval (i.e. a total of n_B bits):

$$b_{is_As_B+q_i}, b_{is_As_B+s_B+q_i}, \ldots, b_{is_As_B+(s_A-1)s_B+q_i}$$

for $i = 0, 1, \ldots, w - 1$.

3. At the end of the agreed time interval it should be clear that A and B will share precisely one bit from each range of bits

$$b_i' = (b_{is_As_B}, b_{is_As_B+1}, \ldots, b_{(i+1)s_As_B-1})$$

for $i = 0, 1, \ldots, w - 1$. I.e. A and B will share precisely w bits.

4. At the end of the agreed time interval (and not before) A sends B its random values $p_0, p_1, \ldots, p_{w-1}$, used to help select which bits from the public channel it has stored during the time interval. Similarly B sends A its secret random values $q_0, q_1, \ldots, q_{w-1}$.

5. A and B can then both very easily determine which key bits they share.

As in the previous section, before attempting to describe the performance of this scheme, we first consider the best strategy for an eavesdropper wishing to find as many key bits as possible using the minimum amount of storage. There would appear to be three obvious strategies for the eavesdropper, C. Firstly C could choose to store a random selection of bits from the channel. Secondly C could store a fixed number of bits from each range, b_t', for $t = 0, 1, \ldots, w - 1$. Thirdly, C could select a number of ranges b_t' for various values of t, $(0 \leq t < w)$, and store all the bits for the selected ranges.

Now, since there is exactly one key bit within any range b_t', $(0 \leq t < w)$, all strategies would appear to yield similar results. In the following simple result, in which we derive the performance of this revised scheme, we therefore arbitrarily assume that the eavesdropper is using the second of the above strategies.

Theorem 3. *Suppose A and B follow Algorithm M, and an eavesdropper C stores n_C bits selected from the public broadcast channel during the agreed time interval. Then the following will hold.*

(i) *The number of bits of key shared by A and B at the end of the process will be exactly $w = n_A n_B / N$.*

(ii) *The expected number of key bits available to C is $w n_C / N = n_A n_B n_C / N^2$.*

Proof. (i) This follows from the discussion given as part of Algorithm M.

(ii) Consider any range: b_t' (for some value of t satisfying $0 \leq t < w$). At the end of the agreed time interval, A and B will store exactly one bit from this range. The probability that C will store this particular bit is equal to n_C / N. Given that there are w such ranges, and given that these probabilities are independent, we see that the expected number of bits held by all of A, B and C at the end of the agreed time interval is equal to $w n_C / N$, as required.

This system then achieves a comparable performance to the system in the previous section. The only disadvantage is the slightly increased communication cost in transferring the values p_i and q_i across the authenticated channel. However, this is a very small cost since the number of these values will only be the same as the number of key bits agreed by A and B, and each value will only require $\log_2 s_A$ or $\log_2 s_B$ bits of storage.

Moreover this variant of the basic scheme has two significant advantages.

- Unlike the first variant (described in Section 4.1), its security is not dependent on the cryptographic properties of a pseudo-random number generator.
- Unlike both the other schemes, it yields a key of guaranteed length to A and B, and not just a varying number of key bits with an associated expected value. The disadvantage of this latter case is that on some occasions the number of key bits provided to A and B may be somewhat less than the expected value, potentially causing problems.

5 Summary and conclusions

We have thus described systems which provide secret key agreement between A and B and whose security rests solely on the following two assumptions.

- The cost of storage remains high relative to the bandwidth of one or more publicly available broadcast channels.
- A and B share an error-free authenticated channel.

It is particularly interesting to note that, with the exception of the scheme described in Section 4.1, the systems' security does not depend on any assumptions regarding the computational difficulty of any problems. In that sense the systems are provably secure (given the two key assumptions listed above).

We now briefly consider possible practical circumstances in which the above two assumptions might be satisfied. There are various ways in which the two legitimate parties might be provided with an authenticated channel (but not with the means to securely agree a key). Two of the more likely are as follows.

- The users may purchase a communications facility which provides an authenticated channel as a premium service (which can be obtained simply by paying the appropriate rate). The communications service provider may, for example, provide this by using digital signatures, MACs or some other type of cryptographic check function. It is certainly conceivable that this could be provided in such a way that the users have no access to the keys used, and hence no direct means to exchange secret keys.
- The users may have access to an implementation of a digital signature function such as DSS, which cannot be used for data encryption. They could then use this digital signature function, in conjunction with authenticated keys for each other, to provide the authenticated channel.

We next briefly describe two possible sources for a high bit rate (say greater than 10 Gbit/sec) public broadcast channel.

- The first is to make use of a high rate public satellite data channel. In this case the bits sent over the channel will not be random—instead they will consist of many data streams intermingled. However, in practice they may be 'random enough' for our purposes (especially if a hashing operation is performed on any agreed set of key bits).
- The second is to employ a purpose-designed high speed random data source, the output of which is made publicly available by some means (e.g. by fibre-optic cable). Although this will now guarantee randomness for the bits, there are obvious problems with this approach if it is simultaneously used by many pairs of parties to agree a secret key (as would almost certainly be necessary to justify the cost of providing such a channel). In such an event there are much greater incentives for third parties to invest resources in storing the channel output, since it could yield many secret keys simultaneously.

Finally, it could be argued that, given that A and B share an authenticated channel, then they can achieve secret key agreement by using the well-known Diffie-Hellman key exchange protocol, (see, for example, [1]). Whilst this is certainly true, there may be situations where users do not wish to take such an approach. For example, users may not choose to trust a system whose security depends entirely on a single mathematical function remaining hard to compute (i.e. the discrete logarithm problem), and they may prefer to trust in arguments about the likely cost of data storage. It is certainly of theoretical interest to observe that key agreement schemes can be devised which rely only on the two assumptions listed above, and which do not require any assumptions about the computational difficulty of certain calculations.

Acknowledgements

The work in this paper is part of the DTI/EPSRC Link Personal Communications Programme project: *Security studies for third generation mobile telecommunications systems (3GS3)*, the colloborating partners in which are Vodafone, GPT and RHUL, and which has been partly funded by the UK EPSRC under Grant Number GR/J17173. The author would like to acknowledge the invaluable advice and encouragement of colleagues in the project.

References

1. Diffie, W., Hellman, M.E.: New directions in cryptography. IEEE Trans. on Information Theory IT-22 (1976) 644–654
2. Maurer, U.M.: Conditionally-perfect secrecy and a provably-secure randomized cipher. J. of Cryptology 5 (1992) 53–66
3. Maurer, U.M.: Secret key agreement by public discussion from common information. IEEE Trans. on Information Theory 39 (1993) 733–742
4. Wyner, A.D.: The wire-tap channel. Bell System Tech. J. 54 (1975) 1355–1387

Soft Decision Decoding of Reed Solomon Codes using the Dorsch Algorithm

H.P. Ho and P. Sweeney

Centre for Satellite Engineering Research
University of Surrey GUILDFORD
Surrey, GU2 5XH, UK.

Abstract This paper describes and evaluates implementations of the Dorsch algorithm for RS codes. This algorithm has previously been applied only to binary codes and requires adaptation for application to multilevel codes. The Dorsch algorithm has also been applied to some Reed Solomon codes which can also be formulated as binary codes in order to compare the two approaches. On the AWGN channel, the binary approach shows better performance for the same complexity than the multilevel approach and it appears that the Dorsch algorithm is not well suited to multilevel codes.

1 Introduction

Reed Solomon codes are multilevel cyclic block codes which are Minimum Distance Separable (MDS). They have found wide applicability because of their efficiency as MDS codes, the burst error correcting capability resulting from their multilevel nature and the existence of several algorithms for high speed decoding. The main general decoding algorithms, however, are not able to make use of soft decision outputs from the demodulator.

Attempts to introduce soft decision capability into RS decoding have centred around two main approaches, namely trellis methods [1] and iterative methods using algebraic decoding, e.g. Forney's GMD algorithm [2] and the Chase algorithm[3]. The Dorsch algorithm [4] is of the latter type and works by erasing n-k symbols of low confidence in a (n,k) block code, generating a small set of test patterns as errors on the k symbols of high confidence, carrying out erasures only decoding on the sequences produced and comparing with the received sequence. It has recently been shown [5] that the Dorsch algorithm is efficient for application to many binary cyclic codes of rate 2/3 or less.

In section 2, the description of the multilevel method of implementing the Dorsch Algorithm is presented and a binary implementation for certain specific RS codes is outlined in section 3. The results can be found in section 4. There is a discussion of the significance of the results in section 5 and, finally, the conclusions are drawn in section 6.

2 Implementation

2.1 The Dorsch Algorithm

The Dorsch algorithm, also known as the Heaviest Weight Information set Algorithm, was first described by Dorsch in 1974. Although falling into the same category of algebraic methods, it differs significantly from the Chase algorithm, where the generated test patterns are used to toggle the most unreliable symbols, and from the GMD approach, which uses erasures with the number of erased symbol varying with different attempts.

The implementation of the Dorsch Algorithm consists of three stages:

Stage 1: The received sequences are sorted out according to the symbol reliability. They are in principle separated into two parts, namely the n-k symbols of low confidence and the k symbols of high confidence. The former is termed the check set and the latter the information set. The encoding and decoding are done in time domain using the generator matrix G and parity check matrix H. By sorting the received sequence according to the reliability of the symbols means that the position of the columns in the parity check matrix has been swapped. For binary codes we need to ensure that the columns corresponding to the check set are linearly independent, although this is not a problem for RS codes; these columns must then be converted into echelon canonical form.

Stage 2: Error patterns or sequences, to be applied to the information set are generated with monotonically increasing weight. Usually the number of patterns chosen is sufficient to cover all single-bit error patterns in the information set

Stage 3: For memoryless channels, e.g. AWGN, maximum-likelihood decoding corresponds to choosing the error pattern which has the minimum weight. After decoding all the generated error patterns, compute the Euclidean distance between the received sequence and each decoded code word, choosing the code word corresponding to the lowest distance.

2.2 Modification to RS code

There are some modifications required when applying the Dorsch Algorithm directly to RS code. Firstly, as mentioned previously, any k symbols can be an information set, so there is no need to check for linear independence of columns of the parity check matrix. Secondly, the concept of selecting reliable and unreliable symbols must be adjusted. In this implementation, the confident level of the symbol is the sum of confidences of all the bits. It is believed that this implementation is suitable for the AWGN channel, but would need to be adjusted for other channels.

3. Binary equivalents of RS codes

It is has been shown [6] that some Reed Solomon codes can be viewed as binary codes. This gives the opportunity to compare the decoding performance of those codes in both the binary and the multilevel implementations. The codes studied in this way are

1. a (7,3) RS code over GF(8), with $g(x)=(x+\alpha)(x+\alpha^2)(x+\alpha^3)(x+\alpha^4)$, and the equivalent (21,9) binary code constructed using the binary generator polynomial $g_b(x) = (x^9+x^3+1)(x^3+x+1)$.
2. a (15,11) RS code over GF(16), with $g(x)=(x+\beta^{14})(x+\beta^0)(x+\beta^1)(x+\beta^2)$ and $\beta=\alpha^4$, and the equivalent (60,44) binary code constructed using the binary generator polynomial $g_b(x) = (x^4+x^3+x^2+x+1)(x+1)(x^4+x+1)(x^2+x+1)^2$.

4. Results

Simulations were carried out using encoders and decoders written in C and an AWGN channel model with 8-level quantization at the receiver. The bit error rate and block error rate performance of various implementation of the (7,3)RS codes are shown in figure 1 and figure 2 respectively.

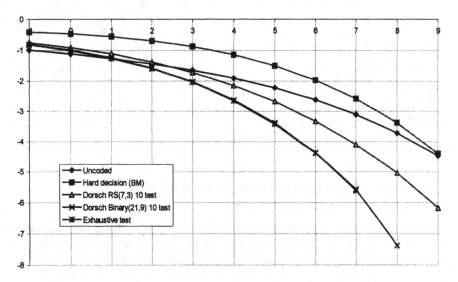

Fig. 1 Bit error rate of RS(7,3) vs Eb/No (dB)

Figure 1 shows the performance for the uncoded channel, hard decision decoding using Berlekamp Massey Algorithm with frequency domain encoding, the Dorsch Algorithm on the (7,3) RS code with 10 test patterns, the Dorsch Algorithm on the equivalent (21,9)binary code with the same test patterns and the maximum likelihood performance obtained using the Dorsch algorithm with 512 test patterns. It can be seen the hard decision decoding produces the worst performance in terms of bit error rate in the range studied. Using the Dorsch algorithm in the binary and multilevel variants shows a distinct difference between the two. The (21,9)binary code is able to produce results close to maximum likelihood while the (7,3) RS code, although showing some improvement over hard decision, is still well short of maximum likelihood performance with this number of test patterns.

In figure 2, an additional curve named as hard decision undetectable error is shown that most uncorrectable errors are, in fact, detected in hard decision decoding. It can be seen that, if only undetected block errors are of importance, the comparison between hard and soft decision decoding is less straightforward, with hard decision actually preferable at low signal to noise ratio.

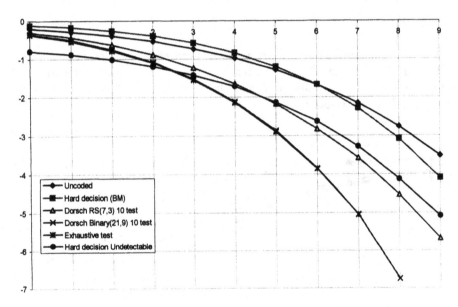

Fig. 2 Block error rate of RS(7,3) vs Eb/No (dB)

Figure 3 shows BER comparisons for the Dorsch algorithm with the same number of test patterns on both the (15,11) RS code and its equivalent (60,44) binary code. Again, it shows that the binary version out-performs the multilevel version.

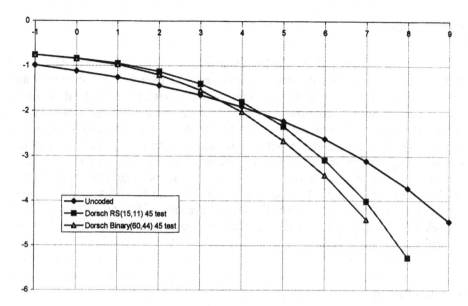

Fig. 3 Bit error rate of RS(15,11) vs Eb/No (dB)

Figure 4 shows the comparison of the bit error performance of a (15,7) RS code using hard decision decoding and using the Dorsch algorithm. It can be seen that using the Dorsch algorithm improves the bit error performance only at low signal to noise ratios.

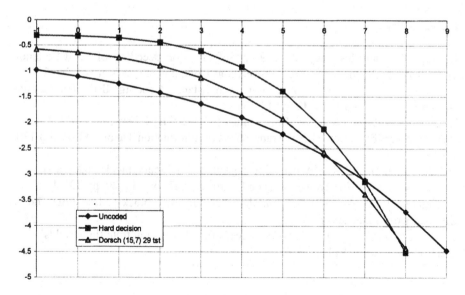

Fig. 4 Bit error rate of RS(15,7) vs Eb/No (dB)

5. Discussion

It appears from these results that the Dorsch algorithm is not well suited to RS codes. Detailed study of data from the simulations suggests that the reason is connected with the multilevel nature of the codes. For binary codes of rate 1/2 or lower, the chosen information set usually consists entirely of high confidence bits, so that there is relatively little need to consider errors in the information set. For multilevel codes, however, errors in the information set are more likely because the low confidence bits cannot be separated from the other bits in the same symbol. Thus we might expect that the more bits per symbol, the less appropriate the Dorsch algorithm will be. The improvements obtained by treating certain codes as binary codes reinforces this conclusion. The coding gains from soft decision decoding of the (60, 44) binary code are relatively modest, but the rate of this code is higher than is ideal for the Dorsch algorithm.

6. Conclusion

A new approach to soft decision decoding of Reed Solomon codes has been introduced in this paper. The Dorsch algorithm did not, however, work well with RS codes on the AWGN channel. This is because several bits are treated as one symbol in RS code and individual bits cannot be separately manipulated. However, if the RS codes can be treated as binary cyclic codes in some special cases, then the performance using the Dorsch algorithm is encouraging.

References

1. Shin, S.K. and Sweeney, P. (1994), 'Soft decision decoding of Reed Solomon Codes using Trellis methods", IEE Proceedings Communications Vol 141. No 5.
2. Forney, G.D.(1966), 'Generalised minimum distance decoding', IEEE Transactions on Information Theory, **IT-12**, pp 125-31
3. Chase D (1972), 'A class of algorithms for decoding block codes with channel measurement information', IEEE Transactions on Information Theory, **IT-18**, 00 170-82.
4. Dorsch, B.G. (1974), 'A decoding algorithm for binary block codes and J-ary output channels', IEEE Transactions on Information Theory, **IT-20**, pp 391-4.
5. Eiguren, J. (1994), 'Soft decision algorithm for linear block codes', PhD Thesis, University of Manchester.
6. Sweeney, P. (1995), 'Cyclic block codes definable over multiple finite fields', Electronics Letters. Vol 31. No 5.

Good Codes based on Very Sparse Matrices

David J.C. MacKay[1] and Radford M. Neal[2]

[1] Cavendish Laboratory, Cambridge, CB3 0HE. United Kindom.
[2] Depts. of Statistics and Computer Science, Univ. of Toronto, M5S 1A1. Canada.

Abstract. We present a new family of error-correcting codes for the binary symmetric channel. These codes are designed to encode a *sparse* source, and are defined in terms of very sparse invertible matrices, in such a way that the decoder can treat the signal and the noise symmetrically. The decoding problem involves only very sparse matrices and sparse vectors, and so is a promising candidate for practical decoding.

It can be proved that these codes are 'very good', in that sequences of codes exist which, when optimally decoded, achieve information rates up to the Shannon limit.

We give experimental results using a free energy minimization algorithm and a belief propagation algorithm for decoding, demonstrating practical performance superior to that of both Bose-Chaudhury-Hocquenghem codes and Reed-Muller codes over a wide range of noise levels.

We regret that lack of space prevents presentation of all our theoretical and experimental results. The full text of this paper may be found elsewhere [6].

1 Background

In 1948, Shannon [14] proved that there exist block codes, for a given memoryless channel, that achieve arbitrarily small probability of error ϵ at any communication rate R up to the capacity C of the channel. We will refer to such code families as 'very good' codes. By 'good' codes we mean code families that achieve arbitrarily small probability of error ϵ at non-zero communication rates R up to some $R_{\max}$ that may be *less than* the capacity C of the given channel. By 'bad' codes we mean code families which can only achieve arbitrarily small probability of error ϵ by decreasing the information rate R to zero. (This does not mean that they are useless for practical purposes.) By 'practical' codes we mean code families which can be encoded and decoded in time and space polynomial in the block length.

Since 1948, few constructive and practical codes that are good have been found, fewer still that are practical, and none at all that are both practical and very good [8]. Goppa's recent algebraic geometry codes (reviewed in [15]) appear to be both practical and good, but we believe that the literature has not established whether they are very good.

In this paper we present a new code family that we call 'MN codes'. These codes have a very sparse structure that shows promise for practical decoding.

At the same time it can be proved that these codes are very good, in that sequences of codes exist which, when optimally decoded, achieve information rates up to the Shannon limit of the binary symmetric channel [6]. In sections 3 and 4 we describe empirical results of computer experiments using first a free energy minimization algorithm [5] and second a 'belief propagation' algorithm for decoding. Our experiments show that practical performance significantly superior to that of BCH and Reed-Muller codes (in terms of information rate for a given probability of decoder error) can be achieved by MN codes.

2 Description of MN codes

We will denote the error probability of the binary symmetric channel (BSC) by f_n, where $f_n < 0.5$, and the binary entropy function by $H_2(f) = f \log_2(1/f) + (1 - f) \log_2(1/(1-f))$. The *weight* of a vector or matrix is the number of 1s in it. We denote the weight of a vector $\mathbf{x}$ by $w(\mathbf{x})$. The *density* of a source of random bits is the expected fraction of 1 bits. A source is *sparse* if its density is less than 0.5. A vector $\mathbf{v}$ is *very sparse* if its density vanishes as its length increases, for example, if a constant number t of its bits are 1s. The capacity $C(f_n)$ of a BSC with noise density f_n is, in bits per cycle, $C(f_n) = 1 - H_2(f_n)$. The rate $R_0(f_n)$ is

$$R_0(f_n) \equiv 1 - \log_2 \left[1 + 2\sqrt{f_n(1 - f_n)} \right]. \tag{1}$$

This is the computational cutoff of sequential decoding for convolutional codes—the rate beyond which the expected cost of achieving vanishing error probability using sequential decoding becomes infinite.

The Gilbert bound $GV(f_n)$ is

$$GV(f_n) = \begin{cases} 1 - H_2(2f_n) & f_n < 1/4 \\ 0 & f_n \geq 1/4. \end{cases} \tag{2}$$

This is the rate at which one can communicate with a code whose codewords satisfy the Gilbert-Varshamov minimum distance bound, assuming bounded distance decoding [7].

2.1 Conventional linear codes, and the ideas behind MN codes

A linear error correcting code can be represented by a N by K binary matrix $\mathbf{G}$ (the generator matrix), such that a binary message $\mathbf{s}$ is encoded as the vector $\mathbf{t} = \mathbf{Gs} \bmod 2$ (figure 1a). (Note that our generator matrices act to the right rather than the left.) The channel adds noise $\mathbf{n}$ to this vector with the resulting received signal $\mathbf{r}$ being given by:

$$(\mathbf{Gs} + \mathbf{n}) \bmod 2 = \mathbf{r}. \tag{3}$$

The decoder's task is to infer $\mathbf{s}$ given the received message $\mathbf{r}$, and the assumed noise properties of the channel. The *optimal decoder* returns the message $\mathbf{s}$ that

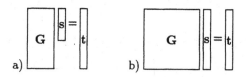

Fig. 1. a) A conventional code. The source vector s, of length K, is dense. The transmitted vector t is of length $N > K$. Here $N = 2K$, so the symbol rate and information rate are both $K/N = 0.5$ bits. b) Square code for a sparse source. The symbol rate is 1, but if the density of the source, f_s, is 0.1 then the information rate is $H_2(0.1) = 0.47$, about the same as that of the conventional code.

maximizes the posterior probability

$$P(\mathbf{s}|\mathbf{r}, \mathbf{G}) = \frac{P(\mathbf{r}|\mathbf{s}, \mathbf{G})P(\mathbf{s})}{P(\mathbf{r}|\mathbf{G})}. \tag{4}$$

It is often not practical to implement the optimal decoder.

It is conventional to define the error correcting code to have $N > K$, and to use signals s of density $f_s = 0.5$. The $(N - K)$ extra bits are parity check bits, which produce redundancy in the transmitted vector t. This redundancy is exploited by the decoding algorithm to infer the noise vector **n**.

MN codes take a different approach. Instead of adding redundancy in the form of parity check bits, we assume that the source itself is redundant, having f_s, the density of s, less than 0.5. Consecutive source symbols are independent and identically distributed. Redundant sources of this type can be produced from other sources by using a variation on arithmetic coding [16, 13]; one simply reverses the role of encoder and decoder in a standard arithmetic coder based on a model corresponding to the sparse messages [6]. Given that the source is already redundant, we are no longer constrained to have $N > K$. In MN codes, N may be less than K, equal to K or greater than K. We distinguish between the 'symbol rate' of the code, K/N, and the 'information rate' of the code, $H_2(f_s)K/N$. Error-free communication may be possible if the information rate is less than the capacity of the channel. For example, consider a BSC having $f_n = 0.1$, and assume that we have a source with density $f_s = 0.1$. Then we might construct a code with $N = K$, i.e., a square linear code with symbol rate 1 (figure 1b). The information rate, 0.47, is less than the channel capacity, 0.53, so it is plausible that we might construct a sequence of codes of this form achieving vanishing probability of error.

The ideas behind MN codes are (1) that we use a sparse source and (2) that we construct the generator matrix in terms of *invertible* matrices, such that the sparse source and the sparse noise can be treated symmetrically in the decoding problem.

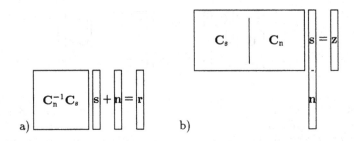

Fig. 2. Pictorial representation of MN Code with $\rho = 1$. a) Encoding, transmission and reception. b) Decoding. The matrices $\mathbf{C_s}$ and $\mathbf{C_n}$ are very sparse. The vectors s and n are sparse. The vector z is given by $\mathbf{z} = \mathbf{C_n r}$.

2.2 Construction of MN codes

The encoder is a linear block code constructed from very sparse matrices as follows. A transmitted block length N and a source block length $K = \rho N$ are selected. Figures 1 and 2 illustrate the case $\rho = 1$. The symbol rate of the code is ρ and the information rate is $\rho H_2(f_s)$. We select a *column weight* t, which is an integer greater than or equal to 3. We create two matrices $\mathbf{C_n}$ and $\mathbf{C_s}$ as follows.

The matrix $\mathbf{C_n}$ is a square $N \times N$ matrix that is very sparse and invertible. It is created randomly with exactly weight t per column and weight t per row. [Such a random sparse matrix is not necessarily invertible, but there is a probability (for large N) of about 0.29 that it is.] The inverse $\mathbf{C_n^{-1}}$ of this matrix is computed. This inverse is likely to be a dense matrix. The inversion takes N^3 time and is performed once only.

The matrix $\mathbf{C_s}$ is a rectangular $N \times K$ matrix that is very sparse. [N rows and K columns.] It is created randomly with exactly weight t per column and a weight per row as uniform as possible. If ρ is chosen to be an appropriate ratio of integers then the number per row can be constrained to be exactly ρt.

We mention three variations on this construction.

1. By slightly relaxing the constraint of weight t per column (by allowing one or two columns to have weight $t + 1$), a random very sparse $\mathbf{C_n}$ may easily be made invertible, by flipping one or two bits.

2. When generating the matrices $\mathbf{C_s}$ and $\mathbf{C_n}$, one can constrain all pairs of columns in the matrix $[\mathbf{C_s C_n}]$ to have an overlap (the number of 1s in common between the two vectors) ≤ 1. This is expected to improve the properties of the ensemble of codes, for reasons explained in [6].

3. One can further constrain the matrix $[\mathbf{C_s C_n}]$ so that the topology of the corresponding belief network does not contain short cycles. This is discussed further in section 3.

2.3 Encoding

A source vector s of length ρN is encoded into a transmitted vector t defined by (figure 2a):

$$\mathbf{t} = \mathbf{C}_n^{-1}\mathbf{C}_s\mathbf{s} \bmod 2. \tag{5}$$

This encoding operation takes time of order $\min\left[\rho Nt + N^2, \rho N^2\right]$.

2.4 The decoding problem

The received vector is

$$\mathbf{r} = \mathbf{t} + \mathbf{n} \bmod 2, \tag{6}$$

where the noise, n, is assumed to be a sparse random vector with independent identically distributed bits, density f_n. The first step of the decoding is to compute:

$$\mathbf{z} = \mathbf{C}_n\mathbf{r}, \tag{7}$$

which takes time of order Nt. Because $\mathbf{z} = \mathbf{C}_n(\mathbf{t}+\mathbf{n}) = \mathbf{C}_s\mathbf{s}+\mathbf{C}_n\mathbf{n}$, the decoding task is then to solve for $\mathbf{x} = \left[\begin{smallmatrix}\mathbf{s}\\\mathbf{n}\end{smallmatrix}\right]$ the equation:

$$\mathbf{A}\mathbf{x} = \mathbf{z}, \tag{8}$$

where A is the N by $(K+N)$ matrix $[\mathbf{C}_s\,\mathbf{C}_n]$ (see figure 2b). The optimal decoder, when $f_s = f_n$, is an algorithm that finds the sparsest vector $\hat{\mathbf{x}}$ that satisfies $\mathbf{A}\hat{\mathbf{x}} = \mathbf{z}$.

We emphasize two properties of equation (8):

1. There is a pleasing symmetry between the sparse source vector s and the sparse noise vector n, especially if $f_s = f_n$.
2. Both the matrix A and the unknown vector x are sparse. The vector x has density f_s or f_n, and the matrix A is very sparse, having only t 1s per column, where t may be much less than N. One might therefore hope that it is practical to solve this decoding problem. The decoding problem is of the type studied by Gallager [4]. However, the sparse parity check codes studied by Gallager are bad. The trick that makes MN codes good is the construction in terms of an invertible matrix.

We now describe theoretical properties that we have proved for MN codes. We then describe empirical results with a practical decoding algorithm.

2.5 Theoretical properties proven for MN codes

In [6] we prove properties of these codes by studying properties of a 'typical set decoder' [3] for the decoding problem $\mathbf{A}\mathbf{x} = \mathbf{z}$, averaging over an ensemble of random matrices A. We prove two theorems (our proofs are computer-aided), whose implications are as follows.

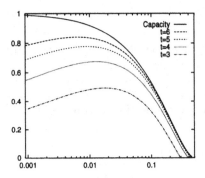

Fig. 3. Main theoretical result. Lower bounds $R^*(f, t)$ on achievable information rate versus noise level f for MN codes with t from 3 to 6. In bits, compared with the channel capacity. The lines are lower bounds on rates achievable by MN codes. As the weight per column t increases the achieveable region rises towards the fundamental limit, the capacity.

1. MN codes with weight per column $t \geq 3$ are *good*, *i.e.*, can achieve error-free transmission up to a non-zero information rate $R^*(f, t)$, if N is made sufficiently large. This rate is plotted numerically in figure 3. This rate is less than the capacity $C(f)$, but for useful values of f, even for t as small as 4, it is not much below the capacity.
2. MN codes are *very good*—if we are allowed to choose t, then we can get arbitrarily close to capacity, still using very sparse matrices with t/N arbitrarily small. The second theorem states:

   Given a density $f < 0.5$, a desired information rate $R < C(f)$, and a desired block error probability $\epsilon > 0$, there exists an integer $t \geq 3$, a symbol rate ρ and an $N_{\min}$ such that for any $N > N_{\min}$, there is a matrix $\mathbf{A}$ having N rows and $K' = N + K = (\rho + 1)N$ columns with weight t or less per column, with the following property: if $\mathbf{x}$ has density f then the optimal decoder from $\mathbf{z} = \mathbf{Ax}$ back to $\hat{\mathbf{x}}$ achieves a probability of error less than ϵ, and the information rate that is achieved is $\geq R$.

3   Practical decoding by free energy minimization

We generated random matrices $\mathbf{A}$ corresponding to symbol rate $\rho = 1$, with uniform weight $t = 4$ per column and $t_r = 8$ per row. We first attempted to solve the decoding problem using a variational free energy minimization algorithm [5]. We found that as the block size N was increased at a constant information rate, the performance improved.

We examined the errors made by the free energy minimization decoder and found that they tended to occur when the vector $\mathbf{x}$ was such that another slightly different typical vector $\mathbf{x}'$ had a similar (but not identical) encoding $\mathbf{z}'$. These errors were attributable to rare topologies in the network corresponding to the $\mathbf{A}$ matrix such as the topology illustrated in figure 4c. We can eliminate the possibility of these errors by modifying the ensemble of random matrices $\mathbf{A}$ so that the corresponding network does not have short cycles in it.

The topological modifications gave codes which were able to communicate at higher rates with a smaller probability of error. The conclusion of these experiments was that MN codes, when decoded by free energy minimization, can be superior to Reed-Muller codes, but not to BCH codes. Significantly better results were obtained when we used the belief net decoder which we now describe.

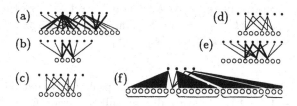

Fig. 4. The vectors $\mathbf{x}$ and $\mathbf{z}$ viewed as nodes in a belief network. White circles denote bits x_k. Black dots denote checks z_n. We illustrate the case $t = 4, t_r = 8$. (a) This figure emphasizes with bold lines the 8 connections to one check and the 4 connections from one bit. Every bit x_k is the parent of 4 checks z_n, and each check z_n is the child of 8 bits. (b-e) Certain topological structures are undesirable in the network defined by the matrix $\mathbf{A}$: in (b) there is a cycle of length 4 in the network; we forbid this topology by saying, equivalently, that the overlap between two columns of $\mathbf{A}$ must not exceed 1; in (c, d, e) more complex topologies are illustrated. Our most successful experiments have used matrices $\mathbf{A}$ in which these topologies are also forbidden [we eliminate bits that are involved in structures like the 'doublet' (e), of which (c) and (d) are hazardous special cases]. This means that every bit's 'friends' (other bits that are parents of its children) consist of t non-overlapping sets of bits as shown in (f).

4 Belief network decoding

We have developed a 'belief net decoder' for the problem $\mathbf{Ax} = \mathbf{z} \bmod 2$, which generalizes the methods of Gallager [4] and Meier and Staffelbach [9] by using methods of belief propagation over networks [11].

We refer to the elements z_n corresponding to each row $n = 1 \dots N$ of $\mathbf{A}$ as checks. We think of the set of bits $\mathbf{x}$ and checks $\mathbf{z}$ as making up a 'belief network', also known as a 'Bayesian network', 'causal network', or 'influence diagram', in which every bit x_k is the parent of t checks z_n, and each check z_n is the child of t_r bits (figure 4). We aim, given the observed checks, to compute the marginal posterior probabilities $P(x_k = 1|\mathbf{z}, \mathbf{A})$ for each k. Algorithms for the computation of such marginal probabilities in belief networks are found in [11]. These computations are expected to be intractable for the belief net corresponding to our problem $\mathbf{A}\mathbf{x} = \mathbf{z} \bmod 2$ because its topology contains many cycles. However, it is interesting to implement the decoding algorithm that would be appropriate if there were no cycles, on the assumption that the errors introduced might be relatively small (c.f. [1]). As the size N of the code is increased, it becomes increasingly easy to produce codes in which there are no cycles of any given length, so we expect that, asymptotically, this algorithm will be an effective algorithm.

4.1   The algorithm

In the following algorithm quantities q_{nk} and r_{nk} associated with each 1 bit in the $\mathbf{A}$ matrix are iteratively updated. We denote the set of bits k that participate in check n by $\mathcal{K}(n) \equiv \{k : A_{nk} = 1\}$. Similarly we define the set of checks in which bit k participates, $\mathcal{N}(k) \equiv \{n : A_{nk} = 1\}$.

Initialization. Let $p_k^0 = P(x_k = 0)$ (the prior probability that bit x_k is 0), and let $p_k^1 = P(x_k = 1) = 1 - p_k^0$. Normally, p_k^1 will be either f_s or f_n, depending on whether bit k is part of the message or the noise. For every (k, n) such that $A_{nk} = 1$ the variables q_{nk}^0 and q_{nk}^1 are initialized to the values p_k^0 and p_k^1 respectively.

Horizontal pass. In the horizontal step of the algorithm, we run through the checks n and compute for each $k \in \mathcal{K}(n)$ two probabilities: the probability of the observed value of z_n arising when $x_k = 0$, given that the other bits $\{x_{k'}, k' \neq k\}$ have a separable distribution given by the probabilities $\{q_{nk'}^0, q_{nk'}^1\}$:

$$r_{nk}^0 = \sum_{\{x_{k'} : k' \neq k\}} P(z_n \mid x_k = 0, \{x_{k'} : k' \neq k\}) \prod_{k' \neq k} q_{nk'}^{x_{k'}} \qquad (9)$$

and the probability of the observed value of z_n arising when $x_k = 1$, r_{nk}^1, defined similarly. These probabilities can be computed efficiently using forward and backward passes (c.f. [5]), in which products of the differences $\delta q_{nk} \equiv q_{nk}^0 - q_{nk}^1$ are computed. We obtain $\delta r_{nk} \equiv r_{nk}^0 - r_{nk}^1$ from the identity:

$$\delta r_{nk} = (-1)^{z_n} \prod_{k' \in \mathcal{K}(n), k' \neq k} \delta q_{nk'}. \qquad (10)$$

Vertical pass. The vertical step takes the computed values of r_{nk}^0 and r_{nk}^1 and updates the values of the probabilities q_{nk}^0 and q_{nk}^1. For each k we compute:

$$q_{nk}^0 = \alpha_{nk} p_k^0 \prod_{n' \in \mathcal{N}(k), n' \neq n} r_{n'k}^0, \quad q_{nk}^1 = \alpha_{nk} p_k^1 \prod_{n' \in \mathcal{N}(k), n' \neq n} r_{n'k}^1, \qquad (11)$$

where α_{nk} is a constant such that $q^0_{nk} + q^1_{nk} = 1$. We can also compute the 'pseudoposterior probabilities' q^0_k and q^1_k at this iteration, given by:

$$q^0_k = \alpha_k\, p^0_k \prod_{n \in \mathcal{N}(k)} r^0_{nk}, \quad q^1_k = \alpha_k\, p^1_k \prod_{n \in \mathcal{N}(k)} r^1_{nk}. \tag{12}$$

At this point, the algorithm repeats from the horizontal pass.

Decoding. Our decoding procedure is to set $\hat{x}_k$ to 1 if $q^1_k > 0.5$ and see if the checks $\mathbf{A}\hat{\mathbf{x}} = \mathbf{z}$ are all satisfied, halting when they are, and declaring a failure if some maximum number of iterations (*e.g.*, 1000) occurs without successful decoding.

4.2   Relationship to Gallager's algorithm

Gallager [4] and Meier and Staffelbach [9] implemented algorithms very similar to this belief net decoder, also studied by Mihaljević and Golić [10]. The main difference in their algorithms is that they did not distinguish between the probabilities q^0_{nk} and q^1_{nk} for different values of n; rather, they computed q^0_k and q^1_k, as given above, and then proceeded with the horizontal pass with all q^0_{nk} set to q^0_k and all q^1_{nk} set to q^1_k.

4.3   Empirical results: belief net decoder

We found the performance of the belief net decoder to be far better than that of the free energy minimization decoder. We found that the results were best for $t = 3$ and became steadily worse as t increased.

In figure 5 we compare two MN codes with BCH codes, which are described in [12] as "the best known constructive codes" for memoryless noisy channels, and with Reed-Muller (RM) codes (block sizes up to 1024). Figure 5 shows the codes' probability of block error versus their rate. All relevant BCH codes listed in [12] are included. To compute the probability of error for BCH codes we evaluated the probability of more than t errors in n bits, as specified in the (n, k, t) description of the code. In principle, it may be possible in some cases to make a BCH decoder that corrects more than t errors, but according to Berlekamp [2], "little is known about... how to go about finding the solutions" and "if there are more than $t + 1$ errors then the situation gets very complicated very quickly." Similarly, for RM codes of minimum distance d, performance was computed assuming that more than $\lfloor d/2 \rfloor$ errors cannot be corrected.

The mean number of iterations of the algorithm to obtain a successful decoding was about 20 for all the experiments reported here. In some cases as many as 800 iterations took place before a successful decoding emerged.

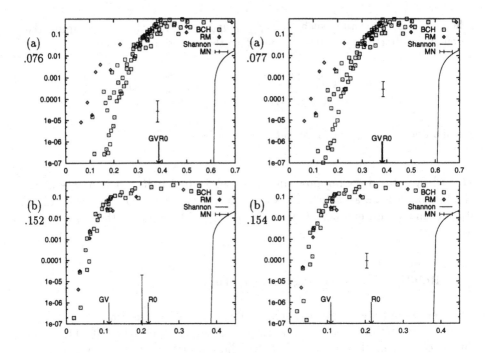

Fig. 5. Comparison of empirical decoding results for MN codes using belief network decoder with calculated performance of Reed-Muller codes and BCH codes, and the Shannon limit. A BSC with (a) $f_n = 0.076$-7 (b) $f_n = 0.152$-4 is assumed. Horizontal axis: information rate R. Vertical axis: block error probability. The best codes are towards the bottom (low error probability) and the right (large rate). Curve: Shannon limit on achievable (rate, bit error probability) values. Arrows show the values of R_0 and $GV(f_n)$ for this channel. Diamonds: Reed-Muller codes. Boxes: BCH codes. MN codes: Empirical results shown are for two topologically modified matrices with $N = 10000$ rows and $N+K$ columns where $K =$ (a) 9839 (b) 3296. The weight per column was $t = 3$. In the case where the error bars extend down to the bottom of the error probability axis, no decoding errors occurred in more than 100,000 trials.

5 Discussion

Our experiments have demonstrated excellent error correction at rates well above the Gilbert bound. In [6] we give an analysis of two practical decoding algorithms. This analysis, and the empirical results we have described, lead us to conjecture that *given a BSC with noise density f, there exist practical decoders for MN codes with any rate R up to $R_0(f)$ which can achieve negligible probability of error, for sufficiently large N.*

The properties of MN codes that we have demonstrated appear to constitute a significant step forward in information theory and coding theory.

The descriptive complexity of these codes is $t(N+K)\log N$, which is much smaller than the complexity of arbitrary linear codes. The set-up time for the code scales as N^3, the encoding time as N^2, and the decoding time as N, where N is the block size.

5.1 Contrasts with convention in coding theory

In a conventional linear (N, K) code, the codewords form a complete linear subspace of $\{0, 1\}^N$. MN codes are only linear in the sense that the transmitted vector t is a linear function of a source vector s. The source is *sparse*, so the codewords that have high probability are only a small subset of a complete linear subspace.

We have obtained the biggest improvement over BCH codes and RM codes by going to high noise levels, *e.g.*, $f_n = 0.15$. Critics might assert that real channels do not have such high noise levels. We would respond that perhaps they ought to—if one increases the clock rate of a channel so that its noise level also increases, there might well be a net increase in capacity. Maybe the main reason that channels with high noise levels are not used is that until now the available codes for error correction have not been good enough.

5.2 Future work

MN codes can also be defined over q-ary alphabets consisting of the elements of $GF(q)$. These codes would be suitable for the q-ary symmetric channel. The decoding algorithms presented here would also generalize. It remains to be established whether our decoders' performance would be any better or worse under this generalization to q-ary alphabets.

We conjecture that as we get closer to the Shannon limit, the decoding problem gets harder. It would be interesting to obtain a convergence proof for the belief net decoding algorithm and to develop ways of reducing the inaccuracies introduced by the approach of ignoring the cycles present in the belief network. The most interesting challenge is to understand whether $R_0(f)$ is indeed the fundamental limit for practical decoding of MN codes.

Acknowledgements

DJCM (mackay@mrao.cam.ac.uk) is grateful to Roger Sewell and David Aldous for helpful discussions and M.D. MacLeod and the Computer Laboratory, Cambridge for kind loans of books. DJCM also thanks Geoff Hinton for generously supporting his visits to the University of Toronto. DJCM was supported by the Royal Society Smithson research fellowship. RMN (radford@stat.toronto.edu) was assisted by a grant from the Natural Sciences and Engineering Research Council of Canada.

References

1. S. Andreassen, M. Woldbye, B. Falck, and S. Andersen. MUNIN - a causal probabilistic network for the interpretation of electromyographic findings. In *Proc. of the 10th National Conf. on AI, AAAI: Menlo Park CA.*, pages 121–123, 1987.

2. E. R. Berlekamp. *Algebraic Coding Theory*. McGraw-Hill, New York, 1968.

3. T. M. Cover and J. A. Thomas. *Elements of Information Theory*. Wiley, New York, 1991.

4. R. G. Gallager. Low density parity check codes. *IRE Trans. Info. Theory*, IT-8:21–28, Jan 1962.

5. D. J. C. MacKay. Free energy minimization algorithm for decoding and cryptanalysis. *Electronics Letters*, 31(6):446–447, 1995.

6. D. J. C. MacKay and R. M. Neal. Good codes based on very sparse matrices. Available from http://131.111.48.24/, 1995.

7. F. J. MacWilliams and N. J. A. Sloane. *The theory of error-correcting codes*. North-Holland, Amsterdam, 1977.

8. R. J. McEliece. *The theory of information and coding: a mathematical framework for communication*. Addison-Wesley, Reading, Mass., 1977.

9. W. Meier and O. Staffelbach. Fast correlation attacks on certain stream ciphers. *J. Cryptology*, 1:159–176, 1989.

10. M. J. Mihaljević and J. D. Golić. Convergence of a Bayesian iterative error-correction procedure on a noisy shift register sequence. In *Advances in Cryptology - EUROCRYPT 92*, volume 658, pages 124–137. Springer-Verlag, 1993.

11. J. Pearl. *Probabilistic reasoning in intelligent systems: networks of plausible inference*. Morgan Kaufmann, San Mateo, 1988.

12. W. W. Peterson and E. J. Weldon, Jr. *Error-Correcting Codes*. MIT Press, Cambridge, Massachusetts, 2nd edition, 1972.

13. J. Rissanen and G. G. Langdon. Arithmetic coding. *IBM Journal of Research and Development*, 23:149–162, 1979.

14. C. E. Shannon. A mathematical theory of communication. *Bell Sys. Tech. J.*, 27:379–423, 623–656, 1948.

15. M. A. Tsfasman. Algebraic-geometric codes and asymptotic problems. *Discrete Applied Mathematics*, 33(1-3):241–256, 1991.

16. I. H. Witten, R. M. Neal, and J. G. Cleary. Arithmetic coding for data compression. *Communications of the ACM*, 30(6):520–540, 1987.

Quantum Cryptography : Protecting our Future Networks with Quantum Mechanics

Simon J. D. Phoenix   and   Paul D. Townsend

BT Laboratories, Martlesham Heath, Ipswich, IP5 7RE. United Kingdom

In a series of recent experiments a radical new technique has been demonstrated that could have far-reaching consequences for the provision of security on our communications networks. This technique, known as quantum cryptography, is the result of a synthesis of ideas from fundamental quantum physics and classical encryption. We review the developments in this rapidly-growing field.

1.   Introduction

1.1   Quantum Cryptography - a Practical Reality

In 1989 a collaboration between IBM and the University of Montreal performed an experiment that could have far-reaching consequences for the provision of security on communications networks [1]. In essence, this deceptively simple experiment used single photons, and a clever protocol that exploited their quantum properties, to establish an identical random sequence of bits at two locations 30cm apart. The transmission was performed in such a way that only the transmitter and receiver could know this sequence. The secrecy of the random bit string was guaranteed by the quantum properties of the single photons used for transmission. The technique, known as 'quantum cryptography' had become a practical reality. Ideas that had first arisen [2] early in the 1970s had now reached fruition and resulted in the remarkable IBM-Montreal experiment.

In the last few years progress has been rapid and several experiments have demonstrated the feasibility of quantum cryptography by establishing identical secret random bit strings over a range of distances and wavelengths in optical fibre using readily-available telecommunications components [3,4]. A working prototype has been built at BT Laboratories to establish secret keys between two users over distances of up to 30km in optical fibre using quantum cryptography at potential key transfer rates of 20kBit/s [5]. This prototype operates at a wavelength of 1.3 microns and exploits the properties of single photons using a phase coding scheme. Other prototype quantum cryptography systems have also been developed to operate at shorter wavelengths using polarisation coding schemes over distances of up to 1km in optical fibre [6] and the group at Los Alamos has demonstrated a different quantum key transmission protocol over a distance of 14km [7]. Although still very much a laboratory-based demonstrator requiring further work before commercial exploitation, quantum cryptography is now a practical reality.

If quantum cryptography is, however, to succeed as a viable method for protecting data on the communication systems of the future it must be capable of implementation on optical networks. In other words, the focus of its applicability, if it is to be widespread, must shift from *point-to-point* links to distributed communications networks. We have developed several techniques for achieving this aim. After having developed the fundamental quantum mechanics necessary for understanding the first quantum cryptography protocol to be invented we shall describe the current BT optical fibre-based demonstrator. We shall briefly explore some of the other quantum cryptography protocols to have been invented before we discuss one of the proposed network implementations of quantum cryptography. We shall conclude with a brief look at other future possibilities for processing data at a quantum level. In particular, recent developments in quantum computing have seriously brought into question the long-term security of certain widely-used encryption techniques. If the inherent potentialities of quantum computing are fully utilised, quantum cryptography may well be the only defence against the quantum code breakers of the future !

1.2   Quantum Mechanical Preliminaries

The first complete protocol for exchanging keys in secret using quantum cryptography was published in 1984 [8]. This protocol is now known as the BB84 protocol. BB84 is a development of the earlier ideas for using quantum mechanics to protect data [2]. The protocol works by exploiting some of the features of quantum mechanics that cannot be explained within the framework of classical physics. Along with the recent advances in quantum computing, quantum cryptography has been instrumental in changing the way in which we view quantum and classical information processing.

Perhaps the most important difference between classical and quantum descriptions of the world can be summarised by the word "complementarity". In simple terms the essence of complementarity is that measurement of a quantum system disturbs it. This complementarity finds its rigorous expression in the incompatibility of quantum observables; by which we mean that if $\hat{A}$ and $\hat{B}$ are the quantum operators representing two physical observables of a quantum system then these observables are said to be incompatible if

$$[\hat{A}, \hat{B}] = \hat{A}\hat{B} - \hat{B}\hat{A} \neq 0 \qquad (1.1)$$

that is, they do not commute.

The properties associated with incompatible observables will, to a greater or lesser extent, be complementary and measurement of one will disturb the other. One of the consequences of (2.1) is the Heisenberg uncertainty relation which connects the variances of two operators to the commutator (2.1) by the following inequality:

$$\left\langle (\Delta \hat{A})^2 \right\rangle \left\langle (\Delta \hat{B})^2 \right\rangle \geq \frac{1}{4} \left| \left\langle [\hat{A}, \hat{B}] \right\rangle \right|^2 \tag{1.2}$$

Precise measurement of one property will result in a 'fuzziness' associated with any complementary property. This is often phrased as an impossibility to know, with arbitrary accuracy, both of the incompatible properties represented by $\hat{A}$ and $\hat{B}$. It is the ingenious exploitation of the properties of complementary observables that enables the BB84 protocol, and indeed some of the other protocols that have been invented, to offer guaranteed secure key distribution.

In quantum mechanics a physical system (a photon, or electron, for example) is described by a state vector in Hilbert space. Each of the physical properties (position, momentum, for example) of the system is represented by an operator whose eigenstates form an orthonormal basis for the Hilbert space. Any state vector of the system can be expanded in terms of these basis states. So a general state, usually written as $|\psi\rangle$, can be expanded in terms of the eigenstates of an operator $\hat{A}$ as follows

$$|\psi\rangle = \sum_j |\alpha_j\rangle\langle\alpha_j|\psi\rangle \tag{1.3}$$

where the eigenvalue relationship is $\hat{A}|\alpha_j\rangle = \alpha_j|\alpha_j\rangle$. Measurement of the physical quantity represented by $\hat{A}$ will yield the value α_j with probability $|\langle\alpha_j|\psi\rangle|^2$ and project the measured system into the *new* state $|\alpha_j\rangle$ with the same probability. It is important to note that if the system were to be prepared in the state $|\alpha_j\rangle$ a measurement of the operator $\hat{A}$ would yield the result α_j with unit probability and the system would remain in the state $|\alpha_j\rangle$.

1.3   Coding Information on Quantum Systems

Information can be encoded on quantum systems in various ways but the simplest and most natural is probably to associate a particular symbol with a particular quantum state. Thus if the operator $\hat{A}$ generates a basis of N states we can associate up to N distinct symbols with this basis. For simplicity let us consider a space spanned by just two states. An example of a quantum system that can be described in this way is a spin-1/2 particle such as an electron which for a given spin direction has just two spin states 'up' and 'down'. We can imagine sending a binary message with a sequence of such particles as depicted in figure 1.

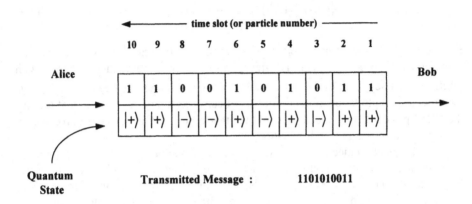

Fig. 1 Alice sends a 10 bit binary message to Bob. Each bit is coded onto a different particle. Alice and Bob use the same quantum basis for coding and reading the message.

In this scenario Alice codes her information using the same basis that Bob uses to read the message. In the example shown in figure 1 Alice has chosen some spin direction characterised by the spin operator in that direction, which we label as $\hat{A}$. This operator has two eigenstates labelled by $|\pm\rangle_A$ and the eigenvalues $\pm 1/2$ associated with these eigenstates. Alice and Bob have chosen to code a '1' on the $|+\rangle_A$ state and a '0' on the $|-\rangle_A$ state. In order to send the message Alice prepares a particle for each time slot in the state indicated by the coding scheme and sends them in sequence. To read the message Bob measures the spin of each particle in the chosen direction, that is, he makes a measurement of the physical quantity represented by the operator $\hat{A}$ for each particle. Bob will record the result for each time slot , either $\pm 1/2$, and can unambiguously associate his result with the quantum state that Alice sent in that time slot. The information transmitted per particle is clearly ln 2.

Let us suppose now that Bob makes a mistake and aligns his apparatus to measure spin along a *different* direction, represented by the operator $\hat{B}$. In quantum mechanical terms spins along different directions do not commute and do not therefore share the same eigenbasis. Because Bob is measuring along the spin direction represented by $\hat{B}$ his measurement will project the particle into one of the states $|\pm\rangle_B$. From equation (1.3) we see that in the basis that Bob is using Alice's states can be expanded as

$$|\pm\rangle_A = {}_B\langle+|\pm\rangle_A|+\rangle_B + {}_B\langle-|\pm\rangle_A|-\rangle_B \qquad (1.4)$$

If, for example, Alice sends the state $|+\rangle_A$ then Bob will read the correct bit, that is, he will obtain the result $|+\rangle_B$, with probability $\left|{}_B\langle+|+\rangle_A\right|^2$. The information transmission rate between Alice and Bob is therefore now dependent on the overlaps between the quantum states and is only maximal if Alice and Bob code and read in the same basis. It is possible for Alice and Bob to choose bases for which no information transmission occurs in the sense that for a given input state of Alice, Bob is equally likely to obtain any of his output states. Bases with this property are said to be 'conjugate' and the operators representing the observable quantities can be thought of in some sense as maximally non-commuting.

   If Alice transmits a random binary sequence to Bob he has no way of knowing whether the sequence he has received is the correct one until some sort of check is made. Furthermore, his measurements have projected the particles into new quantum states and it is *impossible in principle* to recover the original quantum state. Bob might be tempted to try and copy the particles so that he could measure different properties on the identical copies. However, in order to copy the particles he would need to know the basis in which they have been prepared. As we shall see in the next section quantum key distribution works because Alice does not reveal the bases she has chosen to code her bits. It is a general quantum mechanical result that the state of a single particle cannot be determined, and therefore copied, unless the 'coding' basis is known. This "no cloning" property of single quanta is a consequence of the linearity of quantum mechanics [9] and is a rather important result for quantum cryptography.

2.   Quantum Key Distribution over Optical Fibre

2.1   An Interferometric Binary Quantum Communication System

   There are other quantum systems and properties that can be used to encode binary information. The prototype quantum key distribution system we shall examine here, uses a phase-coding scheme for single photons. This scheme is based on the properties of interferometers and the coding is effected by changing the relative phase between the internal arms of the interferometer. We shall describe in this section the implementation of the BB84 4-state protocol on interferometric systems. Such quantum interferometric systems have been used to securely transmit keys over optical fibre up to distances of 30km [5]. A simple communication scheme based on a Mach-Zehnder interferometer is sketched in figure 2 where Alice and Bob each control one half of the interferometer.

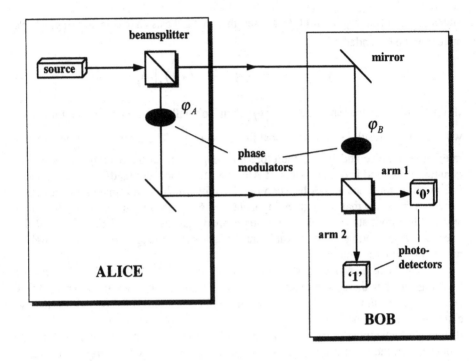

Fig. 2 A simple interferometric binary communication system. Alice and Bob each control one half of a balanced interferometer. If the detector in arm 1 fires Bob records a bit value of '0'. If the detector in arm 2 fires, Bob records a bit value of '1'.

Initially we shall assume that Alice sends a pulse with many photons into the input of her interferometer. The pulse is split at the first beamsplitter, the splitting fraction being determined by the reflection and transmission coefficients. The properties at Bob's output depend on the interference of the two pulses when they recombine at the second beamsplitter. It is the phase difference generated by the relative settings of Alice and Bob's phase modulators that determines the interference behaviour. Depending on the relative phases chosen and the beamsplitter coefficients the pulse exits the device in either of the output arms, or both. For balanced interferometers, that is, where the reflection and transmission coefficients are equal, the pulse is split in half at the first beamsplitter. If Bob keeps his phase set at zero and Alice switches hers between zero and 180° then Bob will see the pulse emerge in arms 1 and 2, respectively. Thus by choosing the coding scheme phase shift (0) $\equiv$ '0', phase shift (180°) $\equiv$ '1', arm 1 $\equiv$ '0' and arm 2 $\equiv$ '1', Alice can communicate a binary message to Bob.

Suppose now that Alice reduces her input intensity to just one photon per pulse. In simple terms, a single photon incident on a beamsplitter cannot go both ways, in other words it cannot be split in the sense that photodetectors placed in both output ports of the first beamsplitter will not simultaneously register a count. At a beamsplitter a photon will randomly 'choose' one output port or another, the probabilities being determined by the reflection and transmission coefficients of the beamsplitter. This probabilistic nature of photons at beamsplitters has been used to generate random sequences. However, the two paths recombine at Bob's beamsplitter and so we must consider the interference between the probability amplitudes for the different paths the photon can take to determine the output characteristics of the entire device. The output characteristics are as before except that, of course, only a single photon emerges from the device. Alice and Bob can use precisely the same coding scheme as for the multi-photon case.

ALICE		BOB	
Phase Setting	**Bit Value**	**Phase Setting**	**Bit Value**
0°	0	0°	0 (arm 1)
180°	1		1 (arm 2)
0°	0	90°	probabilistic
180°	1		outcome
90°	0	0°	probabilistic
270°	1		outcome
90°	0	90°	0 (arm 1)
270°	1		1 (arm 2)

Table 1   Outcomes of various phase settings for the balanced interferometer. We see that it is only when Alice and Bob choose the same coding basis that there is a transmission of information.

The 0°/180° coding is not the only one that will work. There are, in fact, an infinite number of phase settings that can be chosen to convey binary information. If Bob chooses a phase shift of 90°, for example, and Alice modulates between 90° and 270° they can convey a binary message as before. These coding schemes are listed in the table. The operators representing these two distinct coding schemes are maximally non-commuting and the bases are in fact conjugate. Thus if Alice chooses a 0°/180° scheme and Bob sets his modulator at 90° the output at Bob's detectors will be uncorrelated with Alice's input and no information will be transmitted.

2.2 The BB84 Protocol

We now have the basic tools for understanding how the BB84 protocol works on an interferometric optical fibre system. The steps are as follows

❂ Alice and Bob agree on a coding scheme. Alice generates a random binary sequence and for each bit randomly chooses whether to code it on a single photon using the 0°/180° basis or the 90°/270° basis.

❂ Alice sends the sequence of single photons to Bob who, for each incoming photon, chooses randomly and independently of Alice to measure in the 0°/180° basis or the 90°/270° basis.

❂ Alice and Bob now enter into an authenticated public discussion. Bob reveals publicly which basis he chose for each measurement, and in which time slots he did not receive any photon. He does not disclose the bit value he obtained.

❂ Alice publicly compares Bob's list with hers and any time slots where different bases were used or no photon was received are discarded. Alice and Bob should now be left with an identical random bit sequence.

❂ Alice and Bob now publicly choose a random subset of this data and compare the bit values. If there are any errors they infer the presence of an eavesdropper, Eve. If there are no errors they can use the remaining data as a secret key. The test bits are, of course, discarded.

These steps require some further explanation. Why is the final key secret ? Let us put ourselves in the position of Eve. She has access to the channel over which the single photons are transmitted and can monitor, but not alter, all of the subsequent public discussion between Alice and Bob. The quantum and public channels need not be physically distinct. We shall suppose that Eve wishes to obtain all of the secret key so that she intercepts each photon and listens to all of the public discussion. Because Alice is randomly choosing between two incompatible coding schemes for each bit, Eve does not know the correct basis in which to read the information. Furthermore, if she guesses incorrectly she cannot, at this stage, know she has made a mistake. She is also in the unfortunate position of having to retransmit a photon to Bob for each time slot. If Bob receives nothing, that time slot is discarded and forms no part of any secret between Alice and Bob. The problem for Eve is to decide on which basis to use for her retransmission. Bob will randomly, and independently of Alice, choose between two incompatible coding schemes for each incoming photon. Of course this means that about 50% of the data

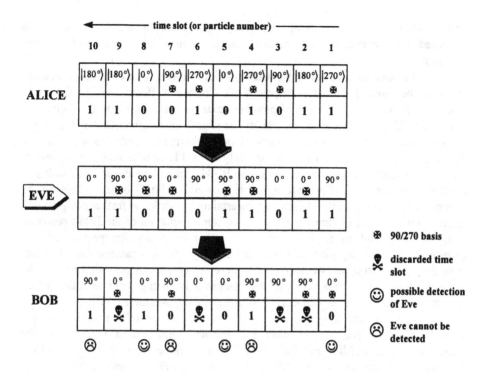

Fig. 3 An illustration of a 10 bit quantum transmission between Alice and Bob which Eve has attempted to eavesdrop. The symbol ⌖ has been used whenever Alice, Bob or Eve have chosen to use the 90/270 coding basis. We see that in time slots 2, 3, 6 and 9 Alice and Bob have chosen incompatible coding schemes and these time slots, marked by the symbol ☠, are discarded. In time slots 1, 5 and 8 Alice and Bob have chosen compatible bases but Eve has not. These instances are marked with the symbol ☺ and will lead to an error in Alice and Bob's data in around 50% of the cases. We see here that only time slots 1 and 8 have lead to an error. On the remaining time slots, marked with the symbol ⊗ Eve cannot be detected as she has chosen the correct basis.

is rejected and forms no part of the key. In the absence of any further information, such as the random sequences used by Alice or Bob, we will assume that Eve faithfully retransmits the quantum state she *thinks* she has measured. As we can see in figure 3 there are instances where Alice and Bob have chosen the same coding scheme but Eve has guessed incorrectly and the result of her measurement has been to modify the initial state sent by Alice, and, in effect to change the coding scheme. This intervention by Eve will lead to a measurable error rate when Alice and Bob publicly test a random subset of their data. If Alice and Bob are using conjugate coding schemes this kind of strategy adopted by Eve will lead to an error rate of 25%. In other words Eve's chance of escaping detection, per tested bit, is 3/4. If just 70 bits are tested, for example, Eve has, approximately, a 1 in a billion chance of escaping detection.

This kind of eavesdropping strategy is known as an intercept/resend strategy and it can be shown [10] that strategies of this kind are ineffective against the BB84 protocol. There is a vast range of possible strategies that are allowed by quantum mechanics. These strategies may, in theory, be slightly more effective than the simple intercept/resend strategy although many of these alternative strategies are currently technically infeasible. *All strategies*, currently infeasible or otherwise, are ultimately doomed to failure because the security of the BB84 protocol resides in the inability to recover a bit from one of two incompatible quantum coding schemes. This can be seen as a consequence of the Heisenberg uncertainty principle. A further complication arises when we come to consider the practical implementation of the BB84 protocol. Real detectors and modulators inevitably introduce errors irrespective of any eavesdropping. for any kind of guaranteed security we must assume that all errors arise from eavesdropping whether or not there is actually an eavesdropper on the channel. For real systems the problem changes from being mere detection of an unwarranted interception to that of rendering the interception ineffective.

This is achieved by the final stage of the full protocol for practical systems. These final steps are reconciliation and privacy amplification [1,11]. In the reconciliation phase Alice and Bob enter into a public error correction procedure. The secrecy of the data is maintained by discarding one undisclosed bit for each publicly disclosed parity check bit. At the end of this process Alice and Bob should have an identical, but only partially secret, random binary sequence. The final stage of the protocol, privacy amplification, takes this partial secret and distils from it, by a process of hashing, a shorter but highly secret binary random sequence which can then be used as a secret key by Alice and Bob. The secrecy of the final key is guaranteed to an astonishingly high confidence level. In the very first quantum cryptography experiment a key of 105 bits was exchanged such that Eve's expected information about the key was estimated to be about 6×10^{-171} bits [1].

2.3   The BT Prototype System

It is currently possible to transmit keys securely over distances of at least 30km of optical fibre using quantum cryptography. With improvements in detector technology this distance should increase to around 100km or more. At BT it has been our aim to develop a usable quantum key distribution system that can operate over significant distances in standard telecommunications fibre. This has been achieved and the first BT prototype [5] implements the BB84 protocol operating at a wavelength of $1.3\mu m$ utilising an interferometric system that is capable of securely transmitting keys over distances of 30km at data rates of around 1kBit/s. This system has also demonstrated the feasibility of achieving key rates of 20kBit/s and greater, although technology limitations have forced us to run the prototype at the slower speed. We have also developed, and are in the process of developing, other prototype systems operating at various wavelengths which exploit different protocols and coding

122

schemes. These systems will enable us to explore some of the technology options for quantum key distribution. All of these systems are laboratory-based systems where the propagation takes place over optical fibre wound on a drum. It is, however, a trivial exercise to let the transmission take place over an installed fibre link. One advantage of working with laboratory fibre systems is that they tend to be less stable than their cabled, installed counterparts so that they can represent a "worst case" scenario.

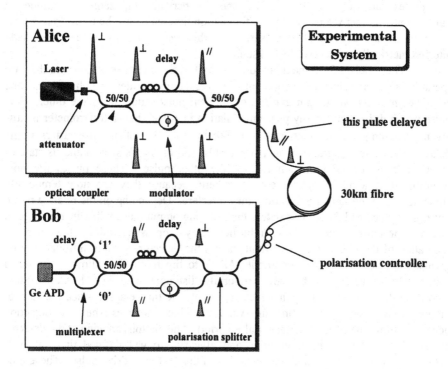

Fig. 4 A schematic diagram of the 30km phase-coded BT prototype interferometric quantum key distribution system. The system consists of a 30km long time- and polarisation-division interferometer in which Alice and Bob each control one half of the device.

The experimental prototype system [5] shown in figure 4 consists of a 30km-long fibre-based Mach-Zehnder interferometer which operates at a wavelength of 1.3μm. The laser source is a standard 1.3μm wavelength semiconductor laser. The pulses from this device are heavily attenuated to a level where the intensity at the input of the transmission fibre is equivalent to 0.1 photons per pulse pair, on average. This attenuated pulse enters an optical coupler, the optical fibre equivalent of a beamsplitter, where the pulse splits and one pulse travels through a lithium niobate phase modulator and experiences a phase shift chosen randomly from one of the four possibilities, 0°, 90°, 180°, 270°. The other travels through a polarisation controller set to act as a half wave plate and a delay loop. The half wave plate rotates the state of polarisation of the pulse to its orthogonal state. These two pulses, now with orthogonal polarisations, enter another optical coupler the output of which is fed directly into a 30km-long length of standard telecommunications fibre which is single-moded at a wavelength of 1.3μm. Because of the delay imposed on one pulse these pulses now travel a few nanoseconds apart in this fibre. The time delay between the two pulses must be set so that they both experience the same environmental fluctuations, in other words the typical fluctuation timescale must be much longer than the time delay between the pulses. In this way the device can be made interferometrically stable over long distances.

These pulses form the input to Bob's half of the interferometer where they are spatially separated by the action of a polarisation splitter which directs one polarisation along one output and the orthogonal polarisation along the other. The pulse which did not suffer any phase modulation in Alice's half-interferometer is now given a random phase modulation of 0° or 90° by Bob. The other pulse suffers a time delay of the same magnitude as its partner pulse did in Alice's system and its state of polarisation is rotated to match that of the other pulse. These pulses are now recombined at a 50/50 optical coupler where, because they are now temporally coincident with the same polarisation, they interfere. Depending on the relative phase settings of Alice and Bob's modulators the resulting output pulse will either emerge in one arm or both. One arm has a further delay loop which allows the temporal separation of the bits so that a '0' value causes the detector, a liquid nitrogen cooled germanium (Ge) avalanche photodiode (APD), to fire first. This allows us to use a single detector where the bits are temporally distinguished. The detector system records each event as a data pair which represents the time elapsed since the start of the key transmission and the time-interval value which indicates where the detection occurred within the concurrent laser pulse period. The photons arriving at the detector are synchronised with the laser drive and give rise to well-defined time intervals between the detector event and the beginning of the next laser drive pulse. These time interval values lie within two windows centred at around 614ns for the '1' output port with the longer path and around 620ns for the '0' output port. These windows are around 1ns wide due to detector timing jitter. In contrast, detector noise mechanisms such as dark counts and after-pulses give rise to counts randomly distributed in time and consequently give rise to random time interval values. After the data transmission is complete the receiver can use the recorded time interval values to classify each detection event as either a '1', a '0', or a noise count. This is shown in figure 5 where an actual data set from a typical key transmission is plotted. In this figure the transmission has been separated into two parts for clarity. In figure 5 we have plotted on separate graphs Bob's results for the time slots where a '1' was transmitted and the result of those time slots where a '0' was transmitted. By separating the data in this way we can see the extremely low error rates achieved in the experiment.

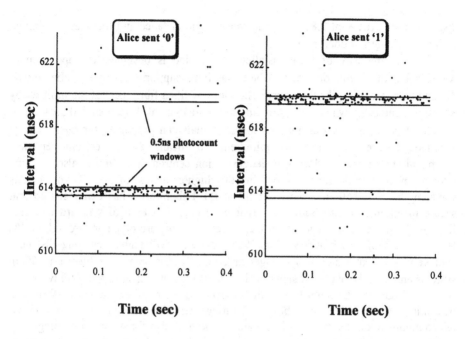

Fig. 5 A typical data set from a key transmission using the BT prototype. The data is taken from a transmission over 10km. The error rate is about 1% and the plots have been separated to emphasise this low count.

Typical error rates achieved on this system are around 4% for distances of 30km and 1.5% for distances of 10km. These errors arise mainly from the relatively high detector dark count and the less than unity fringe visibility which is of the order of 0.99. However, errors of this magnitude are perfectly acceptable and can be accommodated within the error correction and privacy amplification procedures. As the system length increases and more of the single photons are removed by loss processes in the fibre, the relative proportion of dark counts to data points increases giving rise to a higher received error rate. With improved detectors and phase modulators both the error rates and distance of transmission can be improved significantly.

There are a variety of commercially available detectors that can operate in single-photon counting mode for various regions of the spectrum. In general, Silicon (Si) APDs, designed to operate at wavelengths of around $0.8\mu m$, are at an advanced stage of development whereas Ge and Indium Gallium Arsenide (InGaAs) devices, designed for operation at wavelengths of $1.3\mu m$ or $1.5\mu m$ are less advanced. Si APDs for single photon counting have a number of attractive features. They have a high quantum efficiency of around 30%, they have a low dark count rate of less than 100 counts per second and they can be thermo-electrically cooled. However, fibre loss and dispersion are greater at $0.8\mu m$ so that transmission distance and bit rates are limited at this wavelength of operation.

The majority of fibre installed world-wide is designed for operation at wavelengths of $1.3\mu m$ or $1.5\mu m$ where the fibre supports a single spatial mode. Around these wavelengths the transmission loss of silica fibre reaches the limit set by Rayleigh scattering which is approximately 0.3dB/km at $1.3\mu m$ and 0.2dB/km at $1.5\mu m$. With such low transmission losses, signals can propagate for many tens of kilometres before regeneration is required either by means of opto-electronic repeaters or optical amplifiers. A third wavelength region centred on $0.8\mu m$ is also of some interest in telecommunications despite the higher losses of ~ 2dB/km at this wavelength. The 850nm prototype built at BT Laboratories has demonstrated that secure quantum key distribution is possible over at least 8km of standard telecoms fibre at this wavelength. The intrinsic error rates for this prototype are around the 1% level. It is unlikely, however, that 850nm systems will match the long distances achieved by $1.3\mu m$ systems. At longer distances loss becomes a problem for 850nm systems and the advantages of superior detectors at this wavelength are quickly lost.

Trade-offs between factors that depend on the required system performance determine the choice of 'best' operating wavelength for a conventional telecommunications system. This is likely to be true also for quantum cryptography systems using the same fibre transmission medium. Consequently, if quantum cryptography is to achieve a wide customer base, it is important to investigate prototype systems at one or all of these wavelengths to ensure compatibility with existing networks.

3. Future Directions

3.1 Other Quantum Key Distribution Protocols

The test for an eavesdropper in a quantum key distribution system amounts to a test of the integrity of the transmitted quantum state. In the BB84 protocol we can allow Eve an *unlimited* technical sophistication consistent with quantum mechanics and still be confident that she will be unable to successfully eavesdrop on the quantum channel without causing errors. Other quantum key distribution systems have been proposed, all of which are designed to force Eve into causing errors on the quantum channel if she attempts to eavesdrop.

The only other quantum key distribution protocol to have been experimentally demonstrated is the 2-state protocol invented by Charles Bennett [12] and termed the B92 protocol. This can be thought of as "half" of the BB84 protocol in that just one state from two incompatible bases are chosen. To guarantee security using this protocol requires a little more effort than in the BB84 protocol because the eavesdropper knows whether her result is correct or inconclusive. She can therefore adopt a "suppression" attack in which she sends on only those results she knows to be correct and compensates for the decrease in intensity. This attack can be detected but the security of the proposed optical implementation resides, in part, in the existence of a reference pulse of higher intensity which is susceptible to substitution by Eve. Nevertheless, the group at Los Alamos have demonstrated secure key distribution using this protocol over 14km of optical fibre [7].

A potentially useful class of protocols which can be adopted independently of, or in conjunction with, protocols of the BB84 type are the "rejected-data" protocols [13]. These protocols are designed so that an eavesdropper unavoidably disturbs the statistics on the data that would normally be rejected in the key distribution protocol. However, because these tests for an eavesdropper only involve data that would normally be discarded they can be used in conjunction with the standard protocols which use some potential key bits in their test for an eavesdropper. Information about the eavesdropping can be obtained from the normally discarded data. This information can be useful in the privacy amplification procedure where an estimate of the possible information leakage to an eavesdropper is required.

Another class of protocol involving a seemingly different physical phenomenon has also been proposed. This protocol, the EPR protocol, is an ingenious application of the Bell inequality to eavesdropper detection [14]. In this scheme Alice and Bob each receive a particle from a correlated particle pair. The essence of the idea is that correlation in quantum mechanics is of a different character to that in classical mechanics and that an active interception by an eavesdropper disturbs this fragile quantum correlation and reduces it to one of classical character. The Bell inequality [15] can be used as a statistical test for this quantum character in that violation of the inequality implies the existence of a quantum correlation. Experimental systems based on the EPR protocol have been proposed and progress is being made towards a full demonstration but as of yet even the feasibility of an EPR quantum key distribution system has not been experimentally demonstrated. This is in part due to the difficulty of manipulating correlated particles in the laboratory and the more stringent requirements on detector technology to observe violations of the Bell inequality. However, it has been shown that the EPR protocol can be adapted to work with *single* particles [16] achieving precisely the same degree of security. Thus experimental systems operating an EPR protocol are a real feasibility with small adaptations of the existing single-particle implementations of the BB84 protocol, for example.

3.2   Quantum Key Distribution on Optical Networks

Quantum key distribution is an intriguing and exciting possibility. However, its implementation on point-to-point links, as we have so far discussed, is only of limited applicability. If quantum key distribution can be made to work on the next generation of optical networks its potential impact could be considerable.

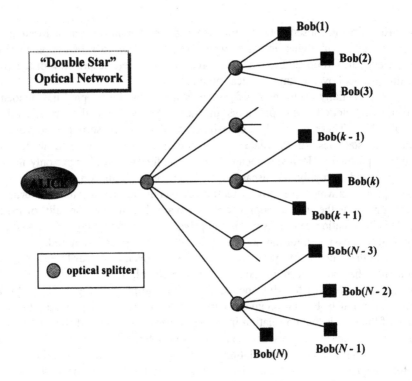

Fig. 6 A schematic illustration of a "double star" optical network. Alice plays the role of broadcaster/gatherer and each of the Bobs is connected to Alice via two layers of optical splitter. A classical signal is split at each layer and an identical copy is received by each Bob. A signal transmitted on single photons, however, behaves very differently

At BT we have developed several techniques that will allow the implementation of these quantum techniques on optical networks [17]. For the purposes of brevity we shall consider here only one network configuration, that of a branched or tree configuration network [17]. A schematic illustration of such a network is shown in figure 6. Alice now plays the role of broadcaster/gatherer and there are now N Bobs, labelled Bob(1) to Bob(N), who can receive downstream signals from Alice and send messages in the upstream direction. The network configuration that we have sketched in figure 6 is known as a "double star" and it has two layers of optical splitters. A classical multi-photon signal from Alice will be split at these points and a copy of the signal will travel along each emergent path. Eventually each Bob will receive a copy of the original signal transmitted by Alice. Single photons, as we have seen, behave in a very different way at optical splitters.

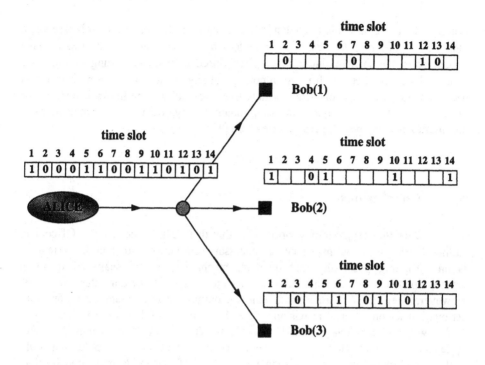

Fig. 7 A simplified version of the kind of optical network sketched in figure 6. Here there are only 3 Bobs. Alice transmits a sequence of single photons onto the network and each Bob receives a unique, but random, subset of this sequence. Which photon eventually arrives at a given Bob cannot be predicted in advance. Alice has, in effect, set up 3 independent point-to-point quantum key distribution links on the network.

A single photon sent by Alice cannot be split or copied so that at each splitting layer it will be found in one, and one only, of the possible outputs. The consequence of this is that any single photon input by Alice at the head end will be received by one, *and one only*, of the Bobs. Which Bob receives any given single photon is purely a matter of probability. Thus, in order to establish secret and individual keys with each of the Bobs, Alice sends a randomly coded sequence of single photons as before for the point-to-point quantum key distribution scheme. Each photon in this initial sequence percolates through the network and reaches one of the Bobs. Which Bob receives a given photon is indeterministic and the sequence of time slots for which a given Bob receives a photon will differ for each separate key transmission by Alice. Therefore a random, and unique, subset of Alice's transmission is received by each Bob. This procedure is equivalent to setting up N distinct point-to-point quantum cryptography links between Alice and each of the Bobs. On average, therefore, assuming equal splitting ratios at each layer, each Bob receives a binary string of length D/N where D is the length of Alice's initial transmission. This procedure is sketched in figure 7 for a simplified network consisting of only one splitting layer and 3 Bobs. It is worth noting that any of the quantum key distribution protocols, including the correlated particle EPR protocol, can be implemented on this optical network. Techniques for using quantum cryptography on other network configurations have been discussed elsewhere [17].

4.   Conclusions

Quantum cryptography works - of this there can be no doubt. Of course, further development and investment are necessary to produce a commercial system. It is an intriguing possibility that quantum mechanics, in the shape of quantum cryptography, will have a significant impact on the security of our future networks. However, if recent ideas prove practicable, quantum mechanics could have a far more serious impact on security provision. Peter Shor of AT&T Laboratories has shown [18] how quantum mechanics can be used to attack some of the most popular cipher systems in use today. He has devised an algorithm for a quantum computer that could perform the operation of factorisation *many* orders of magnitude more quickly than the equivalent calculation on a classical machine. Although we are still some years away from having the technical capacity to build a quantum computer sufficiently complex for factoring numbers of a hundred decimal digits, for example, public key cipher systems are clearly not future-proofed against quantum computers. Surprisingly, perhaps, quantum key distribution systems are *invulnerable* to this threat from quantum computers. Even a quantum computer can't beat the uncertainty principle. It is our guess that as technology edges nearer the capability of building a quantum computer we shall see a renewed interest in secret key cipher systems supported by the invulnerability of quantum key distribution.

References

1.   C. H. Bennett, F. Bessette, G. Brassard, L. Salvail and J. Smolin, "Experimental Quantum Cryptography", *Journal of Cryptology*, **5** 3-28 (1992).

2.   S. Wiesner, "Conjugate Coding", *Sigact News*, **15** 78-88 (1983), original manuscript written circa 1970.

3.   P. D. Townsend, J. G. Rarity and P. R. Tapster, "Single-Photon Interference in a 10km Long Optical Fibre Interferometer", *Electronics Letters*, **29** 634-635 (1993); P. D. Townsend, J. G. Rarity and P. R. Tapster, "Enhanced Single-Photon Fringe Visibility in a 10km Long Prototype Quantum Cryptography Channel", *Electronics Letters*, **29** 1292-1293 (1993); P. D Townsend and I. Thompson, "A

Quantum Key Distribution Channel Based on Optical Fibre", *Journal of Modern Optics*, **41** 2425-2434 (1994).

4.   A. Muller, J. Breguet and N. Gisin, "Experimental Demonstration of Quantum Cryptography using Polarised Photons in Optical Fibre over more than 1km", *Europhysics Letters*, **23** 383-388 (1993).

5.   P. D. Townsend, "Secure Key Distribution System Based on Quantum Cryptography", *Electronics Letters*, **30** 809-810 (1994); C. Marand and P. D. Townsend, "Quantum Key Distribution Over Distances as long as 30km", *Optics Letters*, **20** 1695-1697 (1995).

6.   J. D. Franson and H. Ilves, "Quantum Cryptography using Optical Fibres", *Applied Optics*, **33** 2949-2954 (1994); J. D. Franson and H. Ilves, "Quantum Cryptography using Polarisation Feedback", *Journal of Modern Optics*, **41** 2391-2396 (1994).

7.   R. J. Hughes, G. G. Luther, G. L. Morgan and C. Simmons "Quantum Cryptography over 14km of Installed Optical Fibre", to be published in "*Proceedings of the Seventh Rochester Conference on Coherence and Quantum Optics*".

8.   C. H. Bennett and G. Brassard, "Quantum Cryptography : Public-Key Distribution and Coin Tossing", in *Proceedings of IEEE International Conference on Computers, Systems and Signal Processing*, Bangalore, India, 175-179 (1984).

9.   W. K. Wootters and W. H. Zurek, "A Single Quantum Cannot be Cloned", *Nature*, **299** 802-803 (1982).

10.   S. J. D. Phoenix, "Quantum Cryptography without Conjugate Coding", *Physical Review A*, **48** 96-102 (1993).

11.   C. H. Bennett, G. Brassard and J.-M. Robert, "Privacy Amplification by Public Discussion", *SIAM Journal on Computing*, **17** 210-229 (1988). The procedure is also outlined in reference [1]; C. H. Bennett, G. Brassard, C. Crepeau and U. M. Maurer, "Generalised Privacy Amplification", *IEEE Transactions on Information Theory*, in press.

12.   C. H. Bennett, "Quantum Cryptography Using Any Two Non-Orthogonal States", *Physical Review Letters*, **68** 3121-3124 (1992).

13.   S. M. Barnett and S. J. D. Phoenix, Information-Theoretic Limits to Quantum Cryptography", *Physical Review A*, **48** R5-R8 (1993); S. M. Barnett, B. Huttner and S. J. D. Phoenix, "Eavesdropping Strategies and Rejected-Data Protocols in Quantum Cryptography", *Journal of Modern Optics*, **40** 2501-2513 (1993).

14.   A. K. Ekert, "Quantum Cryptography Based on Bell's Theorem", *Physical Review Letters*, **67** 661-663 (1991).

15.   J. S. Bell, "On the Einstein-Podolsky-Rosen Paradox", *Physics*, **1** 195-200 (1964).

16.   S. M. Barnett and S. J. D. Phoenix, "Bell's Inequality and Rejected-Data Protocols for Quantum Cryptography", *Journal of Modern Optics*, **40** 1443-1448 (1993).

17.   P. D. Townsend, S. J. D. Phoenix, K. J. Blow and S. M. Barnett, "Design of Quantum Cryptography Systems for Passive Optical Networks", *Electronics Letters*, **30** 1875-1877 (1994); S. J. D. Phoenix, S. M. Barnett, P. D. Townsend and K. J. Blow, "Multi-User Quantum Cryptography on Optical Networks", *Journal of Modern Optics*, **42** 1155-1163 (1995).

18.   P. Shor, "Algorithms for Quantum Computation : Discrete Logarithms and Factoring" in *Proceedings of the 35th Annual Symposium on Foundations of Computer Science* (IEEE Computer Society Press, 1994) pp.124-134.

Prepaid Electronic Cheques Using Public-Key Certificates

Cristian Radu, René Govaerts and Joos Vandewalle

Katholieke Universiteit Leuven, Laboratorium ESAT
Kardinaal Mercierlaan 94, B-3001 Heverlee, Belgium

Abstract. An electronic payment system which emulates the payments relying on classical cheques is presented. Its main features are a high computational efficiency and low storage requirements for the tamper resistant smart card implementing the purse of the user. This is achieved using a counter-based solution in combination with public-key certificates that represent the electronic cheques. To simplify both the withdrawal and the deposit stage we renounce to provide the untraceability feature. Our payment system is built around the Guillou-Quisquater identification/signature scheme in an RSA-group.

1 Introduction

During the last few years, the trend of the cryptographic research in relation with off-line electronic payment systems is to provide electronic money which offers at least the properties that real money offers [4, 5, 1, 2, 3]. This means that two goals have to be achieved simultaneously by the payment system designer [6].

The first goal is to provide integrity for all the participants in the system. Basically, this means the unforgeability and uncopiability of the electronic money. The best solution is to guarantee that this feature is independent of the physical requirements related to the electronic purse of the user. This can be achieved using a coin-based system. However, a first drawback is that each coin denomination has to be authenticated by the bank. Another important inconvenience is that the computation and communication complexity for paying are dependent of the specific amount. Therefore, we have opted for a counter-based solution. A tamper-resistant smart card, issued to a user by the bank, holds a counter that represents the amount of electronic money which can be spent by the user, playing the role of a payer during the payment transactions. This implies that, beside the cryptographic assumptions used in the system, the integrity of the bank also relies on a non-cryptographic assumption: *the cost of breaking a smart card in practice will significantly exceed the expected financial profit.*

The second goal is that electronic money should be untraceable. This feature is related to the privacy of the user. In order to provide this feature, two cryptographic primitives are required. The first is a blind signature scheme which guarantees that the identity of the user is encoded in each piece of electronic money, either coins or cheques. The second primitive is a double-spending detection mechanism, which reveals the identity of a dishonest user who attempts to copy electronic money and to use it several times. This involves the management of a large database, where the bank keeps all the transcripts of the payment transactions during a fixed period of time. This is an expensive operation that can exceed the profit which the bank would obtain by providing the "retail" electronic money service.

Therefore, in our paper we are going to present the design of an efficient *off-line traceable counter-based* payment system. The solution relies on the combination between a counter representing the amount of money the user can spend and a set of public key certificates released by the bank, which are used as electronic cheques. Their authenticity is guaranteed by the cryptographic assumption of the intractability of computing RSA-roots. In Section 2 we present the participants in the system and the basic concepts. In the third section we detail the Guillou-Quisquater identification scheme adapted to the RSA-groups and the basic digital signature scheme derived from it. In Section 4 the setup of the system is presented. In the last section we design the protocols of the main transactions: withdrawal, payment and deposit. Finally, our conclusions are presented.

2 A General View of the Payment System

This section provides an overview of the payment system. The participants are the bank $\mathcal{B}$ and its clients, either users $\mathcal{U}_i$ or shops $\mathcal{S}_j$. In order to simplify the system, we do not deal with users and shops that are clients of different banks. During the transactions, the purse of the user $\mathcal{U}_i$ is represented by an electronic tamper-resistant device $\mathcal{T}_i$, usually a smart card. At registration stage, the bank issues this device to the user. It also generates the *user's secret identity i* and *user's public identifier ID_i*, which are kept within the non-volatile memory of $\mathcal{T}_i$.

A *certified key pair, ckp*, is a tuple consisting of a secret key sk, a public key pk, the user's public identifier and a certificate of the bank on the last two items $\sigma_S(pk, ID_i)$. A *certified public key, CPK*, is a triple consisting of the public key, the user's public identifier and the certificate of the bank, $\sigma_S(pk, ID_i)$. The participants to the payment system use electronic cheques. These are represented by certified public keys CPK. Each electronic cheque C of a user is represented by a certified public key CPK.

During withdrawal, the bank signs (pk, ID_i), by releasing the certificate $\sigma_S(pk, ID_i)$. To this end, the bank uses a signature scheme, derived from the Guillou-Quisquater identification scheme [8, 9]. We denote by (S, P) the secret key and the corresponding public key used by the bank to construct the certified public keys for its users.

The certificate issued by the bank, $\sigma_S(pk, ID_i)$, allows the shop to verify the authenticity of a certified public key CPK, in relation with $\mathcal{U}_i$, and its subsequent acceptability at the deposit stage. A payment using a CPK consists of a signature executed by $\mathcal{T}_i$, with respect to the secret key sk (corresponding to the public key pk in the certified key pair ckp), on a message m that specifies the transaction. We refer to this signature as *the payment signature* and we denote it by $\Sigma_{sk}(m)$. During the deposit stage, S_j forwards a transcript of the payment transaction to the bank. This transcript includes the certified public key, CPK, and the pair $(m, \Sigma_{sk}(m))$, which specifies the payment transaction. All the transcripts are recorded by the bank in the transcripts database. Every CPK can only be used once with respect to any user and any payment transcript.

3 The Basic Signature Scheme

In the design of the protocols describing the withdrawal, payment and deposit transactions, we use a signature scheme derived from the Guillou-Quisquater identification scheme. Therefore, we briefly outline this cryptographic primitive, considering the appropriate instance for our payment system. The prover $\mathcal{P}$ generates a secret key (p, q, S). The numbers p and q are primes such that the composite modulus $n = pq$ is of a given length. We denote $\mathbf{Z}_n = \{0, 1, 2, \ldots, n-1\}$ and $\mathbf{Z}_n^* = \{x \in \mathbf{Z}_n | \gcd(x, n) = 1\}$. The secret parameter S is chosen at random from $\mathbf{Z}_n^*$. All the arithmetical operations are performed in the RSA-group $(\mathbf{Z}_n^*, \cdot, 1)$, where the group operation is the multiplication modulo n, and the unit of the group is denoted by 1. It is important to mention that in the RSA-group there exist efficient polynomial-time algorithms for multiplication, inversion, exponentiation, selecting random elements, determining equality of elements, and testing group membership. The prover computes the corresponding public key, (n, v, P), where n is the modulus, v is an RSA-exponent, with $\gcd(v, \varphi(n)) = 1$, and $P \in \mathbf{Z}_n^*$ satisfies the equation $P \cdot S^v = 1$ in the RSA-group.

The GQ identification scheme is a three-step challenge-response protocol. It allows $\mathcal{P}$ to prove the knowledge of its secret key to the verifier $\mathcal{V}$. The following protocol is a version of [9], restricted to the RSA-group $(\mathbf{Z}_n^*, \cdot, 1)$:

1. $\mathcal{P}$ generates at random a commitment r in $\mathbf{Z}_n^*$ and sends the initial witness $T \leftarrow r^v$ to $\mathcal{V}$.
2. $\mathcal{V}$ generates a challenge $d \in \mathbf{Z}_v$, and sends it to $\mathcal{P}$.
3. $\mathcal{P}$ generates the response $t \leftarrow r \cdot S^d$ and sends it to $\mathcal{V}$.

The verifier computes the final witness $T' \leftarrow t^v \cdot P^d$ and accepts the proof of knowledge if and only if $T' \stackrel{?}{=} T$.

From here on a composition of two elements will be denoted by ab instead of $a \cdot b$, each time that it is clear that the composition holds in the RSA-group. If $\mathcal{V}$ generates its challenge according to a uniform probability distribution, the protocol provides a "zero-knowledge" proof. The Guillou-Quisquater identification scheme is presented in Figure 1.

Using a technique described in [7], the GQ identification scheme can be transformed in a signature scheme. In order to sign a message m, the challenge d is computed by the signer S as $\mathcal{H}(m, T)$, where $\mathcal{H}(\cdot)$ is a collision resistant hash function [10], mapping arbitrary long inputs to a fixed length output in $\mathbf{Z}_v$. Thus, the interaction between S and $\mathcal{V}$ is replaced and the signature scheme becomes a

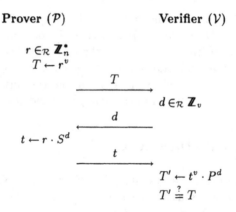

Fig. 1. The GQ identification scheme restricted to the RSA-group

one-step protocol. S generates at random the commitment r in $\mathbb{Z}_n^*$ and computes the initial witness $T \leftarrow r^v$ together with the initial challenge $d \leftarrow \mathcal{H}(m, T)$ and the corresponding response $t \leftarrow rS^d$. The signature, consisting of $m, (d, t)$, is sent to the verifier. V computes the final witness $T' \leftarrow t^v P^d$ and the final challenge $d' \leftarrow (m, T')$. V accepts the signature if the final challenge is equal to the initial challenge. The digital signature scheme derived from the Guillou-Quisquater identification scheme is presented in Figure 2.

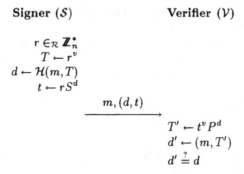

Fig. 2. The digital signature scheme derived from the GQ identification scheme

4 The General Setup of the System

Let $\mathcal{H}(\cdot)$ be a collision resistant hash function mapping arbitrary long inputs to an output of fixed length $|\mathcal{H}|$. The bank generates a secret set of parameters (p, q, S), where p and q are primes selected such that the modulus $n = pq$ is of a given length. The secret parameter S is chosen at random in $\mathbf{Z}_n^*$. The bank publishes a set of parameters $(n, v, \mathcal{H}(\cdot), P)$, where $n = pq$ is the composite modulus, v is an RSA-exponent such that $|v| > |\mathcal{H}|$ and P satisfies the equation $PS^v = 1$.

All the participants in the system can do arithmetics in the RSA-group $(\mathbf{Z}_n^*, \cdot, 1)$. The bank $\mathcal{B}$ also manages two databases, one dealing with information related to the users – the account database – and the other with transcripts of the payment transactions – the transcripts database.

When a person becomes a user $\mathcal{U}_i$ of the payment system, he opens an account with the bank $\mathcal{B}$. The bank generates an appropriate entry in the account database and issues to him the tamper-resistant device T_i. The bank also generates the secret identity of the user, such that $i \in_{\mathcal{R}} \mathbf{Z}_n^*$ and computes $I = i^{-v}$, which is stored in the account database together with the public identifier of $\mathcal{U}_i$ denoted by ID_i.

The shops that subscribe to the payment system, do not need a tamper-resistant device, but only have to be clients of $\mathcal{B}$. The bank also grants a public identifier ID_S for each shop.

5 The Protocols of the Main Transactions

The main transactions of the system are withdrawal, payment and deposit. The protocols of these transactions are detailed in the next three subsections.

5.1 The Withdrawal Protocol

The withdrawal protocol is fulfilled between the smart card of the user and the bank. It consists of two stages. In the first stage, the counter kept with T_i, representing the balance of the smart card, is increased according to the amount claimed by the user. To this end, the user and the bank follow a mutual authentication protocol. The protocol describing this stage is presented in Figure 3.

First, the bank releases a signature (d_a, t_a), with respect to its key pair (S, P), on the challenge m_a of the user. Including this challenge in the message signed by the bank, protection against replay attacks is provided. Therefore, no attacker is able to repeat the identification of the bank, thereby deceiving the card. If the signature verification holds true, T_i is convinced about the authenticity of the bank.

In the next step, the smart card T_i signs, using its key pair (i, I), a message including the current value of its balance bal and the number n of the electronic cheques, which are still available in the card. The message also includes the item t_a of the bank's signature. This inclusion prevents a replay attack, which would enable a malicious party to take advantage of the user's account. This signature is denoted (d, t). It serves to identify the user and the smart card to the bank and to give the bank authentic information about the current value of the card's balance. The latter prevents the smart card from exceeding the maximum value max_bal of the balance.

If the verification of the card's signature is successful, the bank links the user to his account, considering the request for money if the desired amount does not exceed the balance of his account. If so, the bank signs a message, again with respect to the key pair (S, P), entitling T_i to increase its balance by the amount required. The message includes the card's response t from the previous signature. This guarantees that the signature of the bank, denoted (d_u, t_u), is directed to the card that just successfully identified itself. Moreover, no replay attacks that use older signatures of the bank can deceive the card to increase its balance. The account of the user is correspondingly decreased.

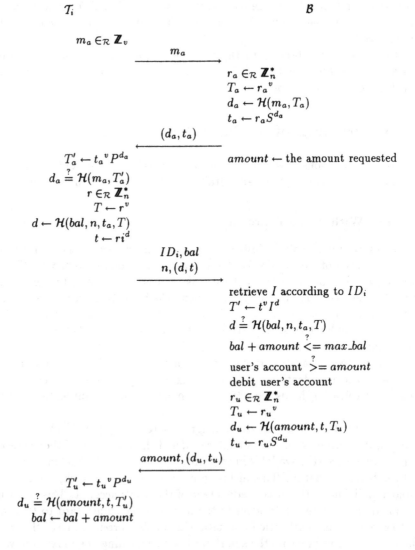

Fig. 3. The first stage of the withdrawal protocol

In the second stage of the withdrawal protocol, the bank issues the electronic cheques for $\mathcal{U}_i$. Using the terminology introduced in Section 2, an electronic cheque C is a certified public key CPK

$$C = CPK = (pk, ID_i, \sigma_S(pk, ID_i)),$$

where $\sigma_S(pk, ID_i) = (d_b, t_b)$ is the certificate such that:

$$d_b = \mathcal{H}(pk, ID_i, t_b{}^v P^{d_b}).$$

During the withdrawal protocol, a certified key pair

$$ckp = (sk, C) = (sk, pk, ID_i, \sigma_S(pk, ID_i))$$

is issued such that $pk \cdot sk^{-v} = 1$ and $(pk, ID_i, \sigma_S(pk, ID_i))$ is a certified public key. The protocol presented in Figure 4 is repetitively executed until the maximum number of cheques is stored in the non-volatile memory of T_i.

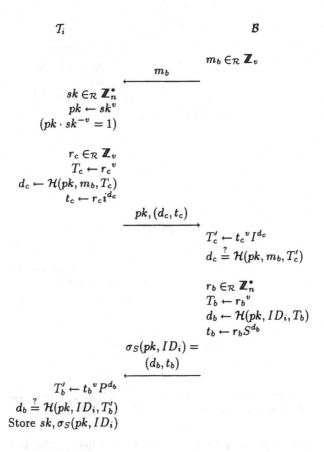

Fig. 4. The second stage of the withdrawal protocol - the withdrawal of electronic cheques

5.2 The Payment Protocol

A payment is carried out between the smart card and the Point-of-Sale (POS) terminal of the payee S_j. The smart card is inserted in the POS and the amount to be paid *amount* is entered by the payee and displayed for the payer. If the payer agrees with this amount, the payment protocol starts. The protocol is presented in Figure 5.

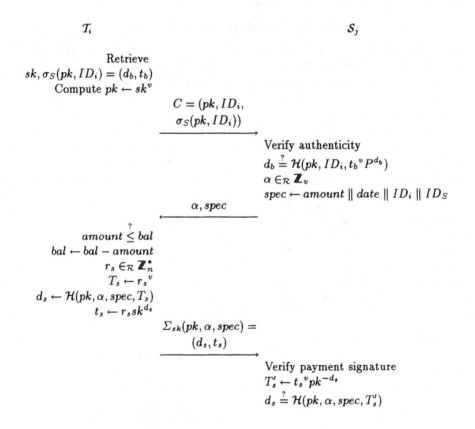

Fig. 5. The payment protocol

The card retrieves $sk, \sigma_S(pk, ID_i) = (d_b, t_b)$ and recomputes pk. Subsequently, it sends the corresponding cheque $C = CKP = (pk, ID_i, \sigma_S(pk, ID_i))$ to the payee.

The payee verifies the authenticity of the certified public key representing the cheque he just received and specifies the message *spec* that describes the current payment transaction. This message includes *amount* and the date *date* when the transaction takes place. To prevent anyone else than the payee from depositing a cheque, the identity of the payee ID_S is included in the message *spec* too. Moreover, measures against replay attacks have to be taken such that the payee does not accept the same cheque filled in with the same amount in different transactions. Therefore, the message that the smart card signs has to include a random number α chosen by the payee for each payment. The payee sends α and *spec* to the card.

The smart card makes a signature on these two items with respect to the secret key sk, corresponding to the certified public key CKP, implementing the current cheque. This signature is denoted $\Sigma_{sk}(pk, \alpha, specs)$ and is referred to as the *payment signature*. The card sends this signature to the payee.

The correctness of the payment signature is verified by the payee, who forwards the purchases to the payer.

5.3 The Deposit Protocol

In order to get his account credited by the bank, the shop deposits the transcript of the transaction together with the electronic cheque used during the payment. The bank performs the same verifications as the payee did during payment. Thus, the authenticity of the cheque and the correctness of writing it out are verified. Moreover, the bank checks whether this cheque has not been deposited before. If not, the payee is credited the amount specified in the transcript of the payment transaction. The protocol to achieve this, is depicted in Figure 6.

$$\mathcal{S}_j \hspace{6cm} \mathcal{B}$$

$$\text{Retrieve}$$
$$C = (pk, ID_i, \sigma_S(pk, ID_i))$$
$$\alpha, spec$$
$$\Sigma_{sk}(pk, \alpha, spec) = (d_s, t_s)$$

$$C, \alpha, spec,$$
$$\Sigma_{sk}(pk, \alpha, spec)$$
$$\xrightarrow{\hspace{3cm}}$$

$$\text{Verify authenticity}$$
$$d_b \stackrel{?}{=} \mathcal{H}(pk, ID_i, t_b{}^v P^{d_b})$$
$$\text{Verify double-deposit of } C$$
$$\text{Verify correctness of } C$$
$$T_s' \leftarrow t_s{}^v pk^{-d_s}$$
$$d_s \stackrel{?}{=} \mathcal{H}(pk, \alpha, spec, T_s')$$
$$\text{Verify double-spending with } \mathcal{U}_i$$

Fig. 6. The deposit protocol

The bank indexes the cheques according to the identity of the user and verifies that no cheque is spent more than once by a user. If so, the smart card of the user is blacklisted. No withdrawals or payments involving this card are furthermore accepted.

6 Conclusions

The off-line payment system we propose is efficient and simple. The solution is suitable for small tamper resistant devices like smart cards. It is developed in the framework of RSA-groups, around the Guillou-Quisquater identification scheme, using the public-key certificate technique. Future work will analyze the possibility to provide untraceability.

References

1. S. Brands, *Off-line cash transfer by smart cards*, Report CS-R9323, Centrum voor Wiskunde en Informatica, March 1993.
2. S. Brands, "Untraceable off-line cash in wallet with observers," *Advances in Cryptology, Proc. Crypto'93, LNCS 773*, D. Stinson, Ed., Springer-Verlag, 1994, pp. 302–318.
3. S. Brands, "Off-line cash transfer by smart cards," *Proc. First Smart Card Research and Advanced Application Conference*, V. Cordonnier and J.-J. Quisquater, Eds., Lille (F), 1994, pp. 101–117.
4. D. Chaum, A. Fiat, M. Naor, "Untraceable electronic cash," *Advances in Cryptology, Proc. Crypto'88, LNCS 403*, S. Goldwasser, Ed., Springer-Verlag, 1990, pp. 319–327.
5. D. Chaum, B. den Boer, E. van Heijst, S. Mjølsnes, and A. Steenbeek, "Efficient off-line electronic checks," *Advances in Cryptology, Proc. Eurocrypt'89, LNCS 434*, J.-J. Quisquater and J. Vandewalle, Eds., Springer-Verlag, 1990, pp. 294–301.
6. D. Chaum, "Achieving electronic privacy," *Scientific American*, Vol. 267, No. 2, 1992, pp. 96–101.
7. A. Fiat and A. Shamir, "How to prove yourself: Practical solutions to identification and signature problems," *Advances in Cryptology, Proc. Crypto'86, LNCS 263*, A.M. Odlyzko, Ed., Springer-Verlag, 1987, pp. 186–194.
8. L.C. Guillou and J.-J. Quisquater, "A 'paradoxical' identity-based signature scheme resulting from zero-knowledge," *Advances in Cryptology, Proc. Crypto'88, LNCS 403*, S. Goldwasser, Ed., Springer-Verlag, 1990, pp. 216–231.
9. L.C. Guillou, M. Ugon, and J.-J. Quisquater, "The smart Card: A standardized security device dedicated to public cryptology, " in *"Contemporary Cryptology: The Science of Information Integrity,"* G.J. Simmons, Ed., IEEE Press, 1991, pp. 561–613.
10. B. Preneel, *Analysis and design of cryptographic hash functions*, PhD thesis, Katholieke Universiteit Leuven, ESAT Department, Leuven (B), January 1993.

How Traveling Salespersons Prove Their Identity

Stefan Lucks

Institut für Numerische und Angewandte Mathematik
Georg–August–Universität Göttingen
Lotzestr. 16–18, D–37083 Göttingen, Germany
(email: lucks@namu01.gwdg.de)

Abstract. In this paper a new identification protocol is proposed. Its security is based on the *Exact Traveling Salesperson Problem* (XTSP). The XTSP is a close relative of the famous TSP and consists of finding a Hamiltonian circuit of a given length, given a complete directed graph and the distances between all vertices. Thus, the set of tools for use in public-key cryptography is enlarged.

1   Introduction

In public-key cryptography it is common to base the security of a cryptosystem on the hardness of number theoretical problems. This remains true for zero-knowledge identification schemes. Motivations to consider other problems are:

- Cryptosystems based on number theory tend to be only moderately efficient, since they typically depend on multiplying large numbers.
- It is dangerous to have all eggs in one basket, i.e. to depend completely on the same source of problems.

In 1989 Shamir [10] published an identification scheme based on an NP-hard algebraic problem, the *Permuted Kernel Problem*. For results about its hardness see [8] and references therein. Two more recent approaches came from Stern. He considered a coding problem, *Syndrome Decoding* [11] and a combinatorial problem, *Constrained Linear Equations* [12]. Both are NP-hard. This year, Pointcheval [9] proposed another identification scheme, based on the NP-hard *permuted Perceptrons Problem* from the theory of learning machines.

All four identification schemes are provably secure—if it is infeasible to solve the underlying problem.

Such an identification scheme's security is lost if the underlying problem is tractable. This motivates us to consider the *Traveling Salesperson Problem* (TSP), one of the oldest and most prominent problems in algorithm and computational complexity theory. It is NP-hard, but has been studied long before the theory of NP-hardness was developed, see [4], chapter 1.

This paper introduces a new scheme, based on the XTSP (*Exact TSP*), a close relative of the TSP where one looks for a Hamiltonian circuit of a given length—instead of the shortest one. Recently this problem was considered for secret-key cryptography, and constructions for one-way hash functions and pseudorandom bit generators based on the XTSP were given [5].

2 The Underlying Problem

In the following $A = (a_{i,j})$ is an $n \times n$-matrix with $a_{i,i} = 0$. By "randomly chosen" we always mean "chosen at random according to the uniform probability distribution". For $i \neq j$, $a_{i,j}$ is randomly chosen from $\{0, \ldots, 2^{l(n)} - 1\}$. We think of A as the distance matrix for distances in the complete directed Graph G_n with n vertices. Therefore the XTSP is actually a family of problems depending on the parameter $l(n)$.

In this paper we only deal with asymmetric XTSPs (i.e. in general $a_{i,j} \neq a_{j,i}$), but our scheme can easily be adopted to symmetric XTSPs.

Any Hamiltonian cycle X for G_n can be coded as an integer with

$$\lceil \log_2((n-1)!) \rceil \quad \text{Bits.}$$

By $\text{Length}_A(X)$ we mean the length of X with respect to A. Given a number B, the XTSP is to find a Hamiltonian cycle X with

$$\text{Length}_A(X) = B.$$

Choosing $l(n) \approx \log_2((n-1)!)$ leads to at least as hard XTSPs as any other choice for $l(n)$ (see [5]), so $l(n) \approx \log_2((n-1)!)$ will be used for our identification scheme.

It is easy to prove the NP-hardness of the XTSP and the NP-completeness of the corresponding existence problem. But since the theory of NP-completeness only deals with worst-case complexity, we need a stronger assumption than simply NP$\neq$P.

Let $l(n) \approx \log_2((n-1)!)$. Given a $n \times n$ distance matrix $A = (a_{i,j})$ with randomly chosen elements $a_{i,j} \in \{0, \ldots, 2^{l(n)}\}$ and the length B of a randomly chosen secret Hamiltoninan cycle X, the *XTSP intractability assumption* states that it is infeasible to find a Y such that $\text{Length}_A(Y)$ holds with non negligible probability.

As usual, a problem is said to be *infeasible*, if there is no efficient algorithm to solve it. We say, a (possibly probabilistic) algorithm is *efficient*, if it computes a correct solution in polynomial time (in n) with *non negligible* probability, i.e. with a probability bounded from below by $1/P(n)$ for a polynomial P.

Note that we could as well restrict ourselves to the modular XTSP, where the length of a Hamiltonian cycle is measured modulo $2^{l(n)}$, cf. [5].

3   The Identification Scheme

Alice wants to prove her identity. To archive this, Alice runs through our protocol to demonstrate her possession of a secret knowledge. Our scheme is zero-knowledge, i.e. the verifier learns nothing about the secret, except that the prover is in its possession.

We need a collision resistant hash function $H : \{0,1\}^* \longrightarrow \{0,1\}^l$, where given a hash output h, it is infeasible to compute statistical information about a hash input x with $h = H(x)$. In the following we assume that H has these properties—in addition to the XTSP intractability assumption. Shamir [10] and Stern [11], [12] make similar use of a hash function.

If the hash function H were not collision resistant, our scheme would be vulnerable to attacks similar to those described in [2]. Shamir [10] and Stern [11] were not aware of that problem.

Let π be a permutation over $\{1, \ldots, n\}$. (W. l. o. g. we may as well define π to be a permutation over $\{2, \ldots, n\}$ and set $\pi(1) = 1$.) Then $\pi(A)$ corresponds to the distance matrix where the vertices are permuted according to π, i.e. $\pi(A) = \pi((a_{i,j})) = (a_{\pi(i)\pi(j)})$. Similarly we define $\pi(X)$ for a Hamiltonian cycle X, thus

$$\text{Length}_{\pi(A)}(\pi(X)) = \text{Length}_A(X).$$

Given two distance matrixes $A = (a_{i,j})$ and $A' = (a'_{i,j})$, their sum $A + A' = (a''_{i,j})$ is defined component-wise, i.e. $a''_{i,j} = a_{i,j} + a'_{i,j}$.

Let the matrix A and the length B be known to the public. Alice's secret is a Hamiltonian cycle X with $\text{Length}_A(X) = B$. Alice proves her identity to the verifyer Bob as follows.

1. Alice chooses at random a matrix A^* and a permutation π over $\{1, \ldots, n\}$.
   She computes $B^* = \text{Length}_{A^*}(X')$, $h_0 = H(A^*)$,
   $$\begin{aligned} X' &= \pi(X), & h_1 &= H(X'), \\ A' &= \pi(A) \quad \text{and} & h_2 &= H(A^* + A') \end{aligned}$$
   and sends B^*, h_0, h_1 and h_2 to Bob.
2. Bob chooses at random $c \in \{0,1,2\}$ and sends c to Alice.
3. In the case   Alice responds
   $$\begin{aligned} c &= 0: & A^* &\text{ and } X', \\ c &= 1: & A^* + A' &\text{ and } X', \\ c &= 2: & A^* &\text{ and } \pi \end{aligned}$$
4. For Bob, Alice's response is a pair (M, N) resp. (M, σ), where M is a matrix, N a Hamiltonian cycle and σ a permutation over $\{1 \ldots, n\}$.
   In the case   Bob checks
   $$\begin{aligned} c &= 0: & H(M) &= h_0, \; H(X') = h_1 \text{ and } \text{Length}_M(N) = B^*. \\ c &= 1: & H(M) &= h_2, \; H(X') = h_1 \text{ and } \text{Length}_M(N) = B + B^*. \\ c &= 2: & H(M) &= h_1 \text{ and } H(M + \sigma(A)) = h_2. \end{aligned}$$

We could imagine the following variant of the above scheme. Assume we know N users $U_1, \ldots, U_N$. Let B_i be an identification string of length $l(n)$ for the user A_i. B_i might be generated as collision resistant hash of U_i's name, address and similar information. For every user U we choose a random Hamiltonian circuit X_i and *then* we generate a matrix A such that for all $i \in \{1, \ldots, N\}$ Length$_A(X_i) = B_i$.

This way, we have made the above scheme identity-based. But it is identity-based in a limited sense, since we can't add new users to the system—except by providing each new user U_j ($j \leq N$) with a pseudonym P_j and the corresponding keys $B_j = H(P_j)$ and X_j—.

4 How Secure is the Scheme?

It is easy to see that Alice (and everyone else who knows X) will always pass the test. What, if Carla, without knowing X, wants to mask herself as Alice?

Due to the collision resistance of H, Carla has to fix the matrices A^*, $A^* + A'$ and the Hamiltonian cycle X' in the first step.

If Carla wants to be able to give the correct reply in the case $c = 2$, she must know a permutation π with $\pi(A) = A' = (A^* + A') - A^*$. (We anticipate that the $a_{i,j}$ are pairwise distinct—then π is unique. Argumenting in a birthday paradox style, we see that for $l(n) \approx \log_2((n-1)!)$, $n(n-1)$ independent random values $a_{i,j} \in \{0, \ldots, 2^{l(n)} - 1\}$ and except for very small n the chance to find two pairs $(i, j) \neq (i', j')$ with $a_{i,j} = a_{i',j'}$ is negligible.)

If Length$_{A^*}(X') \neq B^*$, Carla is lost in the case $c = 0$. For X' the condition

$$\text{Length}_{A'+A^*}(X') = \text{Length}_{A'}(X') + \text{Length}_{A^*}(X') = B + B^*$$

must hold, else Carla is lost in the case $c = 1$.

Even if $X' \neq \pi(X)$, we have

$$\text{Length}_{A'}(X') = B = \text{Length}_A(\pi^{-1}(X')).$$

Thus, if Carla is able to give the correct answer in the cases $c = 0$, $c = 1$ *and* $c = 2$, either she can solve the XTSP or break the hash function H.

If our assumptions hold, the probability for Carla to successfully mask herself as Alice is at most $2/3$. If the identification procedure is repeated r times and Bob's challenges c are chosen independently at random, Carla's chances to remain undetected are bounded by $(2/3)^r$.

What about our claim that the scheme is zero-knowledge? I.e. what does Bob learn about the secret X, if Alice honestly proves her identity to him?

Given h_0, h_1 and h_2, Bob can't compute statistical information about A^*, X' or $A^* + A'$. The only things which come out of the protocol are a random matrix M and either a random Hamiltonian cycle X' or a random permutation π. B^* is the length of X', or the length of a randomly chosen Hamiltonian cycle if X' is unknown. Thus if Bob chooses the challenge c *before* the first step, he can simulate the protocol without the help of Alice, and the probability distributions of the generated random values are the same with or without Alice's cooperation.

Only the ability to give the expected answer for all possible c is genuine for Alice.

5   The Performance of the Scheme

We anticipate that the matrices A and A^* are generated pseudorandomly. (Note that in the case of the identity-based variant A can't be generated pseudorandomly since A depends on the users' identities.) E.g. let $E : \{0,1\}^k \times \{0,1\}^{l(n)} \longrightarrow \{0,1\}^{l(n)}$ be a (secret key) block cipher encryption function. The key size k is a security parameter. We choose a secret key K_A for the matrix A and generate the $a_{i,j}$ by

$$a_{i,j} = E\big(K_A, \ (i-1) + n(j-1)\big).$$

Here $2^{l(n)} \geq n^2$ and "$(i-1) + n(j-1)$" is treated as $l(n)$-bit string.

Similarly the matrix A^* can be generated depending on a secret key K_{A^*}. In other words, we only need $2k$ bits storage for K_A and K_{A^*} to compute all values of A, $A' = \pi(A)$, A^* and $A^* + A'$ efficiently on demand. The complete identification scheme can be implemented using small storage space.

The same idea helps to limit communication requirements. In two of three rounds (on the average), Alice sends the matrix A^* to Bob. It saves bandwidth if Alice sends the key K_{A^*} and Bob generates the values $a_{i,j}^*$ on his own.

It is not that easy to save bandwidth in the case $c = 1$, where Alice sends the matrix $A^* + A'$. It seems impossible to enable Bob to generate the values of that matrix on his own without leaking information about A^*.

Our solution is based on the simple observation that in order to check $\text{Length}_{A^*+A'}(X')$, Bob only needs to know the n distances defined by X'. On the other hand it would be easy for Carla to cheat, if Bob couldn't verify that the distances he sees are the same as the corresponding values of a matrix M with $h_2 = H(M)$.

The following construction is similar to the parallelized version of Damgård's hash function [1] and Merkle's tree authentication [6]. To the author's knowledge, it is the first use of tree authentication in an identification scheme.

The basic idea is to use the given hash function H for small inputs only. Let $(x_1, \ldots, x_n)$ with $x_i \in \{0,1\}^{l(n)}$ be a large input; instead of $H(x_1, \ldots, x_n)$ we compute $H^*(x_1, \ldots, x_n)$ with

$$H^*(x_i, \ldots, x_j) = H\left(H^*\left(x_i, \ldots, x_{i-1+\lceil(j-i)/2\rceil}\right), H^*\left(x_{i+\lceil(j-i)/2\rceil}, \ldots, x_j\right)\right)$$

for $i < j$ and $H^*(x_i) = H(x_i)$. We regard this as a binary computation tree with $H^*(x_1, \ldots, x_m)$ as root and $x_1, \ldots, x_m$ as leafs.

Let $h = H^*(x_1, \ldots, x_m)$ and x be public knowledge. Given $x \in \{0,1\}^{l(n)}$, in order to prove $x = x_i$ we only need to reveal i and at most $\lceil \log_2 m \rceil$ intermediate results $H^*(\ldots)$, following the tree from the leaf x_i to the root.

Our case is slightly more complicated. For simplicity we write $a_{i,j}^{+} = a_{i,j}^{*} + a_{i,j}'$. We modify our scheme such that no longer $h_2 = H(A^* + A')$, but

$$h_2 = H\left(H^*\left(a_{1,1}^{+}, \ldots, a_{1,n}^{+}\right), \ldots, H^*\left(a_{n,1}^{+}, \ldots, a_{n,n}^{+}\right) \right).$$

(The values $a_{i,i}^{+}$ could be left out and should be 0 by definition. Here, their use helps to keep the above expression simple.)

Alice has to reveal n values $a_{i,j}^{+}$. More precise, she has to reveal exactly one value a_{1,j_1}^{+}, exactly one value a_{2,j_2}^{+}, ... and exactly one value a_{n,j_n}^{+}. To prove the correctness of a_{i,j_i}^{+} she reveals about $\log_2(n)$ intermediate results—instead of $n - 1$ leafs.

When computing h_2, Alice knows in advance, which values $a_{i,j}^{+}$ to reveal if Bob sends her $c = 1$. In order to limit her computing load, Alice can store the necessary intermediate results.

If storage space is short, Alice can resort to a time-space tradeoff. She only stores intermediate results, which are close to the root and thus require a high effort to be computed again. The remaining intermediate results actually have to be computed again—if Bob's challenge for Alice is $c = 1$.

The heaviest part of Alice's computing load is to compute the hashes h_0 and h_2. Since Alice's computations may actually take place on a portable device with limited computation power, hashing large distance matrices may be troublesome. But note that these hashes do not depend on what Bob does, i.e. can be computed during the idle time of Alice's device.

6   The Security Parameter n

It is difficult to recommend a minimum size for n. A complete enumeration of all Hamiltonian circuits will solve the XTSP in $(n - 1)!$ steps—but what about more efficient algorithms?

There are good heuristics for the "normal" TSP, especially if the triangle inequality holds. Padberg and Rinaldi [7] solved TSPs with $n =$ "several thousand". But all successful approaches on large-scale TSPs are based on branch–and–bound or (as in the case of Padberg and Rinaldi) on branch–and–cut. The basic branch–and–... principle can roughly be described as follows:

- Start with solution space S, and divide it into subspaces $S_1, S_2, \ldots, S_k$.
- Compute lower and/or upper bounds for all solutions in S_i.
- For all S_i do:
  - If the lower bound is too large (or the upper bound too low) discard S_i
  - else branch, i.e. apply branch–and–... on the solution space S_i.

In the case of the XTSP, this leads to excessive branching because the solution space S we start with (a set of $(n - 1)!$ Hamiltonian cycles) and most subspaces can be expected to contain both too long and too short Hamiltonian cycles.

Empiric evidence supports this reasoning. Thienel [13] wrote a TSP-solver for TSPs under constraints, based on the TSP-code of [3]. He also made some experiments with XTSPs. The condition that the solution is a Hamiltonian circuit with a fixed length can be viewed as such a constraint.

The program was run on a Sun Sparcstation 10 with 60 Mhz and ran out of memory when more than about 30000 branch–and–bound nodes had to be stored. It was not designed for such large branch–and–bound trees. Also it was written for symmetric TSPs. The following distance matrices were considered:

- gr17, a symmetric 17*17 matrix from the TSPLIB and
- pi$x$-1Ey, symmetric $x \times x$ matrices with random distances $\in \{0, \ldots, 10^y - 1\}$ (derived from the digits of π) and

$$(x, y) \in \{(13, 8), (14, 8), (18, 3), (18, 4), (18, 5), (18, 8)\}.$$

For every distance matrix, five random Hamiltonian circuits were chosen and their lengths computed. Among these 35 XTSPs, all gr17-based, four pi-18E3-based and two pi-18E4-based problems were solved, the remaining ones could not be solved due to lack of memory. The required resources can be found in table 1; the time is given in [min:sec]. The corresponding unconstrained TSPs were easy to solve, requiring one branch-and-bound node and < 1 sec time.

Table 1.

matrix	nodes	time
gr17	323	0:13
	333	0:14
	1575	1:06
	1658	0:47
	1193	0:47
pi18–1E3	6881	8:05
	2933	3:00
	5515	5:49
	1575	1:43
	1 unsolved	
pi18–1E4	4361	4:29
	157	0:07
	3 unsolved	
pi18–1E5	5 unsolved	
pi13–1E8	5 unsolved	
pi14–1E8	5 unsolved	
pi18–1E8	5 unsolved	

XTSPs based on pi18-1E3, pi18-1E4 and pi18-1E5 are underconstrained, i.e. in general there is more than one Hamiltonain circuit with a given length.

As could be expected, the program's success rate drops fast when the number length increases.

Even non-heuristic algorithms for the "normal" TSP like the $O(n^2 2^n)$ dynamic programming algorithm (see [4], chapter 3) don't seem to be adaptable to the XTSP.

It appears reasonable to estimate that XTSPs with $n = 21$ are hard, but can be solved with present technology ($20! \approx 2^{61}$), while XTSPs with $n = 41$ can not. When $n = 41$, we have to deal with 160-bit numbers since $l(n) = 160 \approx \log_2(40!)$.

It is not recommended to adopt a cryptographic scheme for actual use too early. This remains true for the scheme proposed here. Experts in cryptanalysis and/or combinatorics are invited to search for efficient algorithms to solve the XTSP.

Acknowledgements

I am grateful to an anonymous referee of [5] for pointing me towards identification schemes and to Stefan Thienel for providing the empiric results.

References

1. I. B. Damgård, *A Design Principle for Hash Functions*, in: Proc Crypto '89, Springer LNCS 435, 416–427.
2. M. Girault, J. Stern, *On the length of cryptographic hash-values used in identification schemes*, in: Proc. Crypto '94, Springer LNCS 839, 202-215.
3. M. Jünger, G. Reinelt, S. Thienel, *Provably good solutions for the traveling salesman problem*, in: Zeitschrift für Operations Research 40 (1994), 183–217.
4. E. L. Lawler, J. K. Lenstra, A. H. G. Rinnoy Kan, D. B. Shmoys (eds.), *The Traveling Salesman Problem*, Wiley, 1985.
5. S. Lucks, *How to Exploit the Intractability of Exact TSP for Cryptography*, to appear in: Proc. Fast Software Encryption (1994), Springer LNCS.
6. R. C. Merkle, *A Certified Digital Signature (That Antique Paper from 1979)*, in: Crypto '89, Springer LNCS 435, 218–238.
7. M. Padberg, G. Rinaldi, *A Branch–and–Cut Algorithm for the Resolution of Large-Scale Symmetric Traveling Salesman Problems*, in: Siam Review 33, No. 1 (1991), 60-100.
8. J. Patarin, P. Chauvaud, *Improved Algorithms for the Permuted Kernel Problem*, in: Proc. Crypto '93, Springer LNCS 773, 391-402.
9. D. Pointcheval, *A New Identification Scheme Based on the Perceptrons Problem*, in: Proc. Eurocrypt '95, Springer LNCS 921, 319-328.
10. A. Shamir, *An Identification Scheme based on Permuted Kernels*, in: Proc. Crypto '89, Springer LNCS 435, 606-609.
11. J. Stern, *A new identification scheme based on syndrome decoding*, in: Proc. Crypto '93, Springer LNCS 773, 13-20.
12. J. Stern, *Designing Identification Schemes with Keys of Short Size*, in: Proc. Crypto '94, Springer LNCS 839, 164-173.
13. S. Thienel, private communication, July 1995.

An Elliptic Curve Analogue of McCurley's Key Agreement Scheme

Andrew Smith and Colin Boyd

Communications Research Group, Electrical Engineering Laboratories, University of Manchester, Manchester M13 9PL, UK

Abstract. McCurley's key agreement scheme is a variation on the well known Diffie-Hellman scheme with enhanced security. In McCurley's scheme a successful attacker must be able to break the ordinary Diffie-Hellman scheme and also factorise large numbers. This paper presents an analogue of McCurley's scheme using elliptic curves. A consequence is that a method to break ordinary Diffie-Hellman would not be applicable to our scheme. An advantage of our scheme over McCurley's is that much smaller key lengths can be used.

1 Introduction

Diffie and Hellman introduced their celebrated key exchange scheme in 1976 [2]. This allows two users to exchange a secret value by an interactive protocol over an insecure channel. The scheme relies for its security on the difficulty of the discrete logarithm problem in the field of integers modulo a prime number p. This is the problem of finding x given $y = a^x \bmod p$ where a is a generator of the non-zero integers modulo p. It is widely believed that the problems of breaking the Diffie-Hellman scheme, and of finding discrete logarithms are equivalent [6], although it remains unproven in general.

In 1988 McCurley [7] published an analogue of the Diffie-Hellman scheme [2] that derives its security from the difficulty of both the discrete logarithm and the integer factorisation problems. More precisely, if an attacker were able to break the ordinary Diffie-Hellman scheme, or solve the integer factorisation problem (but not both) then the scheme remains secure. The basic idea of McCurley's scheme is to work in the ring of integers modulo a composite modulus m. The modulus m is the product of two large primes chosen in a way that is believed to make its factorisation difficult. Careful choice of the primes and the base point a allow the security properties to be proven.

In this paper we present an elliptic curve analogue of McCurley's system. This uses elliptic curves over a composite modulus. The difficulty of breaking the system is proven to be as hard as both breaking the Diffie-Hellman scheme over elliptic curves, and factorising the modulus. In some ways it is to be expected that such an analogue is possible, since it was shown by Lenstra [5] that there is a powerful factorisation method using elliptic curves. However, the details have to be considered carefully in order to make the scheme work. The usefulness of elliptic curves in this context stems particularly from the following observations.

If the size of the prime p used in an instance of the discrete logarithm problem is increased, it is to be expected that the difficulty of the problem increases. Similarly, as the size of the composite m increases, the difficulty of factorising it increases. It turns out that, with the best currently known algorithms, the two problems are of roughly comparable difficulty when p and m are of about the same size [10]. This presents a dilemma when implementing McCurley's system; its security is based on the difficulty of factorising m and taking discrete logarithms mod p and mod q, but in the scheme $m = pq$. Thus if if m is to be chosen to make factorisation of a particular complexity, then the discrete logarithms mod p and mod q will be much easier. On the other hand, if p and q are chosen to make the discrete logarithm problem to be of a particular difficulty then m will have to be taken to be twice this length, making the computations slower, and size of the public key twice as large.

This problem is greatly relieved in our version because the size of numbers over which elliptic curves must be defined is much smaller than those which give comparable security in the analogue schemes in finite fields [8]. For this reason, the size of the composite modulus can be chosen to be whatever size is required to make the factorisation problem of a desired complexity.

The remainder of this paper is organised as follows. Section 2 reviews elliptic curves over prime fields and their generalisation to rings using a composite modulus. Section 3 covers the connection between elliptic curves and factorising in preparation for the description of the scheme in section 4. Implementation issues and possible extensions to the basic scheme are considered in section 5.

2 Elliptic Curves over Rings

Elliptic curves are algebraic equations which may be defined over any field. We first define elliptic curves over prime fields $\mathbf{Z}_p$ where p is a prime larger than 3. An elliptic curve $E_p(a, b)$ consists of the points (x, y) in $\mathbf{Z}_p \times \mathbf{Z}_p$ which are solutions of

$$y^2 = x^3 + ax + b \tag{1}$$

If we define a special point $\mathcal{O}_p$, called the *point at infinity*, then the points on the elliptic curve become an abelian group with a special addition operation defined as follows.

1. $\mathcal{O}_p$ is the identity element: for any $P \in E_p(a, b)$

$$\mathcal{O}_p + P = P$$

2. The inverse of a point $P = (x, y) \in E_p(a, b)$ is the point $-P = (x, -y)$.

$$P + -P = \mathcal{O}_p$$

3. Given any two points $P = (x_1, y_1)$ and $Q = (x_2, y_2)$ on $E_p(a, b)$ with $x_1 \neq x_2$ then define[1]

---

[1] Note that if $x_1 = x_2$ then either $Q = -P$ (previous case) or $Q = P$ (next case).

$$\lambda = \frac{y_2 - y_1}{x_2 - x_1}. \tag{2}$$

Then $P + Q$ is the point (x_3, y_3) with

$$x_3 = \lambda^2 - x_1 - x_2$$
$$y_3 = \lambda(x_1 - x_3) - y_1$$

4. Finally consider any point $P = (x_1, y_1)$ on $E_p(a, b)$ and define

$$\lambda = \frac{3x_1^2 + a}{2y_1}$$

Then $2P$ is the point (x_3, y_3) with

$$x_3 = \lambda^2 - 2x_1$$
$$y_3 = \lambda(x_1 - x_3) - y_1$$

Elliptic curves over rings have been studied previously in order to produce analogues of the RSA cryptosystem [3, 1]. The definition (1) for an elliptic curve over a prime field $\mathbf{Z}_p$ can be easily extended to a ring of integers $\mathbf{Z}_n$. The curve $E_n(a, b)$ consists of points $(x, y) \in \mathbf{Z}_n \times \mathbf{Z}_n$ which are solutions to the equation

$$y^2 = x^3 + ax + b$$

together with the point an infinity $\mathcal{O}_n$. Unfortunately things do not run quite so smoothly when we try to extend the addition operation, since if $(x_2 - x_1, n) > 1$ then λ will not be defined in (2). We are only interested here in the case where $n = pq$ for some large primes p and q and so we will assume this from now on. Following Menezes [8] we may define the product group

$$\tilde{E} = E_p(a, b) \times E_q(a, b)$$

of pairs of points in the curves defined over $\mathbf{Z}_p$ and $\mathbf{Z}_q$.

It is useful to introduce some notation. By the Chinese remainder theorem, if x is any value modulo n, there is a unique pair of values x_p and x_q such that $x \bmod p = x_p$ and $x \bmod q = x_q$. We will denote this equivalence by $x = [x_p, x_q]$. Similarly, if $P_p = (x_p, y_p)$ and $Q_q = (x_q, y_q)$ are points on the curves $E_p(a, b)$ and $E_q(a, b)$ then $[P_p, Q_q]$ represents the unique point on $E_n(a, b) = (x_n, y_n)$ with $x_n \bmod p = x_p$ etc. It may easily be verified that if $P = [P_p, P_q]$ and $Q = [Q_p, Q_q]$ then, as long as the addition is defined, $P + Q = [P_p + Q_p, P_q + Q_q]$. Hence for any integer γ we have $\gamma P_n = [\gamma P_p, \gamma P_q]$. Thus any pair of points in $\tilde{E}$ corresponds to a unique point in $E_n(a, b)$ as long as neither of the points in $\tilde{E}$ is the point at infinity. The group $\tilde{E}$ consists of all points of $E_n(a, b)$ plus a number of points of the form $[P_p, \mathcal{O}_q]$ or $[\mathcal{O}_p, P_q]$. These points cannot be the result of adding any two points on $E_n(a, b)$ and so we call them *non-realisable* points. The point $[\mathcal{O}_p, \mathcal{O}_q]$ is equivalent to the point $\mathcal{O}_n$ on $E_n(a, b)$.

If the addition of two different points is not defined on $E_n(a, b)$ the attempt to perform the addition will result in a non-trivial factor of n. This is because if $x_1 \neq x_2$ the addition operation will fail only if $(x_2 - x_1, n) > 1$. On the other hand, if $x_1 = x_2$ then $y_1^2 = y_2^2$. But since $P \neq Q$ we have $y_1 \neq y_2$, and also $y_1 \neq -y_2$ since this would mean that $P = -Q$ when the addition *is* well-defined. Thus $(y_2 - y_1, n)$ is now a non-trivial factor of n.

Since it is believed that factorisation is a hard problem, it can confidently be asserted that undefined addition is unlikely to occur in practical use of a system[2]. The result is summarised as follows.

Lemma 1. *Suppose* $P = (x_1, y_1)$ *and* $Q = (x_2, y_2)$ *are two different points on* $E_n(a, b)$ *whose addition is not defined. Then knowledge of* P *and* Q *is sufficient to factorise* n.

3 Half Points and Factorisation

In McCurley's scheme, the proof of security results from showing that an attacker who can find the Diffie-Hellman key is able to find two integers x and y in $\mathbf{Z}_n$ with $x^2 \equiv y^2 \bmod n$ but $(x/y)^2 \bmod n \neq 1$. This implies that $gcd(x - y, n)$ is a non-trivial factor of n.

Since addition on elliptic curves is the analogue of multiplication in finite fields, the analogue of a square root in a finite field is a *half point* on an elliptic curve. The idea of obtaining an analogue of McCurley's scheme is to show that finding points on the elliptic curve which have two half points enables factorisation of n. Much of the analysis may be done in arbitrary cyclic groups. The following result is elementary.

Lemma 2. *Let* G *and* H *be two additive cyclic groups of even orders* 2α *and* 2β, *with generators* g *and* h *and identities* 0_g *and* 0_h *respectively. Let* $K = G \times H$. *If* (x_1, y_1) *and* (x_2, y_2) *are elements of* K *such that* $2(x_1, y_1) = 2(x_2, y_2)$ *then* $(x_1, y_1) - (x_2, y_2)$ *is one of the four values* $(0_g, 0_h), (\alpha, 0_h), (0_g, \beta), (\alpha, \beta)$.

Applying this result to elliptic curve groups leads to the following.

Lemma 3. *Suppose* $E_p(a, b)$ *and* $E_q(a, b)$ *are two elliptic curve groups which are cyclic and of even order. If* P *and* Q *are two points on* $E_n(a, b)$ *with* $2P = 2Q$ *but* $P \neq Q$ *and* $P - Q \neq (x, 0)$ *for some* x, *then knowledge of* P *and* Q *is sufficient to factorise* n.

The proof follows almost immediately from the lemmas 1 and 2 with $G = E_p(a, b)$ and $H = E_q(a, b)$. The points $(\alpha, 0_h)$ and $(0_g, \beta)$ correspond to non-realisable points of $E_n(a, b)$. Hence from Lemma 1 calculation of $P - Q$ will

---

[2] Note that, contrary to a remark of Koyama *et. al.* ([3] p.255), it is *not* the case that addition on $E_n(a, b)$ is undefined only when the resulting point, as a point of $\tilde{E}$, is of the form $[P_p, \mathcal{O}_q]$ or $[\mathcal{O}_p, P_q]$. If $P_p = Q_p$ while $x_1 \neq x_2 \bmod q$ then $P + Q$ is undefined, but as points on $\tilde{E}$ the sum may not be the infinity point in either component.

result in factorisation of n if the result in the product group is $(\alpha, 0)$ or $(0, \beta)$. But since $P \neq Q$ the only other possibility would be $P - Q = (\alpha, \beta)$. However this is ruled out by the condition $P - Q \neq (x, 0)$. This is because (α, β) is clearly a point of order two. But if R is any point of order two on an elliptic curve then $2R = \mathcal{O}_n$ so $R = -R$ and so its y component is zero; thus $R = (x, 0)$ for some x.

$\square$

4 The Elliptic Curve Scheme

4.1 Choosing the Curves

There are a number of constraints on the elliptic curves $E_p(a, b)$ and $E_q(a, b)$ that must be satisfied. For the moment we will ignore the problem of how these properties may be achieved in practice, but return to this issue in the following section. The following properties are required.

1. Integers p and q must be generated such that their product $n = pq$ is hard to factorise.
2. $E_p(a, b)$ and $E_q(a, b)$ must each have a large cyclic subgroup of even order, say $2r$ and $2s$. Furthermore, r and s must be odd (it would be best for security if they are prime). Two points are chosen: F_p is a generator of the cyclic subgroup on $E_p(a, b)$ and so has order $2r$, while F_q is twice a generator of the cyclic subgroup of $E_q(a, b)$ and so has order s.
3. The base point B_n is chosen to be

$$B_n = 4F_n$$

where $F_n = [F_p, F_q]$.

Public Parameters

- The composite modulus n.
- The parameters a and b defining the curve

$$y^2 = x^3 + ax + b$$

in the ring $\mathbf{Z}_n$.
- The base point B_n.

4.2 Using the Scheme

Use of the scheme is a direct analogue of the Diffie-Hellman key agreement scheme with exponentiation in the finite field replaced by multiplication in the elliptic curve group. Suppose Alice and Bob wish to agree a secret value using an insecure channel.

1. Alice chooses a random value $k_A \in \mathbf{Z}_n$. She calculates $C_A = k_A B_n$ and sends C_A to Bob over the insecure channel.
2. Bob chooses a random value $k_B \in \mathbf{Z}_n$. He calculates $C_B = k_B B_n$ and sends C_B to Alice.
3. Alice calculates $k_{AB} = k_A C_B$ and Bob calculates the same value by $k_{AB} = k_B C_A$.

4.3   Proof of Security

We show that the scheme remains secure even if an attacker is able to factorise the modulus n, or solve the Diffie-Hellman problem in the groups $E_p(a, b)$ and $E_q(a, b)$, but not do both. To do this we assume the existence of an efficient algorithm $ECDH(n, B_n, a, b, C_A, C_B)$ to calculate k_{AB} from knowledge of C_A and C_B (together with the public values). We then show that this algorithm can be used to solve the other two problems. The equivalence to the Diffie-Hellman problem in the separate groups is easily dealt with first.

Lemma 4. $ECDH(n, B_n, a, b, C_A, C_B)$ may be used to break efficiently the Diffie-Hellman scheme in $E_p(a, b)$.

Proof Suppose that B_p is a base point for the Diffie-Hellman scheme in $E_p(a, b)$ and $C_A = k_A B_p$ and $C_B = k_B B_p$ are recorded by the attacker. Then he may choose a large prime q and find $n = pq$, any point B_q on the curve $E_q(a, b)$ and the point $B_n = (B_q, B_p)$. By presenting the values n, B_n, a, b, C_A, C_B to $ECDH$ he obtains the point $k_A k_B B_n$. Using the Chinese remainder theorem he can split this value as $k_A k_B B_n = [k_A k_B B_p, k_A k_B B_q]$. The required value $k_A k_B B_p$ has thus been found.

$\square$

Next we show that access to the algorithm $ECDH$ is sufficient to factorise n efficiently. The basic idea is to use Lemma 3 of section 3. The choice of base point allows us to do this. The algorithm to factorise n is as follows.

1. Compute
$$J = ECDH(n, B_n, a, b, 2F_n, 2F_n)$$

2. Write $J = (x_j, y_j)$ and $F_n = (x_f, y_f)$.
3. If $x_j \neq x_f$ then $gcd(n, x_j - x_f)$ is a non-trivial factor of n. Otherwise $(n, y_j - y_f)$ is such a factor.

By Lemma 2 , to prove that the algorithm works it is necessary only to establish the following two properties.

$$2F_n = 2J_n \tag{3}$$
$$F_n \neq J_n \text{ and } F_n - J_n \neq (x, 0) \tag{4}$$

Recall that $F_n = [F_p, F_q]$ where F_p and F_q generate cyclic subgroups of order $2r$ and s respectively. Since r and s are both odd, it follows that F_n generates a subgroup $\langle F_n \rangle$ of order $2rs$ while the subgroups $\langle 2F_n \rangle$ and $\langle B_n \rangle$ are both equal and of order rs.

Therefore we may write $2F_n = \alpha B_n$ for some integer α and so by the definition of the Diffie-Hellman key we have

$$J_n = \alpha^2 B_n = 2\alpha F_n$$

and so

$$2J_n = 4\alpha F_n = \alpha B_n = 2F_n$$

and so requirement (3) is satisfied. Now since $\langle F_n \rangle$ and $\langle 2F_n \rangle$ are different subgroups and $J_n = 2\alpha F_n$ it is clear that $F_n \neq J_n$.

Finally we need to show that $F_n - J_n \neq (x, 0)$. Now $(x, 0)$ is a point of order two as discussed earlier. But $\langle F_n \rangle$ has been constructed to be of odd order so it has no points of order two. Hence we have shown condition (4) and the equivalence to factorisation is proven.

Lemma 5. *The algorithm ECDH may be used to efficiently factorise the modulus* n.

5 Implementing the Scheme

In contrast to McCurley's scheme in a ring of integers, implementing our scheme on elliptic curves is not so straightforward. The main problem is to find elliptic curves of the required structure. It is well known [8] that the structure of an elliptic curve group of order N, must be of the form

$$E_p(a, b) \cong \mathbf{Z}_{n_1} \oplus \mathbf{Z}_{n_2}$$

where $n_2 | n_1$ and $n_2 | p - 1$. If $n_2 = 1$ then $E_p(a, b)$ is cyclic so for our purposes it is sufficient to choose $N = 2r$ for some prime r and then it follows that $n_2 = 1$. In order to avoid the known cases where the elliptic curve discrete logarithm collapses to the discrete logarithms problem over prime fields [9], it is also necessary to ensure that the curves chosen are not *supersingular*. For curves defined over a prime field it is necessary only that the order is not $p+1$ to ensure that the curve is not super-singular.

A practical algorithm for selecting curves of a chosen order has been designed by Lay and Zimmer [4]. This is ideal for use in implementation of our scheme. A suitable procedure to find each of the two curves might be as follows.

1. Choose a large prime p to be used as one factor of the modulus.
2. Examine the even numbers from $p + 3$ upwards until one of the form $2r$, for some prime r, is found. Check that $2r < p + 1 + 2\sqrt{p}$, which ensures that an elliptic curve of this order exists; in the unlikely event that this is not the case return to step 1.
3. Use the algorithm of Lay and Zimmer to generate a curve of order $2r$.

Acknowledgements

This work is partly sponsored by a grant from British Telecommunications plc. The first author is also supported by a Research Studentship from the UK Engineering and Physical Sciences Research Council. We are grateful to Alfred Menezes for a number of constructive comments.

References

1. N.Demytko, "A New Elliptic curve Based Analogue of RSA", *Advances in Cryptology - EUROCRYPT 93*, pp.40-49, Springer-Verlag, 1994.
2. W. Diffie and M. Hellman, "New Directions in Cryptography", *IEEE Transactions on Information Theory*, 22, pp. 644-654, (1976).
3. K. Koyama, U.M. Maurer, T. Okamoto, S.A. Vanstone, "New Public-Key Schemes Based on Elliptic Curves over the Ring Z_n", *Proceedings of Crypto 91*, pp.252-266, Springer-Verlag, 1992.
4. G. Lay and H. Zimmer, "Constructing elliptic curves with given group order over large finite fields", *Algorithmic Number Theory: First International Symposium*, Lecture Notes in Computer Science, 877, Springer-Verlag, pp.250-263.
5. H.W.Lenstra, Jr., "Factoring with Elliptic Curves", *Annals of Mathematics*, 126, pp.649-673, 1987.
6. U.M. Maurer, "Towards the Equivalence of Breaking the Diffie-Hellman Protocol and Computing Discrete Logarithms", Proceedings of Crypto '94, pp.271-281, Springer-Verlag, 1994.
7. K. S. McCurley, "A Key Distribution Scheme Equivalent to Factoring", *Journal of Cryptology*, Vol.1, No.2 (1988).
8. A. Menezes, *Elliptic Curve Public Key Cryptosystems*, Kluwer Academic Publishers (1993).
9. A. Menezes, T. Okamoto and S.A. Vanstone, "Reducing Elliptic Curve Logarithms to Logarithms in a Finite Field", *IEEE Transactions on Information Theory*, (39), 1993, pp.1639-1646.
10. P.C. van Oorschot, "A Comparison of Practical Public Key Cryptosystems Based on Integer Factorization and Discrete Logarithms", in G.J.Simmons (Ed.), *Contemporary Cryptology*, IEEE Press 1992.

Multi-Dimensional Ring TCM Codes for Fading Channels

M.Ahmadian-Attari and P.G.Farrell

Communications Research Group, School of Engineering,
University of Manchester, Manchester, M13 9PL, UK.

Abstract A multi-dimensional non-binary ring trellis coded modulation (R-TCM) scheme suitable for fading channels which is superior to conventional binary TCM is considered.

1 Introduction

Ring TCM codes are an expansion of well known conventional TCM codes [1]-[2] based on the Galois field *GF(2)* to codes based on finite rings of integers. These codes were investigated independently by Massey and Mittelholzer [3] and Baldini and Farrell [4],[14] at the same time. The former authors have demonstrated that due to the similarities between the M-ary PSK signal set and the algebraic structure of the ring of integers modulo-M, Z_M, TCM codes based on modulo-M rings are the natural linear codes for M-PSK modulation. They also established a monotonic relationship between the Hamming distance (HD) of ring TCM codes and the Euclidean distance (ED) among M-PSK signal points in the two-dimensional space. Such an interesting relationship does not exist between conventional TCM and M-PSK constellations. The latter authors obtained some ring trellis and block M-PSK codes suitable for AWGN channels. They also realized that the construction of linear codes over rings with phase invariant properties is simpler than in the case of field TCM codes. Later, Lopez, et al [5] reported some modulo-4 ring TCM codes suitable for rectangular M-QAM constellations. Based on the new design criteria for TCM codes in a non-Gaussian channel which were formulated by Divsalar and Simon [6]-[7], Lopez [8] also investigated new fading-optimized modulo-4 ring TCM codes suitable for rectangular M-QAM signal sets.

In this paper we develop multi-dimensional ring-TCM schemes and show that the construction of these codes to be transparent (and thus phase invariant) is much simpler than for their conventional TCM counterparts and particularly that they are more suitable for fading channels.

2 Codes over Rings of Integers

Most of codes proposed so far are constructed with a binary rate $r=(m-1)/m$ convolutional encoder (CE) combined with the M-level ($M=2^m$) modulation (see [9] and[10] for a quick survey of TCM codes). Employing modulo-M arithmetics instead of binary arithmetics is more advantageous and practical. Expansion of the CE from *GF(2)* to *GF(M)* requires M to be a prime integer which is not adapted to most practical multilevel constellations with a composite number of signal points. Since construction of the ring of integers modulo-M with a nonprime positive integer number, M, is suitable for most practical signal sets, it has attracted some researchers to investigate the possible application of ring codes on band-limited channels. Let us consider a ring with $M=2^m$ elements $\{0,1,2,...,M-1\}$ under modulo-M operations. A multilevel convolutional encoder (MCE) with symbols from the modulo-M ring which is shown in Fig. 1 has a generator matrix of the form

$$G(D) = \left[I_m \mid P(D) \right] = \begin{bmatrix} 1 & 0 & 0 & \cdots & 0 & g^{(1)}(D)/f(D) \\ 0 & 1 & 0 & \cdots & 0 & g^{(2)}(D)/f(D) \\ \vdots & \vdots & \vdots & & \vdots & \vdots \\ 0 & 0 & 0 & \cdots & 1 & g^{(m)}(D)/f(D) \end{bmatrix} \quad (1)$$

where $g^{(i)}(D)$ $(i=1,...,m)$ and $f(D)$ are polynomials related to the feedforward and feedback connections, respectively, with coefficients belonging to Z_M. The codewords can be generated by the equation

$$Y(D) = X(D)G(D) \qquad (2)$$

where $X(D)$ and $Y(D)$ are polynomials which represent the nonbinary input and output signals, respectively.

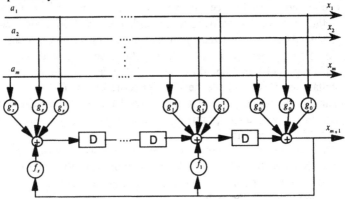

Fig.1. General structure of MCE

3 System Model

In this section we briefly describe the approach taken in implementing the simulation. Fig.ure 2 shows a block diagram of the overall system being simulated. The objective of the simulation is to evaluate the performance of the ring coded modulation technique in the presence of fading and Gaussian noise.

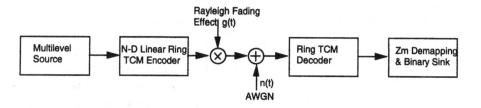

Fig. 2. Multi-dimensional Ring TCM Model over Fading Channel

The multilevel source combines a binary source and a Z_M mapper which generates random signals belonging to Z_M from m parallel random binary inputs. The MCE is fed by m-1 of these signals at a time to produce m encoded outputs belonging to Z_M. Next each encoded output symbol is mapped into one of the M-level N-dimensional constellations according to set partitioning rules pioneered by Ungerboeck [1].To simplify, we suppose that the fading and Gaussian processes are independent and identically distributed. Amongst the fading channel models we consider the Rayleigh fading model which is the worst case representation of fading due to the multipath effect. This means that there is no line of sight (LOS) between the transmitter and receiver and all received signals are reflected versions of the transmitted signal with different delays.We also assume that the receiver performs coherent detection, and hence the channel phase shift is compensated by the receiver. At the receiver, the received faded signal is to be detected as one of the valid signals by the Viterbi decoder, to recover a group of m-1 Z_M decoded symbols. Finally, a Z_M symbol-to-bit demapping process retrieves m output bits. These bits are compared with the original information bits and the ratio of number of bits in error to the total number of the transmitted bits is used to evaluate the system.

4 Multidimensional Modulation

In most conventional TCM schemes, 2-D constellations are employed. To improve the performance of the code more states are required. However, the coding gain increases more slowly if the size of the 2-D constellation is doubled by increasing the constraint length by one. An alternative solution is using a multidimensional signal set to reduce the rapid expansion of the constellation, because fewer redundant bits are added for each dimension. Theoretically, multi-dimensional signaling achieves a significant improvement because of providing more room for the signal points and increasing the ED between the nearest neighbors in the signal set [11]. In general, higher dimensional TCM schemes exhibit larger asymptotic coding gain than 2-D schemes. However, this improvement is compromised by a large number of nearest neighbors in the multi-dimensional space.

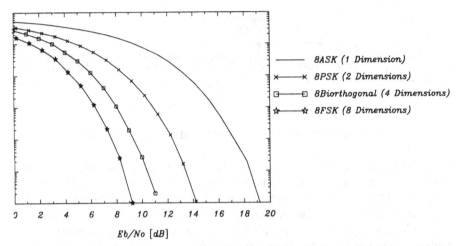

Fig. 3. The effect of dimensionality on the error performance of uncoded modulation schemes

Four and 8-dimensional TCM schemes achieve larger real coding gains than 2-D schemes. However, increasing the number of dimensions further does not improve the performance significantly. This fact is illustrated in Fig.ure 3 for uncoded modulation schemes. In fading channels, denser constellations generally have better performance because they offer the opportunity for design of codes with greater merge distances and more signal redundancy [12].

5 Rotationally Invariant Codes

In general, all classes of TCM codes suffer the phase ambiguities of the expanded signal set, caused by noisy channels which may result in a rotation of the received signals compared to the transmitted signals. Designing TCM codes to be transparent to signal rotations is one approach to compensating the phase ambiguities in the receiver. In this method, every coded sequence has a rotated version under all phase ambiguities of the modulation scheme [13]. Thus, all the rotated versions of a coded sequence are decoded to the same input sequence. It has been proved that a ring TCM code is transparent if and only if its trellis has the all-one sequence as a coded sequence for a self transition (a transition from a state to itself). This is achieved when the modulo-M summation of all coefficients in the MCE is equal to one [14]. Such a code is rotational invariant (RI) to all multiples of $360/M$ degree phase rotations. This simple approach to designing a transparent ring-TCM code has not been reported for other types of TCM codes.

6 Simulation Results

Four and sixteen-state ring-TCM codes over the Gaussian and Rayleigh fading channels are considered. The modulation is a 16-level constellation from 2, 4, and 8-

dimensional space. Since fading channels are sensitive to amplitude fluctuations, we employ constant envelope modulation schemes. Although uniform symbol spacing has been the traditional approach, it is claimed that d^2_{free} can be increased by using nonuniform spacing of the signals [15]. However, after considering several multiamplitude signal constellations, we preferred symmetric and uniform spacing of signal points because for a given average power the minimum distance between distinct symbols is maximised. Also the peak-to-average power ratio, which is an important factor in constellation shaping theory is reduced to its minimum possible value, i.e. one. The minimum phase rotation between adjacent signals in 16-level hypercube and 16-level biorthogonal signalling is 60 and 90 degrees, respectively. So the problem of phase ambiguity can be better treated. Although SED is the best parameter for evaluating a TCM code, especially in the AWGN channel, two other parameters should be considered when we compare schemes:

1) N_{free} : the number of paths of length L that have a competing path at d^2_{free}. Among codes with the same d^2_{free}, the one with smaller N_{free} has a better performance in the Gaussian channel. This conclusion comes from well known lower bound for the error event probability

$$P_r(e) \geq N_{free} Q\left(\frac{d_{free}}{2\sigma}\right) \qquad (3)$$

Error performance of codes 12/2 and 30/2 with the same minimum SED and different N_{free} employing 2(QPSK) modulation scheme is depicted in Fig. 4 to demonstrate the importance of N_{free}. Notice that both codes are transparent. Fig. 5 illustrates that because of having $N_{free}=1$, the code 212/01 performs as good as optimum code 212/31, despite a big difference between their minimum SEDs.

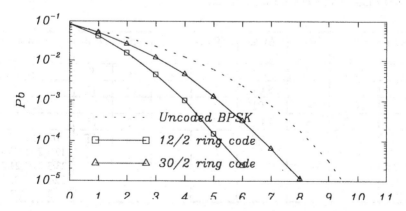

Fig. 4. Error performance of 4-state R-TCM with the same d^2_{free}

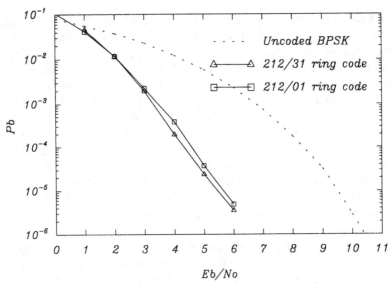

Fig. 5. Error performance of best 16-state R-TCM codes

2) d^2_{next} : the second smallest SED between two paths forming an error event. If this is very close to d^2_{free}, more errors are introduced and this degrades the error performance of the code.

  Thus, the primary objective in the design of optimum codes for the AWGN channel is to maximize the minimum SED and for a given d^2_{free} to minimise N_{free} and finally to maximize the difference d^2_{next}-d^2_{free}. On the other hand, over fading channels, the primary objectives are maximizing the diversity factor of the code, L, then among codes with the same L, minimizing N_{free}. The secondary objective is maximization of the product SED of branches along the paths with diversity factor L. Maximization of d^2_{free} has the least importance in the fading environment. In Table 1, 4-state ring-TCM codes are summarized.

Group	Code	SED	ACG	Nfree	L	spd	Rot	Similar Codes
1	12/2	8	3.01	1	2	16	T	--
2	32/2	8	3.01	1	2	16	NT	--
3	30/2	8	3.01	5	2	12	T	32/0
4	10/2	8	3.01	5	2	12	NT	10/0
5	21/1	10	3.98	4	4	24	NT	11/2,13/2,12/3, 12/1,23/3,23/1, 23/3,23/1,21/3, 32/3,32/1,33/2, 31/2,12/3

Table 1. Summary of 4-state Ring-TCM codes

From the above discussion we expect that codes in Group 5 will have a better performance than those in Group 4 on fading channels because these codes have a diversity factor of 4 which is the maximum possible value. This is justified in Fig. 6. Also the error performances of these codes on the AWGN channel and code 32/2 from Group 2 are depicted in Fig. 7. Notice that in the fading channel a shorter error event path, i.e. L=2, is compromised by the minimum possible value of N_{free}.

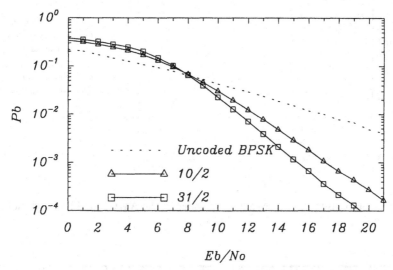

Fig.6. Comparison of two 4-state ring codes on the Rayleigh fading channel

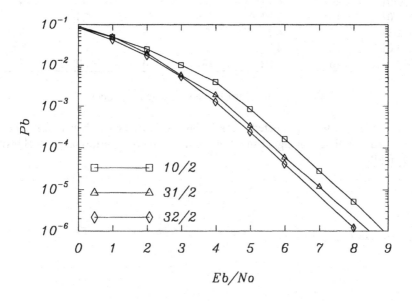

Fig.7. Comparison of three 4-state ring codes on the AWGN channel

Whereas code 010/01 is a poor code on the Gaussian channel, it is seen in Fig. 8 that important parameters, L=3 and N_{free}=2 compensate the performance of the code on the fading channel. Table 2 compares important features of some 16-state ring codes.

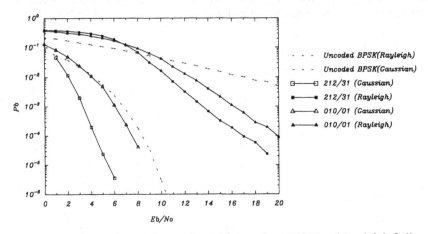

Fig.8. Comparison of two 16-state ring codes on the AWGN and Rayleigh fading channels

Code	SED	ACG	Nfree	L	spd	Rot
010/01	6	1.77	2	3	8	NT
012/01	10	3.98	2	4	24	NT
012/31	12	4.77	5	5	64	NT
212/01	12	4.77	1	3	64	NT
212/31	16	6.02	14	4	128	T

Table 2. Features of some 16-state ring codes

As theoretically was predicted, employing 4-dimensional hypercube and 8-dimensional biorthogonal signal sets can improve the error performance as shown in Fig.9 and Fig.10.

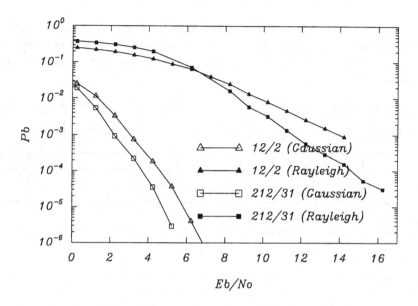

Fig.9. Error performance of 4 and 16-state ring-TCM codes in 4 -D space

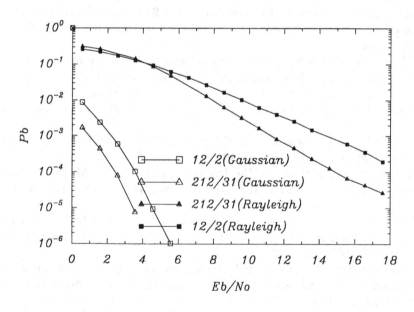

Fig.10. Error performance of 4 and 16-state ring-TCM codes in 8-D space

7 Conclusions

The R-TCM strategy has been shown to provide good performance over both Gaussian and fading channels.Multidimensional modulation can improve the error performance of the code moderately. Investigation of the structure of optimum multidimensional constellation will be continued to clarify the role of symmetry and uniformity in suitable signal sets for fading channels. Codes introduced in this paper are summarized in Table3.

Number of dimensions	Constellation	4-state Gaussian at $p_b = 10^{-5}$	16-state Gaussian at $p_b = 10^{-5}$	4-state Rayleigh at $p_b = 10^{-5}$	16-state Rayleigh at $p_b = 10^{-5}$
4	2(QPSK)	2.8	3.3	11	12.2
4	hypercube	3.7	4.8	12	14.2
8	biorthogonal	5	6	11.2	14.4

Table 3. Advantage of multidimensional R-TCM on the Gaussian and Rayleigh fading channels

References

[1] G.Ungerboeck,"Channel Coding with Multilevel/Phase Signals",IEEE Trans. Inform.Theory, Vol.28, No.1, pp 55-67,Jan. 1982.

[2] G.Ungerboeck,"Trellis-Coded Modulation with Redundant Signal Sets-- Part I: Introduction, and Part II: State of the Art", IEEE Comm.Magazine, Vol.25, No.2, pp 5-21 Feb. 1987.

[3] J.L.Massey and T.Mittelholzer, "Convolutional Codes over Rings", Proc.Fourth Joint Swedish-Soviet Int.Workshop on Inform. Theory, Gotland,1989.

[4] R.Baldini,F. and P.G.Farrell," Coded Modulation Based on Rings of Integers Modulo-q; Part 2: Convolutional Codes", IEE Proc. Comm. Vol.141, No.3, June 1994.

[5] F.J.Lopez, R.A. Carrasco and P.G.Farrell, "Ring TCM Codes for QAM", Electronic Lett., Vol.28, No. 25, Dec. 1992.

[6] D.Divsalar and M.K.Simon,"The Design of Trellis Coded MPSK for Fading Channels: Performance Criteria", IEEE Trans. Comm., Vol. 36, No. 9, pp 1004-1012, Sep.1988.

[7] D.Divsalar and M.K.Simon,"The Design of Trellis Coded MPSK for Fading Channels: Set Partitioning for Optimum Code Design", IEEE Trans. Comm., Vol. 36, No. 9, pp 1013-1021, Sep.1988.

[8] F.J.Lopez,'Optimal Design and Application of Trellis Coded Modulation Techniques Defined over Ring of Integers' Ph.D. Thesis, Staffordshire University, Staffordshire, UK ,1994.

[9] E.Biglieri, D.Divsalar, P.J.McLane, and M.K.Simon, Introduction to Trellis-Coded Modulation with Applications, McMillan, N.Y., 1991.

[10] S.H.Jamali and T.Le-Ngoc, Coded-Modulation Techniques for Fading Channels, Kluwer, USA, 1994.

[11] C.E.Shannon, "Communications in the Presence of Noise", Proc. IRE, Vol. 37, pp 10-21, Jan.1949.

[12] P.K.Ho, J.K.Cavers and J.L.Varaldi,"The Effects of Constellation Density on Trellis Coded Modulation in Fading Channels",IEEE Trans. Vehicular Tech., Vol.42,No.3,pp.318-325, Aug.1993.

[13]L.F.Wei, "Rotationally Invariant Trellis-Coded Modulation with Multidimensional M-PSK", IEEE Journal of Selected Area in Communications, Vol.7, No.9, Dec.1989.

[14] R.Baldini F.,'Coded Modulation Based on Rings of Integers', Ph.D. Thesis, University of Manchester, 1992.

[15] D.Divsalar and M.K.Simon, "Trellis Coding with Asymmetric Modulations", IEEE Trans. Comms.,Vol.35, No.2, Feb.1987.

Authentication Codes:
an Area where Coding and Cryptology Meet

Henk C. A. van Tilborg

Department of Mathematics and Computer Science
Eindhoven University of Technology
P.O. Box 513, 5600 MB Eindhoven
The Netherlands
E-mail: henkvt@win.tue.nl

Abstract. Among many applications of cryptography, the use of *authentication schemes* is of great practical importance. The purpose of authentication schemes [3], [10] is to add proof to a message that the message is authentic, i.e. it was not sent by an imposter and it has not been altered on its way to the receiver. The imposter may replace an authenticated message by another message (substitution) or may just try to send his own message (impersonation). The aspect of secrecy could also be introduced here, but in many cases the receiver just wants to be sure that the message is genuine. Think for instance of offices that are communicating with each other.

An important distinction to be made is that between authentication schemes that are unconditionally secure and schemes that are based on certain complexity theoretic assumptions. It is the first category that will be the main topic of this paper. A common technique here is to append to a message a (relatively short) tail that depends in an essential way on every bit in the message and also on a key that is shared with the legitimate receiver.

Some well-known bounds on the probability of successful substitution and impersonation will be given. Further, a direct connection with the existence of error-correcting codes will be given. (This relation is not a direct one-to-one correspondence!) Interesting results have already been obtained in this way, but there is ample room for improvement. It is the purpose of this paper to make the reader acquainted with this area of research.

1 Introduction

The most important goals that cryptography tries to achieve are:

confidentiality The inability of people that have gained unauthorized access to certain data to understand their meaning.

authentication A proof for the receiver of data that they came from the person that supposedly transmitted them.

integrity A proof for the receiver of data that nothing has been changed in those data.

Although confidentiality (privacy, secrecy) is an important and mathematically interesting issue, for the vast majority of data communication it does not play an essential role. Authentication and integrity on the other hand are almost always essential. Think of receivers of data files, E-mail messages, fax, telex, etc. For instance, banks often use tapes with all the transactions of that day written on it. Violation of the confidentiality does (in general) little harm, but this is not the case if somebody else has been able to tamper with the files. Exactly the same protection is needed in situations were vulnerable data are stored in a computer for later retrieval.

When studying authentication schemes one needs to distinguish the following aspects:

- unconditional security versus computational security,
- trusted parties versus non-trusted parties,
- large data files versus small data files,
- confidentiality versus no confidentiality,
- multiple use versus single use.

Especially the first two distinctions have lead to completely different research areas. The main topic here will be authentication schemes with unconditional security involving large data files between trusting parties for multiple use, while secrecy does not play a role. These schemes are usually called *authentication codes* or simply *A-codes*.

With unconditional security we mean that the security of the scheme does not depend on the computing power of the opponent. In case of computational security it does. We shall give examples of two kinds of computationally secure authentication schemes: a *digital signature scheme* and a *Message Authentication Code* (shortened to MAC).

RSA digital signature system [8]. This is a public key system based on:

Theorem 1 Euler, Fermat. *Let a and n be integers. Then*

$$\gcd(a,n) = 1 \quad \Rightarrow \quad a^{\phi(n)} \equiv 1 \pmod{n}, \tag{1}$$

where ϕ denotes Euler's Totient Function.

Each user U in the system chooses two different primes, say p_U and q_U, of about 100 digits long. Let $n_U = p_U \cdot q_U$. By the definition of ϕ

$$\phi(n_U) = (p_U - 1)(q_U - 1). \tag{2}$$

Secondly, user U chooses an integer $1 < P_U < \phi(n)$ with $\gcd(P_U, \phi(n_U)) = 1$. With Euclid's Algorithm user U computes (in less than $2 \cdot \log_2 \phi(n)$ operations) the integer S_U, satisfying

$$P_U \cdot S_U \equiv 1 \pmod{\phi(n_U)}, \ 1 < S_U < \phi(n_U). \tag{3}$$

Each user U makes the numbers P_U and n_U public but keeps S_U secret. If user A, say Ann, wants to authenticate a message for user B (Bob) she represents her message in any standard way by numbers $0 < m < n_A$. So Ann sends

$$c = (m^{S_A} \bmod n_A), \tag{4}$$

where $(x \bmod n)$ denotes the non-negative remainder of x after division by n. Bob (and anybody else) can recover m from c and check its authenticity and integrity by computing $c^{P_A} \bmod n_A$. Indeed for some integer l one has by (4), (3) and (1)

$$c^{P_A} \equiv m^{P_A S_A} \equiv m^{1+l\phi(n_A)} \equiv m \pmod{n_A}, \tag{5}$$

when $\gcd(m, n_A) = 1$. The reader may easily verify that these relations also hold when $\gcd(m, n_A) \neq 1$. The system is summarized in Table 1.

Public:	P_U and n_U of all users U.
Secret:	S_U of user U.
Property:	$P_U \cdot S_U \equiv 1 \pmod{\phi(n_U)}$.
Messages from A to B:	$0 < m < n_A$.
m with signature by A:	$c = (m^{S_A} \bmod n_A)$.
B computes:	$(c^{P_A} \bmod n_A) = m$.
Signature:	the pair (m, c).

Table 1. The RSA digital signature scheme

The security of the system is based on the fact that only Ann can convert a meaningful message m into a c such that (5) holds, because only Ann knows P_A. Once a third party knows the secret S_A, he can also put Ann's signature on m, i.e. map m to c. He is of course able to compute S_A from (3) with Euclid's Algorithm and the publicly known P_A, once he knows $\phi(n_A)$. To find $\phi(n_A)$ from (2) and the publicly known n_A, he needs to find the factorization of n_A. The computational complexity of factoring a 200-digit long number with the fastest known algorithms is such that this approach is not feasible. Faster techniques to attack this system are not known at the moment.

Note the characteristics of the above system. A third party with umlimited computing power can factor n_A, which would break the system. So the RSA digital signature scheme is computationally secure, not unconditionally. Parties involved do not have to trust each other and the system is very suitable for multiple use. In the description above there is no secrecy but it is easy to add that feature by means of the same RSA system. The scheme is not so suited for large data files, because exponentiation modulo 200-digits long numbers is still relatively slow.

Message Authentication Codes

We shall describe the *Digital Signature Standard* (DSS) as an example of a message authentication code. Its characteristics are: computationally secure, no trusted parties, suitable for multiple use, no secrecy and very suited for large data files. The system is designed by the National Security Agency (NSA) and adopted as standard by the National Institute of Standards and Technology (NIST).

DSS adds two sequences of 160 bits each to the end of a document as guarantee of its authenticity and integrity. To this end, it first applies the *Secure Hash Algorithm* (**SHA**) to the file.

In general an hash function h maps an arbitrary file M to a sequence of certain fixed length (in the case of the SHA this length is 160), such that

H1 $h(M)$ is easy to evaluate.

H2 From a value $h(M)$ it is infeasible to determine another file M', such that $h(M') = h(M)$.

H3 It is infeasible to find two files M and M' with the same hash value.

It follows that the evaluation of $h(M)$ depends in an essential way on every single bit in M. The precise working of the SHA is not relevant here but can be found in [9].

To set up the system the following joint parameters are chosen:

- A prime number p whose binary representation has a word length divisible by 64 and between 512 and 1024,
- a prime factor q of $p-1$ that is 160 bits long,
- a value $g = h^{(p-1)/q} \pmod p$ where h is less than $p-1$ such that g is greater than 1.

Next, as in the logarithm system proposed in [2], each user U chooses a secret m_U, computes $c_U \equiv g^{m_U} \pmod p$, and makes c_U public.

When Ann wants to sign a file M, she first computes its value $h(M)$ under SHA. Next, she chooses a random number $r < q$ and adds as signature to M the numbers u and v, both of length 160, defined by:

$$u = \left((g^r \bmod p) \bmod q\right),$$

$$v = \left((r^{-1}(h(M) + m_A u)) \bmod q\right).$$

A receiver can check the authenticity and integrity of the received message M by evaluating:

$$w \equiv v^{-1} \pmod q,$$

$$x \equiv (h(M) \times w) \pmod q,$$

$$y \equiv (u \times w) \pmod q,$$

$$U = \left((g^x \times c_A^y \bmod p) \bmod q\right).$$

If $u = U$ the document will be accepted as genuine and coming from Ann. By a simple substitution one can verify that the relation $u = U$ indeed should hold. The security of the signature is based on the computational security of the the logarithm cryptosystem. Note that the security of SHA is also of the computational type.

The function of the random number r above is to hide the secret key of A.

No authentication scheme can give an absolute guarantee that an accepted message comes from a particular user A. For instance, there is always a small probability that a (randomly or otherwise) generated sequence could have been made A, but in fact was not. Hence, it is necessary to define and compute the probability of a successful fraud. However, in such computations there is an essential difference between assuming the computational security of certain problems (as we have done above), or not making any further assumptions at all. This last situation will be the topic of the next sections. Interesting connections with the theory of error-correcting codes will come up.

2   Authentication Codes

The authentication schemes that we will discuss here will be unconditionally secure and hence will not be based on a public key cryptosystem. This implies that sender and receiver must trust each other and must have agreed upon a secret key. This is not strictly necessary, but then a trusted third party (like an arbiter) must be introduced.

Let us start with a simple example.

Example 1. Ann wants to send a single bit of information (a yes or a no) to Bob by means of a word of length 2. Ann and Bob have 4 possible keys available. Ann and Bob make use of the following scheme:

$$
\begin{array}{c|cccc}
& \multicolumn{4}{c}{\text{send}} \\
& 00 & 01 & 10 & 11 \\
\hline
1 & 0 & 1 & - & - \\
\text{key } 2 & 1 & - & 0 & - \\
3 & - & 0 & - & 1 \\
4 & - & - & 1 & 0 \\
\end{array}
$$

So, message 1 will be sent as word 11 under the third key.

The probability of succesful impersonation is $1/2$, because only two of the four words are possible as transmitted word under a particular key.

Also, an opponent who tries to replace a transmitted word by another will know that only two keys can possibly have been used, but does not know which one. So, the probability of a succesful substitution is also $1/2$.

The above scheme even gives secrecy, because every transmitted word can come from message 0 with probability $1/2$ and from message 1 with probability $1/2$.

The general definition of an authentication code (we deviate here from the standard notation in the theory of authentication codes in order to avoid confusion with the standard notation in the theory of error-correcting codes) is as follows:

Definition 2. An *authentication code* is a triple $(\mathcal{M}, \mathcal{K}; \mathcal{C})$ and a mapping $f : \mathcal{M} \times \mathcal{K} \to \mathcal{C}$ such that

$$\forall_{k \in K} [\, f(m, k) = f(m', k) \Rightarrow m = m']. \tag{6}$$

The letters $\mathcal{M}, \mathcal{K}$ and $\mathcal{C}$ stand for *message space* resp. *key space* and *code space*. In the example above, $\mathcal{M} = \{0, 1\}$, $\mathcal{K} = \{1, 2, 3, 4\}$ and $\mathcal{C} = \{00, 01, 10, 11\}$. Condition (6) makes f into an injective mapping for each possible key. To make the system practical for each fixed key f should be easily invertible.

The legitimate receiver of a codeword c accepts it as a signed version of m if and only if $f(m, k)$ is equal to c. Note that the key k is also known to the receiver.

Let M and K denote random variables defined on $\mathcal{M}$ and $\mathcal{K}$. They denote the probability that a sender wants to send a particular message $m \in \mathcal{M}$, resp. the probability that a sender and receiver have agreed upon a particular key $k \in \mathcal{K}$. The distributions M and K induce a probability distribution on $\mathcal{C}$ in a natural way.

The authenticator should have the property that possible changes made in m by an opponent result in words that are spread evenly over $\mathcal{C}$, the set of codewords, while the subset of words that the receiver expects knowing the key k, i.e. the range of $f(m, k)$, $m \in \mathcal{M}$, is only a fraction of this set. Thus the aim of an authentication code is that the legitimate receiver of c (but also an arbiter) can check the authenticity of m, but an enemy who knows the authentication code but not the key k and who attempts to impersonate the legitimate sender by sending $\hat{c}$ with $\hat{c} = f(\hat{m}, \hat{k})$ has only a small probability of getting message $\hat{m}$ accepted. The same should be true if the enemy wants to substitute $\hat{c} = f(\hat{m}, k)$ for a transmitted $c = f(m, c)$ where $\hat{m} \neq m$ (now more information on the key in use is available). The maximum probability of a successful substitution will be denoted by P_S.

Since an authentication function f is injective for fixed key k, it follows that at least $|\mathcal{M}|$ codewords must be authentic for any given key. This implies that when M and K have a uniform distribution P_I satisfies

$$P_I \geq \frac{|\mathcal{M}|}{|\mathcal{C}|}. \tag{7}$$

Similarly, when one codeword is observed, at least $|\mathcal{M}| - 1$ of the remaining codewords are still authentic. This shows (under the same condition of uniform distribution) that P_S satisfies

$$P_S \geq \frac{|\mathcal{M}| - 1}{|\mathcal{C}| - 1}. \tag{8}$$

To make an authentication scheme more suitable for large data files one usually prefers authentication codes of a special type.

Definition 3. An authentication code is called *systematic* if the mapping $f : \mathcal{M} \times \mathcal{K} \to \mathcal{C}$ is of the form $f(m, k) = (m, \tau(m, k))$, where the *tail* $\tau(m, k)$ is a mapping from $\mathcal{M} \times \mathcal{K}$ to some set $\mathcal{T}$. The tail $\tau(m, k)$ that is appended to m is called the *authenticator* of m.

Note that the set C of possibly transmitted words c can be now be viewed as a systematic code. This interpretation will lead to some interesting connections that will be discussed in the next section. From now on we restrict our attention to systematic authentication codes.

For the impersonation attack the opponent can never do better than send a pair $(\hat{m}, \hat{t})$ in which $\hat{t}$ is the authenticator of $\hat{m}$ for the largest fraction of the keys.

$$P_I = \max_{m \in \mathcal{M}, t \in \mathcal{T}} \frac{|\{k \in \mathcal{K} \mid \tau(m, k) = t\}|}{|\mathcal{K}|}. \tag{9}$$

If one wants to substitute $(\hat{m}, \hat{t})$ for a transmitted message (m, t) one can do no better than optimize

$$\max_{\hat{m} \in \mathcal{M} \backslash \{m\}, \hat{t} \in \mathcal{T}} \frac{|\{k \in \mathcal{K} \mid \tau(m, k) = t \ \& \ \tau(\hat{m}, k) = \hat{t}\}|}{|\{k \in \mathcal{K} \mid \tau(m, k) = t\}|}.$$

So, the maximum chance of a successful substitution attack is

$$P_S = \max_{m \in \mathcal{M}, t \in \mathcal{T}} \max_{\hat{m} \in \mathcal{M} \backslash \{m\}, \hat{t} \in \mathcal{T}} \frac{|\{k \in \mathcal{K} \mid \tau(m, k) = t \ \& \ \tau(\hat{m}, k) = \hat{t}\}|}{|\{k \in \mathcal{K} \mid \tau(m, k) = t\}|}. \tag{10}$$

The maximum of the two probabilities in (9) and (10) is often called the probability of successful *deception*, so it is defined by

$$P_D = \max\{P_I, \ P_S\}. \tag{11}$$

Without proof we quote the following lower bounds on the probabilities P_I, P_S, and P_D. Proofs and references can be found in [6].

Theorem 4. *Let M, K, and T denote random variables defined on $\mathcal{M}, \mathcal{K}$, resp. $\mathcal{T}$, related by a function $\tau : \mathcal{M} \times \mathcal{K} \to \mathcal{T}$. Further, let $H(X|Y)$ and $I(X;Y)$ denote the usual (conditional) entropy function resp. the mutual information function (see [7] for definitions). Then*

$$P_I \geq 2^{-I(M,T;K)}, \tag{12}$$

$$P_S \geq 2^{-H(K|M,T)}, \tag{13}$$

$$P_D \geq \frac{1}{\sqrt{K}}, \tag{14}$$

The bound in (14) is called the *square root bound*. Authentication codes meeting this bound are called *perfect*. Incidentally, the proof of (14) in [3] is correct only for uniform distributions, while for general random variables it was given only recently in [1]. In [3] the following inequality is discussed. A (more general) proof can be found in [6].

Theorem 5. *A necessary condition for an authentication code to be perfect is that*

$$|\mathcal{M}| \leq \sqrt{|\mathcal{K}|} + 1. \tag{15}$$

Construction 6 (Projective plane construction, [3]) *Let $(\mathcal{P}, \mathcal{L})$ denote the point resp. line set of a projective plane $PG(2, q)$. Select an arbitrary line l from $\mathcal{L}$. To obtain an authentication code take*

1. *The message set $\mathcal{M}$ consists of the points on l.*
2. *The key space $\mathcal{K}$ consists of all points not on l.*
3. *The authenticator set $\mathcal{T}$ consists of all lines in $\mathcal{L}$ except for l.*
4. *The authenticator function τ is defined by $\tau(m, k) = l'$ with l' the unique line through m and k.*

That the above scheme defines an authentication code is easy to check. Its parameters are given in the following theorem.

Theorem 7. *The authentication code described in Construction 6 has parameters*

$$\mathcal{M} = q + 1, \quad \mathcal{K} = q^2, \quad \mathcal{T} = q^2 + q. \tag{16}$$

The probability of success for the impersonation and substitution attack are given by

$$P_I = P_S = \frac{1}{q}.$$

Proof: The parameters in (16) follow directly from the definitions in Construction 6 and the fact that $PG(2, q)$ contains $q^2 + q + 1$ points and lines and that each line contains $q + 1$ points.

To compute P_I we observe that an opponent can do no better than selecting a line l'' as authenticator that contains as many points outside l (keys) as possible. But this number is q independent of the choice of l''. So

$$P_I = \frac{q}{\mathcal{K}} = \frac{q}{q^2} = \frac{1}{q}.$$

Similarly, if the opponent has observed the authenticator l' (not equal to l), there are still q keys (points on l' but not on l) possible. For any message m' not equal to m and any of the q possible keys k' there is a unique line (through them) that is the corresponding authenticator. So

$$P_S = \frac{1}{q}.$$

□

The authentication codes coming from Construction 6 are perfect because P_I, P_S, and P_D are all $1/q$ which is equal to $1/\sqrt{|\mathcal{K}|}$. Moreover, $|\mathcal{M}| = q + 1 = \sqrt{|\mathcal{K}|} + 1$, so its message set is maximal (by (15)).

A construction of authentication codes by means of shift register sequences can be found in [5]. Its implementation is simpler than the projective plane construction above. For large message sets, e.g. data files, the codes discussed in the next section may be more practical.

3 Authentication Codes from Error Correcting Codes

In [4] it is shown how authentication codes can be constructed from error-correcting codes and vice versa. To distinguish between the two types of codes we shall speak of A-codes and EC-codes. In this section we shall show the EC-to-A transformation.

Let C be any $(n, |C|, d_H)$ EC-code over $GF(q)$, i.e. a code of length n, cardinality $|C|$ and minimum Hamming distance d_H. Let C have the additional property that

$$\underline{c} \in C \quad \Rightarrow \quad \forall_{\lambda \in GF(q)}[\underline{c} + \lambda\underline{1} \in C]. \tag{17}$$

For instance, any linear code containing the all-one vector satisfies (17). Note that property (17) implies that q divides $|C|$. If (17) holds when the all-one vector is replaced by any other vector of weight n then code C is equivalent to another $(n, |C|, d)$ code for which property (17) does hold.

The relation $\sim$ defined on C by $\underline{c} \sim \underline{c}'$ if and only if $\underline{c} - \underline{c}' = \lambda\underline{1}$, for some λ in $GF(q)$, is an equivalence relation on C. Let M be a subcode of C, containing one representative from each equivalence class. So M has cardinality $|C|/q$; it will be the set of messages $\mathcal{M}$ in the authentication code that we are constructing. In the remainder of this paper, $\underline{m}$ will denote the word in M, corresponding to the message m in $\mathcal{M}$.

Let $GF(q)$ consist of elements $\alpha_1, \alpha_2, \ldots, \alpha_q$. Define the set V of q-ary vectors of length nq by

$$V = \{\underline{v}^{(m)} = (\underline{m} + \alpha_1\underline{1}, \underline{m} + \alpha_2\underline{1}, \ldots, \underline{m} + \alpha_q\underline{1}) \mid \underline{m} \in M\}. \tag{18}$$

The set of keys $\mathcal{K}$, of the authentication code that we are constructing, will be the coordinate set of the vectors in V. So $|\mathcal{K}| = nq$. The authenticator $\tau(m, k)$ of message m under key k simply is given by the k-th coordinate of $\underline{v}^{(m)}$.

Theorem 8. *The authentication code for the messages m in $\mathcal{M}$ described above by means of a q-ary $(n, |C|, d)$ code C satisfying (17), by relation (18) and by the authenticator function $\tau(m, k) = (v^{(m)})_k$ satisfies*

$$|\mathcal{M}| = |C|/q, \quad |\mathcal{K}| = nq, \quad |\mathcal{T}| = q, \tag{19}$$

$$P_I = 1/q, \quad P_S = 1 - d/n. \tag{20}$$

Proof: The parameters in (19) follow immediately from the construction.

To compute P_I, we note that an opponent who wants to impersonate the sender needs to find the right authenticator for his message m'. However in $(m' + \alpha_1 \underline{1}, m' + \alpha_2 \underline{1}, \ldots, m' + \alpha_q \underline{1})$ each symbol occurs equally often. So, the probability that the opponent will choose the correct authenticator is $1/q$, independent of the choice of the authenticator and independent of the message m' that the opponent tries to transmit. This proves that $P_I = 1/q$.

An opponent who wants to replace the authenticated message (m, t), where $t = (v^{(m)})_k$, by another authenticated message, knows that the key in use is from a set of n possible keys. To be more precise, each of the (coordinate) sets $\{i, i+n, \ldots, i+(q-1)n\}$, $1 \leq i \leq n$, contains exactly one possible key. Let k_i, $1 \leq i \leq n$, denote the key that is congruent to i mod n. Note that $k_1, k_2, \ldots, k_n$ forms a complete residue system modulo n and that on the corresponding coordinates $\underline{v}^{(m)}$ has $t = (v^{(m)})_k$ as entry.

The optimal strategy for the opponent that wants to substitute another authenticated message for (m, t) is to find a message m', $m' \neq m$, such that in $\underline{v}^{(m')}$ on the coordinates indexed by $k_1, k_2, \ldots, k_n$ a symbol t' occurs the maximum number of times and use that symbol as most likely authenticator for m'. It remains to show that t' occurs at most $n - d$ times, so that the probablity of successful substitution is at most $(n - d)/n = 1 - d/n$. Because the possible keys form a complete residue system modulo n, this follows from

$$|\{1 \leq j \leq n \mid (\underline{v}^{(m)})_{k_j} = t \ \& \ (\underline{v}^{(m')})_{k_j} = t'\}| =$$
$$= |\{1 \leq j \leq n \mid ((\underline{m} + \alpha_1 \underline{1}) - (\underline{m}' + \alpha_1 \underline{1}))_{k_j} = t - t'\}| =$$
$$= n - d\left(((\underline{m} + \alpha_1 \underline{1}) - (\underline{m}' + \alpha_1 \underline{1})), (t - t')\underline{1}\right) =$$
$$= n - d(\underline{m} - t\underline{1}, \underline{m}' - t'\underline{1}),$$

which is at most $n - d$, because $\underline{m} - t\underline{1}$ and $\underline{m}' - t'\underline{1}$ are different words in the $(n, |C|, d)$ code C, since $\underline{m}$ and $\underline{m}'$ represent different messages.

$\square$

As an example, if C is the extended q-ary $(q, k+1, q-k)$ Reed Solomon code, then $|\mathcal{M}| = q^k$, $|\mathcal{K}| = q^2$, $P_I = 1/q$ and $P_S = k/q$.

Example 2. Let C be the quarternary $[4, 3, 2]$ EC-code that is the dual code of the repetition code. It clearly meets requirement (17). Now, M contains 4^2 messages, one vector from each set $\{\underline{c}, \underline{c} + \underline{1}, \underline{c} + \alpha \underline{1}, \underline{c} + \alpha^2 \underline{1}\}$, where $\underline{c} \in C$ and $\alpha^2 = 1 + \alpha$. For instance, M can be taken as the set of codewords in C with last coordinate equal to 0. The vectors $\underline{m}$ in M correspond to 16 messages in $\mathcal{M}$.

Taking $\alpha_1 = 0, \alpha_2 = 1, \alpha_3 = \alpha$ and $\alpha_4 = \alpha^2$, we can make an authenticator look-up table, depicted in Table 2. Note that the 16 possible messages can be represented by two quarternary symbols (the two leftmost). The third coordinate in the corresponding word in M is such that the sum of these coordinates is 0. The 4-th coordinate is always 0. This gives a codeword in M, one from each equivalence class defined on C. The authenticator consists of one symbol. For instance, message $(1, \alpha)$ being the first two symbols of the word $(1, \alpha, \alpha^2, 0)$ will get α as authenticator when key 13 is used.

Since each symbol occurs four times in each row of Table 2, we may conclude that $P_I = 1/4$.

Suppose on the other hand that $(1, \alpha, \alpha^2, 0, 0, \alpha)$ is an intercepted codeword. From Table 2 an opponent can conclude that the key that has been used is among keys 2, 7, 12 and 13. Looking at the corresponding columns, we observe that for each message the same authenticator is used at most twice in these columns. So the best the opponent can do is transmit a message with an authenticator that occurs twice and obtain $P_S = 1/2$. For instance, he can send $(1, \alpha^2, \alpha, 0, \alpha^2)$.

m				key number															
				1	2	3	4	5	6	7	8	9	10	11	12	13	14	15	16
m				$\underline{v}^{(m)}$															
0	0	0	0	0	0	0	0	1	1	1	1	α	α	α	α	α^2	α^2	α^2	α^2
0	1	1	0	0	1	1	0	1	0	0	1	α	α^2	α^2	α	α^2	α	α	α^2
0	α	α	0	0	α	α	0	1	α^2	α^2	1	α	0	0	α	α^2	1	1	α^2
0	α^2	α^2	0	0	α^2	α^2	0	1	α	α	1	α	1	1	α	α^2	0	0	α^2
1	0	1	0	1	0	1	0	0	1	0	1	α^2	α	α^2	α	α	α^2	α	α^2
1	1	0	0	1	1	0	0	0	0	1	1	α^2	α^2	α	α	α	α	α^2	α^2
1	α	α^2	0	1	α	α^2	0	0	α^2	α	1	α^2	0	1	α	α	1	0	α^2
1	α^2	α	0	1	α^2	α	0	0	α	α^2	1	α^2	1	0	α	α	0	1	α^2
α	0	α	0	α	0	α	0	α^2	1	α^2	1	0	α	0	α	1	α^2	1	α^2
α	1	α^2	0	α	1	α^2	0	α^2	0	α	1	0	α^2	1	α	1	α	0	α^2
α	α	0	0	α	α	0	0	α^2	α^2	1	1	0	0	α	α	1	1	α^2	α^2
α	α^2	1	0	α	α^2	1	0	α^2	α	0	1	0	1	α^2	α	1	0	α	α^2
α^2	0	α^2	0	α^2	0	α^2	0	α	1	α	1	1	α	1	α	0	α^2	0	α^2
α^2	1	α	0	α^2	1	α	0	α	0	α^2	1	1	α^2	0	α	0	α	1	α^2
α^2	α	1	0	α^2	α	1	0	α	α^2	0	1	1	0	α^2	α	0	1	α	α^2
α^2	α^2	0	0	α^2	α^2	0	0	α	α	1	1	1	1	α	α	0	0	α^2	α^2

Table 2. Authenticator look-up table for Example 2

The method explained in this section is certainly not the only way to make A-codes from EC-codes. However, it does have the additional property that each impersonation has the same probability of success (here $1/q$). Authentication codes with this property are called *I-equitable*. So, I-equitable codes have the property that

$$P_I = \frac{|\{k \in \mathcal{K} \mid \exists_{m \in \mathcal{M}}[\tau(m, k) = t]\}|}{|\mathcal{K}|}, \quad \text{for all } t \in \mathcal{T}. \tag{21}$$

Further the construction above satisfies (7) with equality since the authentication code has size $|M| \times q$ (because every message can have any symbol as authenticator).

4   Error Correcting Codes from Authentication Codes

Consider a systematic A-code described by the triple $(\mathcal{M}, \mathcal{K}; \mathcal{T})$ and the authenticator function $\tau : M \times K \to \mathcal{T}$. Define $n = |\mathcal{K}|$ and $q = |\mathcal{T}|$. Further, let $k_1, k_2, \ldots, k_n$ be a labeling of the keys. Then an EC-code C is defined as the set of all words:

$$\underline{c}^{(m)} = (\tau(m, k_1), \tau(m, k_2), \ldots, \tau(m, k_n)), \quad m \in \mathcal{M}. \tag{22}$$

If the authentication code is I-equitable (see (21)), it follows that the EC-code coming from it has the property that each symbol occurs the same number of times in a codeword. This number is n/q.

The expression in (10) for the maximum probability of successful substitution P_S for an A-code can now also be expressed in terms of the EC-code it defines:

$$P_S = \max_{\substack{\underline{c}, \underline{c}' \in C \\ \underline{c} \neq \underline{c}'}} \max_{\alpha, \beta \in GF(q)} \frac{|\{1 \leq i \leq n \mid c_i = \alpha, c_i' = \beta\}|}{|\{1 \leq i \leq n \mid c_i = \alpha\}|}. \tag{23}$$

Note that for I-equitable codes the denominator in (23) is equal to n/q, independent of the choice of $\underline{c}$ and $\underline{c}'$. On $GF(q)^n$ we define $\delta(\underline{a}, \underline{b})$ by

$$\delta(\underline{a}, \underline{b}) = \max_{\alpha, \beta \in GF(q)} |\{1 \leq i \leq n \mid a_i = \alpha, b_i = \beta\}| \tag{24}$$

The *A-distance* (for authentication distance) between vectors $\underline{a}$ and $\underline{b}$ is defined by

$$d_A(\underline{a}, \underline{b}) = n - q\delta(\underline{a}, \underline{b}). \tag{25}$$

The A-distance is not a proper distance function, because the triangle inequality does not hold. The *minimum A-distance* $d_A(C)$ of an EC-code C is defined by

$$d_A(C) = \min_{\substack{\underline{c}, \underline{c}' \in C \\ \underline{c} \neq \underline{c}'}} d_a(\underline{c}, \underline{c}'). \tag{26}$$

It follows from (23)–(26) that that the following relation holds between an I-equitable code and the EC-code C defined by it:

$$d_A(C) = n(1 - P_S). \tag{27}$$

Example 3. The I-equitable A-code in Example 2 defines an EC-code C consisting of the restriction of the rows in Table 2 to the last 16 columns. It is a quarternary $(16, 16, 8)$ EC-code (to verify the minimum Hamming distance observe that in each of the groups of 4 coordinates words differ in at least 2 places). We know from Example 2 that $P_S = 1/2$. It follows that $d_A(C) = 8$.

In general, $d_A(C)$ and $d_H(C)$ do not have to be equal, but there is a relation. Indeed,

$$
\begin{aligned}
n - d_H(C) &= n - \min_{\substack{c, c' \in C \\ c \neq c'}} d_H(c, c') \\
&= \max_{\substack{c, c' \in C \\ c \neq c'}} |\{1 \leq i \leq n \mid c_i = c_i'\}| \\
&= \max_{\substack{c, c' \in C \\ c \neq c'}} \sum_{\alpha \in GF(q)} |\{1 \leq i \leq n \mid c_i = c_i' = \alpha\}| \\
&\leq \max_{\substack{c, c' \in C \\ c \neq c'}} \sum_{\alpha \in GF(q)} \max_{\beta \in GF(q)} |\{1 \leq i \leq n \mid c_i = \alpha, c_i' = \beta\}| \\
&\leq \max_{\substack{c, c' \in C \\ c \neq c'}} q\delta(c, c') \\
&= n - \min_{\substack{c, c' \in C \\ c \neq c'}} d_A(c, c') \\
&= n - d_A(C).
\end{aligned}
$$

This proves the following proposition.

Proposition 9. *Let C be an EC-code made from an I-equitable A-code in the way described above and let $d_H(C)$ and $d_A(C)$ denote its minimum Hamming rep. authentication distance. Then*

$$d_H(C) \geq d_A(C). \tag{28}$$

5 Some Further Bounds on Authentication Codes

Let $m(n, \epsilon, q)$ denote the cardinality of the largest message set $\mathcal{M}$ in an I-equitable authentication code (see (21)) with $|\mathcal{K}| = n$, $|\mathcal{T}| = q$, and $P_S \leq \epsilon$ (they have $P_I = 1/q$ by definition).

The connection between A-codes and EC-codes makes it possible to derive lower and upper bounds on $m(n, \epsilon, q)$. These connections also lead to new and interesting questions about EC-codes with particular properties.

Let $A_q(n, d_H)$ denote the maximum cardinality of a q-ary EC code of length n with minimum Hamming distance d_H. Let $A_q^*(n, d_H)$ denote the same quantity for codes that satisfy the additional property that in each codeword each symbol occurs equally often. Similarly, let $A_q^\bullet(n, d_H)$ denote the same quantity for codes that satisfy the additional property that with each codeword $\underline{c}$ also $\underline{c} + \lambda \underline{1}$ is in the code for all $\lambda \in GF(q)$ (see (17)).

Theorem 10. *Let q be a prime power and $\epsilon < 1$. Then*

$$m(n, \epsilon, q) \geq \frac{1}{q} A_q^\bullet(n/q, (1 - \epsilon)n/q), \tag{29}$$

$$q(q - 1)m(n, \epsilon, q) \leq A_q^*(n, (1 - \epsilon)n), \tag{30}$$

$$q(q - 1)m(n, \epsilon, q) + q \leq A_q(n, (1 - \epsilon)n). \tag{31}$$

Proof: The lowerbound in (29) follows directly from the construction described in Section 3. Indeed, by Theorem 8, a q-ary code C of length n/q with minimum Hamming distance $d_H = (1 - \epsilon)n/q$ and of size $A_q^\bullet(n/q, d_H)$ yields an A-code with $|\mathcal{M}| = \frac{1}{q} A_q^\bullet(n/q, (1 - \epsilon)n/q)$, $|\mathcal{K}| = n$, $|\mathcal{T}| = q$, $P_I = 1/q$, and $P_S = \epsilon$.

To prove (30), we enlarge the EC-code C constructed from an A-code by means of the definition in Equation (22) to a code C' as follows:

$$C' = \{\alpha\underline{c} + \beta\underline{1} \mid \underline{c} \in C, \alpha, \beta \in GF(q), \alpha \neq 0\}.$$

Let $P_{\alpha,\beta}(\underline{c})$ denote such a word $\alpha\underline{c} + \beta\underline{1}$. We want to say something about the Hamming distance between the words in C'. Let $\underline{u}' = P_{\alpha,\beta}(\underline{u})$ and $\underline{v}' = P_{\hat{\alpha},\hat{\beta}}(\underline{v})$ be two words in C' and γ in $GF(q)$. Then

$$|\{1 \leq i \leq n \mid (P_{\alpha,\beta}(\underline{u}))_i = (P_{\hat{\alpha},\hat{\beta}}(\underline{v}))_i = \gamma\}|$$

$$= |\{1 \leq i \leq n \mid (\underline{u})_i = \frac{\gamma - \beta}{\alpha}, (\underline{v})_i = \frac{\gamma - \hat{\beta}}{\hat{\alpha}}\}|,$$

which is at most $\delta(\underline{u}, \underline{v})$ by (24). It follows from the above and by (27) that

$$d_H(\underline{u}', \underline{v}') = n - \sum_\gamma |\{1 \leq i \leq n \mid (P_{\alpha,\beta}(\underline{u}))_i = (P_{\hat{\alpha},\hat{\beta}}(\underline{v}))_i = \gamma\}|$$

$$\geq n - q\delta(\underline{u}, \underline{v}) = d_A(\underline{u}, \underline{v}) \geq d_A(C) = (1 - \epsilon)n.$$

Since in each word in C' each symbol occurs equally often (as is the case in C) and since $|C'| = q(q - 1)|C|$ we have proved inequality (30).

Adding the vectors $\alpha\underline{1}$, $\alpha \in GF(q)$, to C' gives an EC-code C'' that still has minimum distance $d_A(C) = (1 - \epsilon)n$. This proves (31).

$\square$

To use (29) in conjunction with the the Gilbert-Varshamov bound it is necessary to prove a similar bound on $A_q^{\bullet}(n, d)$. However, in the standard proof of the unrestricted Gilbert-Varshamov bound and also in its linear version, it is possible to choose the all-one vector as first choice of a non-zero codeword. So the Gilbert-Varshamov bound is also a lower bound for $A_q^{\bullet}(n, d)$.

For further bounds on $m(n, \epsilon, q)$ the reader is referred to [4] and [6].

6 Conclusion

Authentication codes can be constructed from error-correcting codes and vice versa. This leads to lower and upper bounds on authentication codes. These connections also give rise to new questions on EC-codes. There may be other ways to transform EC-codes to A-codes and back. Possibly, stronger bounds can be obtained in that way.

References

1. Bassalygo, L.A., *Lower bounds for the probability of successful message deception*, Probl. Inf. Trans., **29**, No. 2, pp. 104–108, 1993.
2. Diffie, W. and M.E. Hellman, *New directions in cryptography*, IEEE Trans. Inf. Theory, **IT–22**, pp. 644–654, Nov. 1976.
3. Gilbert, E.N., F.J. MacWilliams, and N.J.A. Sloane, *Codes which detect deception*, Bell System Technical Journal, Vol. **53**, pp. 405–424, 1974.
4. Johansson, T., G. Kabatianskii, and B. Smeets, *On the relation between A-codes and codes correcting independent errors*, Proceedings of Eurocrypt '93, pp. M1–M10, 1993.
5. Johansson, T., *A shift register of unconditionally secure authentication codes*, Designs, Codes and Cryptography, Vol. **4**, pp. 69–81, 1994.
6. Johansson, T., *Contributions to unconditionally secure authentication*, KF Sigma, Lund, 1994.
7. McEliece, R.J., *The theory of information and coding*, Encyclopedia of Math. and its Applications, Vol. 3, Addison-Wesley Publ. Comp., Reading, Mass., 1977.
8. Rivest, R.L., A. Shamir, and L. Adleman, *A method for obtaining digital signatures and public key cryptosystems*, Comm. ACM, Vol. **21**, pp. 120–126, Febr. 1978.
9. Schneier, B., *Applied Cryptography: Protocols, Algorithms, and Source Code in C*, John Wiley & Sons, New York etc., 1994.
10. Simmons, G.J., *A survey of information authentication*, in Contemporary cryptology: the science of information integrity, G.J. Simmons, Ed., IEEE Press, New York, pp. 379–419, 1992.
11. van Tilborg, H.C.A., *An introduction to cryptology*, Kluwer Academic Publishers, Boston, etc., 1988.
12. van Tilborg, H.C.A., *Coding theory, a first course*, Chartwell Bratt Studentlitteratur, Lund, Sweden, 1993.

Efficient Generation of Binary Words of Given Weight

Nicolas Sendrier

INRIA, Domaine de Voluceau, Rocquencourt, BP 105,
78153 Le Chesnay CEDEX, FRANCE

1 Introduction

In some cryptographic systems [2, 3], it is necessary to put some information in binary words of given weight, say t, and given length, say n.

It is possible to produce an explicit bijection between the set $W_{n,t}$ of these words and the set of integer $\{1, 2, \ldots, \binom{n}{t}\}$. As far as we know, this cannot be achieved efficiently.

We propose here a procedure that converts any binary sequence into elements of $W_{n,t}$ in linear time both for encoding and decoding. The solution we propose is an approximation: all the words of $W_{n,t}$ will not be reached and the words of $W_{n,t}$ are not obtained with uniform probability.

2 Representing Words of Given Weight

Let x be an element of $W_{n,t}$, and let $k_0 < \ldots < k_{t-1}$ denote the positions of the "1"s, the positions are numbered from 1 to n. Let $l_0 = k_0$, and for all i, $0 < i < t$, let $l_i = k_i - k_{i-1}$. The word x will be represented by the t-tuple of integers $(l_0, \ldots, l_{t-1})$.

Reciprocally, any t-tuple of strictly positive integers $(l_0, \ldots, l_{t-1})$ such that $\sum_{i=0}^{t-1} l_i \leq n$ will represent an element of $W_{n,t}$.

Our goal is to encode binary information into such t-tuples.

3 Huffman Code

We consider the memoryless source $X = \{1, \ldots, K\}$ with the law probability $P_X(i) = \lambda \binom{n-i}{t-1}/\binom{n}{t}$, where λ is such that $\sum_{i=1}^{K} P_X(i) = 1$.

When $K = n - t + 1$, then $\lambda = 1$, and the probability $P_X(i)$ is the probability to have a sequence of $i - 1$ "0" between two "1" when we consider the set $W_{n,t}$ with a uniform distribution. We will see that taking values of K smaller than $n - k + 1$ will reduce the amount the memory for the encoder and the decoder, but will not significantly reduce the performance.

Let h_K be a Huffman code of source X [1, Ch. 3]. We have

$$h_K : \{1, \ldots, K\} \longrightarrow \{0, 1\}^* = \cup_{i>0}\{0, 1\}^i$$
$$l_i \longmapsto h_K(l_i)$$

To h_K we can associate the encoder of infinite sequences:

$$H_K : \{1, \ldots, K\}^{\mathbf{N}} \longrightarrow \{0, 1\}^{\mathbf{N}}$$
$$(l_i)_{i \geq 0} \longmapsto (h_K(l_i))_{i \geq 0}$$

Since h_K is a Huffman code, H_K is a bijection. By extension we will also use H_K to denote the image of finite sequences.

4 Encoding a Binary Sequence

Let $\mathbf{a} = (a_i)_{i \geq 0}$ be a binary sequence to be encoded, and let

$$\mathbf{l} = (l_i)_{i \geq 0} = H_K^{-1}(\mathbf{a}).$$

The encoding of the t first letters of $\mathbf{l}$ will give

$$H_K(l_0, \ldots, l_{t-1}) = (a_0, \ldots, a_{m-1})$$

for some m. If $\sum_{i=0}^{t-1} l_i \leq n$, then the t-tuple $(l_0, \ldots, l_{t-1})$ represents a word of $W_{n,t}$ "containing" m bits of information.

The probability that $\sum_{i=0}^{t-1} l_i \leq n$ can be computed by using generating function techniques. We have

$$F(Y, Z) = \left(\sum_{i=1}^{K} Y^i Z^{r_i} \right)^t = \sum_{u,v} A_{u,v} Y^u Z^v$$

where r_i is the length of $h_K(i)$, and $A_{u,v}$ is the number of sequences of X^t which adds to u, and whose image by H_K has a binary length of v.

The probability of the event "$u \leq n$", when the sequence $\mathbf{a}$ is produced by a uniform-ally distributed binary source, is equal to

$$P_{n,t}(K) = \sum_{u \leq n} \frac{A_{u,v}}{2^v} = \sum_{u \leq n} [Y^u] F(Y, \frac{1}{2})$$

and the average value of v given that $u \leq n$ is equal to

$$\mathcal{I}_{n,t}(K) = \frac{1}{P_{n,t}(K)} \sum_{u \leq n} \frac{v A_{u,v}}{2^v} = \frac{1}{P_{n,t}(K)} \sum_{u \leq n} [Y^u] \frac{1}{2} \frac{\partial F}{\partial Z}(Y, \frac{1}{2})$$

This number $\mathcal{I}_{n,t}(K)$ is the average number of information bits contained in one word of $W_{n,t}$ obtained that way.

For $n = 1024$, $t = 50$ and $K = 77$, we have $P_{n,t}(K) = 0.747$ and $\mathcal{I}_{n,t}(K) = 276.0$ while we have $\log_2 \binom{n}{t} = 284.0$. Such a result is practically not acceptable since on average 25% of the binary sequences cannot be encoded.

5 Encoding All Binary Sequences

We consider a family $(c_i)_{i \geq 0}$ of binary patterns (of variable length). For instance we can take for c_i the representation of i in base 2, the length of c_i will be s_i where $2^{s_i - 1} \leq i < 2^{s_i}_i$. The encoding will work as follows, starting with $i = 0$:

1. let $\mathbf{a}^{(i)}$ be the result of a repeated bitwise addition of pattern c_i to $\mathbf{a}$,
2. put "$1 \ldots 10$" with i "1" at the beginning of $\mathbf{a}^{(i)}$,
3. proceed with h_K as described in the previous section,
4. if $\sum_{j=0}^{t-1} l_j \leq n$ then we have a word of $W_{n,t}$, else $i \leftarrow i + 1$ and we repeat steps 1 to 4.

To decode a word of $W_{n,t}$ obtained that way, we first encode the t-tuple according to the Huffman code h_K, and then the number of "1"s in the beginning of the obtained sequence will give us the pattern to be added to the result.

If we assume that the encoding with the different patterns are independent, the average number of try will be $1/P_{n,t}(K)$, and the number of information bits lost (those added in the beginning of $\mathbf{a}$) is also $1/P_{n,t}(K)$. From the simulation we have made, this assumption seems to be true as far as the number of attempts is concerned.

For $n = 1024$, $t = 50$ and $K = 77$, we have $P_{n,t}(K) = 0.747$ and thus the average number of information bits is $\tilde{\mathcal{I}}_{n,t}(K) = \mathcal{I}_{n,t}(K) - 1/P_{n,t}(K) = 274.7$. If we defined the information rate r_K as the ratio between $\tilde{\mathcal{I}}_{n,t}(K)$ and $\log_2 \binom{n}{t}$, we have $r_K = 0.967$. We give in Table 1 the result for other values of K, the information rate ranges from 0.893 to 0.974 when K ranges from 41 to 100. As a measure of the efficiency we will take the ratio $E_{n,t}(K) = \tilde{\mathcal{I}}_{n,t}(K)/\log_2 \binom{n}{k}$.

K	$P_{n,t}(K)$	$\tilde{\mathcal{I}}_{n,t}(K)$	$E_{n,t}(K)$	K	$P_{n,t}(K)$	$\tilde{\mathcal{I}}_{n,t}(K)$	$E_{n,t}(K)$
45	0.9993	257.57	0.907	75	0.8479	272.74	0.960
50	0.9971	261.04	0.919	80	0.7095	275.12	0.969
55	0.9714	266.84	0.939	85	0.6834	275.48	0.970
60	0.9457	268.94	0.947	90	0.6735	275.67	0.971
65	0.9019	270.98	0.954	95	0.5827	276.44	0.973
70	0.8666	272.13	0.958	100	0.5743	276.54	0.974

Table 1. $P_{n,t}(K)$, $\tilde{\mathcal{I}}_{n,t}(K)$ and $E_{n,t}(K)$ for different values of K

6 Complexity

Encoding and decoding with a prefix code can be done in linear time, the number of attempts for the encoding will generaly be small (less that 2 on average), and the amount of memory needed to store the Huffman code tree is $2K \log_2 K$ bits.

7 Conclusion

We thus provide here a linear time encoder/decoder of binary sequences into words of given weight and length. Furthermore, our algorithm is easy to implement and requires a limited amount of memory. Any exact solution, that a bijection between $W_{n,t}$ and $\{1, 2, \ldots, \binom{n}{t}\}$ will require, as far as we know, an algorithm of complexity at least $O(n^2)$, and computations over large integers.

It must be noted however that the statistical properties of the words generated are not very good. The major drawback being that the average length of the last sequence of consecutive "0"s is much larger than the other. For instance with $n = 1024$, $t = 50$ and $K = 80$ the length of this last sequence is 124 instead of 20.

References

1. R. G. Gallager. *Information Theory and Reliable Communication*. John Wiley & Sons, 1968.
2. H. Niederreiter. Knapsack-type cryptosystems and algebraic coding theory. *Prob. Contr. Inform. Theory*, 15(2):157–166, 1986.
3. V.M. Sidelnikov. A public-key cryptosystem based on Reed-Muller codes. *Discrete Mathematics and Applications*, 4(3):191–207, 1994.

Distribution of Recurrent Sequences Modulo Prime Powers
(Abstract)

Richard G.E. Pinch

University of Cambridge, Department of Pure Mathematics and Mathematical Statistics,
16 Mill Lane, Cambridge CB2 1SB, U.K.

Abstract. We study the distribution of linear recurrent sequences modulo p^n for prime p when the auxiliary polynomial is irreducible and the period is maximal. We show that such a sequence takes each possible value equally often up to an error of order $p^{n/2}$.

1 Introduction

In this paper we study the possible periods of linear recurrent sequences modulo p^n for prime p in the case when the the auxiliary polynomial is irreducible and the period is maximal: see [7]. We show that such a sequence takes each possible value equally often up to an error of order $p^{n/2}$, generalising the corresponding result for sequences to a prime modulus, [2], [3].

The result has applications to the generation of pseudo-random numbers [5], [6] and answers a question of Mascagni [4].

We give the results in the case $p = 2$, although they generalise readily to any characteristic.

Lemma 1. *Let x_n be a sequence modulo $M = 2^m$ with period R and let $Z(b)$ be the number of times a value b occurs in one full period of the sequence. Suppose that*

$$\left| \sum_{n=0}^{R-1} \zeta_M^{x_n} \right| < E$$

for ζ_M any M-th root of unity other than 1. Then

$$|Z(b) - R/M| \leq E.$$

So an estimate for the error in an exponential sum can be converted into an estimate for the difference between the distribution of a recurrent sequence and the uniform distribution, and conversely.

Lemma 2. *Let f be a polynomial of degree k, irreducible over $\mathbf{Q}_2$, and let x_n be a recurrent sequence in $\mathbf{Z}_2$ with auxiliary polynomial f. Suppose that x_n has period exactly $2^k - 1$ modulo 2^l but not modulo 2^{l+1}. Then x_n has period exactly $\left(2^k - 1 \right) 2^{m-l}$ modulo 2^m for $m \geq l$.*

We note that $(2^k - 1)2^{m-1}$ is the maximum possible period for such a sequence.

Theorem 3. *Suppose that x_n is a sequence modulo $M = 2^m$, with $m \geq 3$, satisfying a recurrence relation of degree k, with an auxiliary polynomial irreducible over $\mathbf{Q}_2$ and with maximal period $R = (2^k - 1)2^{m-1}$. Then*

$$\left| \sum_{n=0}^{R-1} \zeta_M^{x_n} \right| \leq C_k 2^{m/2}.$$

where ζ_M is an M-th root of unity other than 1.

References

1. M. Ganley (ed.), *Cryptography and coding III*, IMA conference series (n.s.), vol. 45, Institute of Mathematics and its Applications, Oxford University Press, 1993, Proceedings, 3rd IMA conference on cryptography and coding, Cirencester, December 1991.
2. Rudolf Lidl and Harald Niederreiter, *Finite fields*, Encyclopaedia of Mathematics and its applications, vol. 20, Addison–Wesley, Reading Mass., 1983, 0-201-13519-1, Republished, Cambridge University Press, 1984.
3. _____, *Introduction to finite fields and their applications*, second ed., Cambridge University Press, 1994, First edition 1986.
4. M. Mascagni, S.A. Ciccaro, D.V. Pryor, and M.L. Robinson, *A fast, high-quality and reproducible lagged-Fibonacci pseudrandom number generator*, Technical report SRC-TR-94-115, Supercomputing Research Center, IDA, Bowie, MD, U.S.A., Feb 1994.
5. Harald Niederreiter, *Quasi-Monte Carlo methods and pseudo-random numbers*, Bull. Amer. Math. Soc. **84** (1978), no. 6, 957–1041.
6. _____, *Recent trends in random number and random vector generation*, Ann. Oper. Res. **31** (1991), 323–346.
7. R.G.E. Pinch, *Recurrent sequences modulo prime powers*, In Ganley [1], Proceedings, 3rd IMA conference on cryptography and coding, Cirencester, December 1991., pp. 297–310.

On-line Secret Sharing

Christian Cachin

Institute for Theoretical Computer Science
ETH Zürich
CH-8092 Zürich, Switzerland
E-mail: cachin@inf.ethz.ch

Abstract. We propose a new construction for computationally secure secret sharing schemes with general access structures where all shares are as short as the secret. Our scheme provides the capability to share multiple secrets and to dynamically add participants on-line, without having to re-distribute new shares secretly to the current participants. These capabilities are gained by storing additional authentic (but not secret) information at a publicly accessible location.

1   Introduction

Secret sharing is an important and widely studied tool in cryptography and distributed computation. Informally, a secret sharing scheme is a protocol in which a dealer distributes a secret among a set of participants such that only specific subsets of them, defined by the *access structure*, can recover the secret at a later time.

Secret sharing has largely been investigated in the *information-theoretic* security model, requiring that the participants' shares give no information on the secret, i.e. that the respective probability distributions are independent. Called *perfect* secret sharing schemes, they require that for every participant the number of bits needed to represent a share must be at least as large as the number of bits required to describe the secret itself (analogous to Shannon's theorem about key size for a perfectly secure cipher).

If the access structure allows any subset of k or more of the n participants to reconstruct the secret but not $k-1$ or less, the secret sharing scheme is called a *threshold scheme*. It can be implemented with Shamir's construction [14] based on polynomial interpolation. Secret sharing schemes for general monotone access structures are known, based on monotone circuit constructions [1, 9]. The surveys by Stinson [16] and Simmons [15] provide a general description of secret sharing schemes.

For many access structures it can be proved that some shares have to be considerably larger than the secret in perfect schemes [5]. Moreover, there exist families of special access structures on n participants where the size of some shares must grow unboundedly as $n \to \infty$ [6].

In the schemes described so far, the set of participants remains unchanged until the secret is recovered. Blakley *et al.* [2] study threshold schemes with *disenrollment* capabilities, where a participant is free to leave and to give away his share. The dealer then shares a new secret by broadcasting a message over a public channel. For perfect threshold schemes with m-fold disenrollment it can be shown that the size of the initially distributed shares must grow linearly in m [2]. These results are extended to general dynamic access structures by Blundo et al. [3]. Schemes for distributing multiple secrets are examined in [4].

Recently, Krawczyk [10] introduced a construction for *computationally* secure threshold secret sharing schemes where the shares can be shorter than the secret and that uses a secure encryption function. Basically, this protocol works as follows: The (potentially large) secret is encrypted with a symmetric encryption function. The result is distributed among the participants using an information dispersal protocol [13] based on error correcting codes. Any k out of the n participants can reconstruct the encrypted secret. To prevent an unauthorized set of participants from learning anything about the secret, the secret key used for encryption is distributed among the participants using a conventional, unconditionally secure secret sharing scheme (e.g. Shamir's threshold scheme [14]).

Much research in the area of secret sharing has concentrated on the size of the shares. Although the size of the shares is important because the shares have to be transmitted and stored secretly, this is not the only information the participants must know to reconstruct the secret. Additional knowledge needed is, for example, the identity of the participants or the description of the protocol, including the access structure. These parameters are publicly known, but at the same time it is vital that they are authentic, i.e. no malicious participant has changed these descriptions. This is particularly important if the participants are computer systems that receive the descriptions over a potentially insecure communications link.

We propose a novel computationally secure secret sharing scheme for general access structures where all shares are as short as the secret. Our scheme provides the capability to share multiple secrets and to dynamically add participants on-line, without having to re-distribute new shares secretly to the current participants. These capabilities are traded for the need of storing additional authentic (but not secret) information at a publicly accessible location, e.g. on a bulletin board. Alternatively, this information can be broadcast to the participants over a public channel. The protocol gains its security from any one-way function. In particular, our construction has the following properties:

- All shares that must be transmitted and stored secretly once for every participant are as short as the secret.
- Multiple secrets can be shared with different access structures requiring only one share per participant for all secrets. This includes the ability for the dealer to change the secret after the shares have been distributed.
- The dealer can distribute the shares on-line: When a new participant is added and the access structure is changed, already distributed shares remain valid. Apart from the new participant's share that is secretly transmitted to him, only publicly readable information has to be changed.

The scheme is secure given any secure one-way function in the sense that a non-qualified set of participants running a polynomial-time algorithm cannot determine the secret with non-negligible probability. To prevent an attack by exhaustive search, however, the set of possible secrets must not be too small. Our construction solves an open problem of [10], albeit in a somewhat different way than proposed there.

Compared to traditional, unconditionally secure secret sharing schemes the proposed method is very flexible and uses only small shares. The differences lie in the additional use of publicly accessible information and in the security model. As for the use of authentic storage, we note that public information is needed in all traditional secret sharing schemes and that, authenticity usually costs much less than secrecy to implement.

Regarding the security model, computational security is theoretically weaker than information-theoretic or perfect security. On the other hand, for many applications that use a perfectly secure protocol, the cost of generating the needed random bits is prohibitively high and the bits are generated by a computationally secure pseudo random number generator. This makes the perfectly secure protocol vulnerable to adversaries with unlimited computing power.

The proposed scheme has many practical applications in situations where the participants and the access rules or the secret itself frequently change. No new shares have to be distributed secretly when new participants are included or participants leave. Such situations often arise in key management, escrowed [7] and fair [12] encryption systems, to name a few.

Consider, e.g., a high security area in a laboratory or in a bank where employees and managers are not permitted during off-hours. Only groups of one manager and at least two employees may enter and a secret sharing scheme is used to share the access code. If, for example, a manager is fired, he will disclose his share. With our scheme, only the access code and the bulletin board have to be updated—the other managers and employees do not have to be given new shares.

Another example is a group of frequently changing participants and alternating size where always two thirds of the current group members are needed to invoke some action, for example to reconstruct a master key used for escrowing keys of malicious users.

The paper is organized as follows: The basic scheme is presented in Section 3 and extended for sharing multiple secrets in Section 4. On-line secret sharing is then described in Section 5.

2   Preliminaries

We first need to formalize some aspects of a secret sharing scheme. A secret sharing scheme is a protocol between a set of participants $\mathcal{P} = \{P_1, \ldots, P_n\}$ and a dealer D, where $D \notin \mathcal{P}$ is assumed. The *access structure* $\Gamma \subseteq 2^{\mathcal{P}}$ is a family of subsets of $\{P_1, \ldots, P_n\}$ containing the sets of participants qualified to recover the secret. It is natural to require Γ to be monotone, that is, if $X \in \Gamma$ and $X \subseteq X' \subseteq \mathcal{P}$, then $X' \in \Gamma$. A *minimal* qualified subset $Y \in \Gamma$ is a set of participants such that $Y' \notin \Gamma$ for all $Y' \subset Y, Y' \neq Y$. The *basis* of Γ, denoted by Γ_0, is the family of all minimal qualified subsets. Note that Γ_0 uniquely determines Γ and vice versa.

For simplicity, we assume that the secret K is an element of a finite Abelian group $\mathbf{G} =< G, + >$ with $l = \log_2 |G|$. $\mathbf{G}$ could be the set of l-bit strings under bitwise addition modulo 2.

A *computationally secure secret sharing scheme* [10] is a protocol between D and the members of $\mathcal{P}$ to share a secret K, respective to an access structure Γ such that

a) the dealer D transmits a share S_i secretly to participant P_i, for $i = 1, \ldots, n$,
b) all qualified sets of participants $X \in \Gamma$ can efficiently compute K from their set of shares $\{S_i | P_i \in X\}$, and
c) every unqualified subset of participants $X \notin \Gamma$ running any polynomial-time algorithm cannot determine K with non-negligible probability.

To make the definition of security rigorous, we have to resort to asymptotics and consider the family of probability distributions of K indexed by the length l of the secret. Under this definition, for any unqualified subset $X \notin \Gamma$ running any algorithm A to recover K in time polynomial in l, the output of A must be equal to the correct K only with probability less than l^{-c}, for all constants c and suitably chosen $l > l_c$.

We will make use of a one-way function on G, $f : G \to G$ such that $f(x)$ is easy to compute for all $x \in G$ (i.e. can be computed in time polynomial in l) and that it is computationally infeasible, for a given $y \in G$, to find an $x \in G$ such that $f(x) = y$. The notion can be made rigorous analogous to the definition of security.

To achieve reasonable security, the security parameter l and thus the set of possible secrets have to be chosen large enough. Today, many secure one-way functions exist with typical l ranging from 64 to 128.

3 The Basic Scheme

Our protocol uses a publicly accessible location where the dealer can put up non-forgeable information that can be accessed by all the participants. We will refer to this location as the *bulletin board*. Alternatively, if communication and storage were not too expensive, the dealer could broadcast the information to the participants instead of storing it centrally. Implicitly, such a bulletin board is present in all existing secret sharing schemes and contains at least the number of participants n and the access structure Γ.

The basic protocol to share a secret $K \in G$ works as follows:

1. The dealer randomly chooses n Elements $S_1, \ldots, S_n$ from G according to the uniform distribution.
2. For all $i = 1, \ldots, n$, the dealer transmits S_i over a secret channel to P_i.

3. For each minimal qualified subset $X \in \Gamma_0$, the dealer computes

$$T_X = K - f\left(\sum_{x:P_x \in X} S_x \right)$$

and publishes $\mathcal{T} = \{T_X | X \in \Gamma_0\}$ on the bulletin board.

Addition and subtraction are performed in $\mathbf{G}$. To recover the secret K, a qualified set of participants Y proceeds similarly:

1. The members of Y agree on a minimal qualified subset $X \subseteq Y$.
2. The members of X add their shares together to get $V_X = \sum_{x:P_x \in X} S_x$ and apply the one-way function f to the result V_X.
3. They fetch T_X from the bulletin board and compute $K = T_X + f(V_X)$.

One can easily verify the completeness of the protocol: every qualified subset $X \in \Gamma$ can recover K.

Analyzing the security is only slightly more complicated: The relation between K and the shares is given by the $|\Gamma_0|$ equations

$$K = T_X + f(V_X)$$

for all $X \in \Gamma_0$, where the $V_X = \sum_{x:P_x \in X} S_x$ are all computed from different sets of shares. In the following, we denote by V_X the sum $\sum_{x:P_x \in X} S_x$ for any set of participants X. An unqualified subset $U \notin \Gamma$ cannot compute any of the $V_X, X \in \Gamma_0$ directly. So the members of U cannot compute K by exploiting one equation alone. However, they can link several equations through K or through any V_X. Linking two equations via K, one obtains relations of the form

$$T_Y - T_Z = f(V_Y) - f(V_Z)$$

with $Y, Z \in \Gamma_0$, of which the right sides are unknown to the members of U. Except for the unlikely case that $V_Y = V_Z$ which can be recognized on the bulletin board from $T_Y = T_Z$, this if of no use to them.

Linking two equations via V_W with $W \cap U = \emptyset$ and $W \cup U' \in \Gamma_0$, $W \cup U'' \in \Gamma_0$, for $U' \subset U$, $U'' \subset U$, and $U' \neq U''$ yields

$$f^{-1}(K - T_{W \cup U'}) - f^{-1}(K - T_{W \cup U''}) = V_{U'} - V_{U''},$$

thus nothing what the members of U could not have computed by themselves.

The size of $\mathcal{T}$ deserves some consideration. In general, $\mathcal{T}$ and the bulletin board are of size $O(2^n)$. However, note that for almost all general Γ the description of Γ itself is of the same size. This does not apply to threshold schemes that can be described by a list of participants plus two parameters (t, n) and for which $\mathcal{T}$ contains $\binom{n}{t}$ elements. But threshold schemes may be more important in theory than in practice: Reflecting on the way large companies and organizations are structured hierarchically today, it seems unlikely that a threshold scheme with more than several thousands of members will be realized by them.

In case the size of the bulletin board is limited, the authenticity of $\mathcal{T}$ can also be guaranteed by a digital signature of $\mathcal{T}$ by the dealer. If $\mathcal{T}$ is large, then parts of it could be signed such that not the entire table has to be read to validate a single entry.

Only one member of $\mathcal{T}$ has to be accessed to recover the secret. Therefore, the bulletin board could be implemented dynamically as a sever that broadcasts the desired entry upon request, together with a signature.

The shares of the participants in X are the inputs to a computation that ultimately yields K. For the basic scheme where one secret is shared once, the shares do not have to be kept secret during this computation. However, in the next two sections where additional capabilities of the scheme will be introduced, the shares and the result of their addition have to be kept secret. We will then assume that the participants can compute $f(V_X)$ without revealing their shares.

The protocol also allows the dealer to change the secret after the shares have been distributed by modifying $\mathcal{T}$ on the bulletin board.

4   Sharing Multiple Secrets

To share multiple secrets $K^1, K^2, \ldots$ with different access structures $\Gamma^1, \Gamma^2, \ldots$ among the same set of participants $\mathcal{P}$, the dealer distributes the shares S_i only once but prepares $\mathcal{T}^1, \mathcal{T}^2, \ldots$ for each secret. However, straightforward repetition of the basic scheme is not secure. Consider a set of participants X qualified to recover both K^1 and K^2: Any group $Y \in \Gamma^1$ can obtain K^2 as

$$K^2 = T_X^2 + T_Y^1 + f(V_Y) - T_X^1,$$

because $K^1 = T_X^1 + f(V_X) = T_Y^1 + f(V_Y)$ and because $f(V_X)$ is the same for K^1 and K^2. So $K^h - T_X^h$ must not be the same for different secrets.

To remedy this deficiency we replace f by a family $F = \{f_h\}$ of one-way functions so that different one-way functions are employed for different secrets. The following protocol is used to share m secrets K^h with access structures Γ^h for $h = 1, \ldots, m$:

1. The dealer randomly chooses n Elements $S_1, \ldots, S_n$ from G according to the uniform distribution.
2. For all $i = 1, \ldots, n$, the dealer transmits S_i over a secret channel to P_i.
3. For each secret K^h to share (with $h = 1, \ldots, m$) and for each minimal qualified subset $X \in \Gamma_0^h$, the dealer computes

$$T_X^h = K^h - f_h\Big(\sum_{x:P_x \in X} S_x \Big)$$

and publishes $\mathcal{T}^h = \{T_X^h | X \in \Gamma_0^h\}$ on the bulletin board.

To recover some secret K^h, a set of participants $Y \in \Gamma^h$ proceeds similarly:

1. The members of Y agree on a minimal qualified subset $X \subseteq Y$.
2. The members of X compute $V_X = \sum_{x:P_x \in X} S_x$ and apply f_h to V_X.
3. They fetch T_X^h from the bulletin board and compute $K^h = T_X^h + f_h(V_X)$.

The scheme does not demand a particular order for the reconstruction of the secrets. The required family F of one-functions can easily be obtained from f by setting $f_h(x) = f(h + x)$ when h is represented suitably in G.

Because a different one-way function f_h is used for each secret K^h, the information T^h on the bulletin board corresponding to K^h is (computationally) independent of the other $T^{h'}$ and $K^{h'}$ for $h \neq h'$. Thus, the security of the protocol is the same as for the basic protocol.

As noted above, the shares have to be protected from the eyes of other participants during the reconstruction phase. Otherwise, these participants could subsequently recover other secrets they are not allowed to know. We assume therefore that the computation of $f_h(V_X)$ is performed without revealing the set of inputs $\{S_i | P_i \in X\}$. Possible ways of achieving this include the presence of a trusted device to perform the computation or the use of a distributed circuit evaluation protocol [8].

The protocol does not impose any limitation on m except for $|\mathcal{F}|$, the size of the family of hash functions, such that any number of secrets can be distributed via the bulletin board while the shares of the participants remain the same.

5 On-line Secret Sharing

In many situations, the participants of a secret sharing scheme do not remain the same during the entire life-time of the secret. The access structure itself may change, too, if it is adapted to the new constellation of participants. In analogy to the monotonicity of the access structure we will assume that the changes to the access structure are monotone, i.e. participants are only added and qualified subsets remain qualified.

We define a *computationally secure on-line secret sharing scheme* to be a protocol between a dealer D and the members of a sequence of sets of participants $\mathcal{P}(0), \mathcal{P}(1), \ldots$ with $\mathcal{P}(t) \subset \mathcal{P}(t+1)$ for all $t \geq 0$ to share a secret K, respective to a sequence of access structures $\Gamma(0), \Gamma(1), \ldots$ with $\Gamma(t) \subseteq \Gamma(t+1)$ for all $t \geq 0$, such that

a) the shares S_i for $P_i \in \mathcal{P}(0)$ form a computationally secure secret sharing scheme for K respective to $\Gamma(0)$,

b) at time $t > 0$ the dealer D transmits a share S_i secretly to every participant $P_i \in \mathcal{P}(t) \setminus \mathcal{P}(t-1)$,

c) for all $t \geq 0$, every qualified set of participants $X \in \Gamma(t)$ can efficiently compute K from their set of shares $\{S_i | P_i \in X\}$, and

d) for all $t \geq 0$, all unqualified subsets of participants $X \notin \Gamma(t)$ running any polynomial-time algorithm cannot determine K with non-negligible probability.

The basic scheme from Section 3 satisfies the above definition when the dealer operates step-by-step, distributing the shares to the new participants and updating the bulletin board accordingly for every step. In particular, at step $t > 0$, D chooses a random S_i for every $P_i \in \mathcal{P}(t) \setminus \mathcal{P}(t-1)$ and publishes the T_X with

$X \in \Gamma_0(t)$ and $X \notin \Gamma_0(t-1)$. The previously issued shares are not invalidated and no shares have to be retransmitted.

The protocol of Section 4 to share multiple secrets can be extended similarly for on-line sharing of multiple secrets.

6 Extensions

The flexibility and the simplicity of the protocols allow many extensions to handle additional situations. We briefly discuss removing participants as opposed to adding them in on-line schemes and the secrecy of the shares during reconstruction in multi-secret schemes.

If a participant P_i is removed or disenrolled at time t, he will publish his share S_i, eventually enabling an unqualified set $X \notin \Gamma(t')$ to recover K if $X \cup S_i \in \Gamma(t')$ for some $t' \geq t$ if such an X exists. But in contrast to traditional secret sharing schemes, if a new secret is chosen to be shared, the dealer needs only update the bulletin board and no information has to be transmitted secretly. The same situation arises if the sequence of access structures is allowed to be non-monotonic.

The proposed multi-secret sharing protocols are only secure if the members of a qualified set X do not disclose their shares when a secret K^h is reconstructed. Otherwise, some schemes for $K^{h'}, h' \neq h$ could be compromised. If the shares cannot be hidden to carry out this computation, the protocols can be modified as follows: Similar to Lamport's one-time user authentication scheme [11], S_i is replaced by $f^{(N-h)}(S_i)$ for all $i = 1, \ldots, n$ in the h-th scheme (N is a predefined constant and $f^{(h)}(x)$ denotes h-fold repeated application of f to x). The secrets $K^h, h = 1, \ldots, n$ have to be recovered in increasing order. Thus, after the reconstruction of K^h, only values $f^{(N-h')}(S_i)$ for $h' > h$ are needed to reconstruct additional secrets, but these values cannot be computed from $f^{(N-h)}(S_i)$ if the one-way function is secure. The drawback is, apart from the fixed order of reconstruction, that N has to be chosen in advance and poses an upper limit on m, the number of secrets that can be distributed.

References

1. J. BENALOH AND J. LEICHTER, *Generalized secret sharing and monotone functions*, in Advances in Cryptology — CRYPTO '88, S. Goldwasser, ed., vol. 403 of Lecture Notes in Computer Science, Springer-Verlag, 1990, pp. 27–35.
2. B. BLAKLEY, G. R. BLAKLEY, A. H. CHAN, AND J. L. MASSEY, *Threshold schemes with disenrollment*, in Advances in Cryptology — CRYPTO '92, E. F. Brickell, ed., vol. 740 of Lecture Notes in Computer Science, Springer-Verlag, 1993, pp. 540–548.
3. C. BLUNDO, A. CRESTI, A. DE SANTIS, AND U. VACCARO, *Fully dynamic secret sharing schemes*, in Advances in Cryptology — CRYPTO '93, D. R. Stinson, ed., vol. 773 of Lecture Notes in Computer Science, Springer-Verlag, 1994, pp. 110–125.

4. C. BLUNDO, A. DE SANTIS, G. DI CRESCENZO, A. G. GAGGIA, AND U. VACCARO, *Multi-secret sharing schemes*, in Advances in Cryptology — CRYPTO '94, Y. G. Desmedt, ed., vol. 839 of Lecture Notes in Computer Science, Springer-Verlag, 1994, pp. 150–163.

5. R. M. CAPOCELLI, A. D. SANTIS, L. GARGANO, AND U. VACCARO, *On the size of shares for secret sharing schemes*, in Advances in Cryptology — CRYPTO '91, J. Feigenbaum, ed., vol. 576 of Lecture Notes in Computer Science, Springer-Verlag, 1992, pp. 101–113.

6. L. CSIRMAZ, *The size of a share must be large*, in Advances in Cryptology — EUROCRYPT '94, A. De Santis, ed., vol. 950 of Lecture Notes in Computer Science, Springer-Verlag, 1995, pp. 13–22.

7. D. E. DENNING AND M. SMID, *Key escrowing today*, IEEE Communications Magazine, 32 (1994), pp. 58–68.

8. O. GOLDREICH, S. MICALI, AND A. WIGDERSON, *How to play any mental game or a completeness theorem for protocols with honest majority*, in Proc. 19th ACM Symposium on Theory of Computing (STOC), 1987, pp. 218–229.

9. M. ITO, A. SAITO, AND T. NISHIZEKI, *Secret sharing scheme realizing general access structure*, in Proceedings of IEEE Globecom '87, 1987, pp. 99–102.

10. H. KRAWCZYK, *Secret sharing made short*, in Advances in Cryptology — CRYPTO '93, D. R. Stinson, ed., vol. 773 of Lecture Notes in Computer Science, Springer-Verlag, 1994, pp. 136–146.

11. L. LAMPORT, *Password authentication with insecure communication*, Comm. ACM, 24 (1981).

12. S. MICALI, *Fair public-key cryptosystems*, in Advances in Cryptology — CRYPTO '92, E. F. Brickell, ed., vol. 740 of Lecture Notes in Computer Science, Springer-Verlag, 1993, pp. 113–138.

13. M. O. RABIN, *Efficient dispersal of information for security, load balancing, and fault tolerance*, J. Assoc. Comput. Mach., 36 (1989), pp. 335–348.

14. A. SHAMIR, *How to share a secret*, Comm. ACM, 22 (1979), pp. 612–613.

15. G. J. SIMMONS, *An introduction to shared secret and/or shared control schemes and their application*, in Contemporary Cryptology: The Science of Information Integrity, G. J. Simmons, ed., IEEE Press, 1991, pp. 441–497.

16. D. R. STINSON, *An explication of secret sharing schemes*, Designs, Codes and Cryptography, 2 (1992), pp. 357–390.

Church-Rosser Codes

Vladimir A. Oleshchuk

Department of Electrical Engineering and Computer Science
Agder College
N-4890 Grimstad, Norway

Abstract. The notion of code, called *Church-Rosser code*, is proposed and studied. The necessary and sufficient conditions for a finite set of being a Church-Rosser code are presented. It is proved that property of being a Church-Rosser code defined by a *monadic* confluent string-rewriting system is decidable. We also propose decidable sufficient conditions for a finite set of being a Church-Rosser code defined by a *finite* Church-Rosser string-rewriting system.

1   Introduction

Informally, a nonempty set of words C is a code if any product of words from C can be decoded or factorized in a unique way into words of C [4]. In this paper we introduce a new definition of code given with respect to string-rewriting system with Church-Rosser property. The codes are called *Church-Rosser codes* and have been used in public-key cryptosystems proposed in [10]. The main feature of our approach is that each code presents, in general, infinitely many different codes with the same unique decoding.

We introduce and study a notion of codes based on string-rewriting systems. Informally, a string-rewriting system is a set of ordered pair (l, r) of strings over a finite alphabet. The rewriting of a given string w is performed by (non-deterministically) replacing some occurrence of the string l in w by the string r or by replacing some occurrence of string r in w by string l. For a given system T a string x is reduced to a string y if y can be obtained from x by applying a sequence of length-decreasing rules from T. The system T has the Church-Rosser property if for every x and y, they are congruent if and only if there exists z such that both x and y can be reduced to z. The Church-Rosser property of finite string-rewriting systems means that there is a unique normal form for every congruence class in such systems. It is known that such a normal form can be found in linear time [5]. We will use that property as a basis to define codes such that decoding messages are considered as finding normal forms of these messages.

We find necessary and sufficient conditions for a finite set of being a Church-Rosser code and show that the property of being a Church-Rosser code is decidable for monadic Church-Rosser string rewriting systems. We also present sufficient conditions of being Church-Rosser code with respect to any finite Church-Rosser string-rewriting system.

2 Definitions and Notations

Here we provide formal definitions of string-rewriting systems and related notions. For additional information and comments regarding the various notions introduced, the reader is asked to consult [6, 7].

If Σ is a finite alphabet, then Σ^* is a set of all finite words in alphabet Σ with empty word λ, and Σ^+ denotes a set $\Sigma^* - \{\lambda\}$. If $w \in \Sigma^*$, then the *length* of a word w is denoted by $|w|$ where $|\lambda| = 0, |a| = 1$ for $a \in \Sigma$, and $|wa| = |w| + 1$ for $w \in \Sigma^*$, $a \in \Sigma$. For $w \in \Sigma^*$, $w = a_1 a_2 ... a_n$, the word w^R is defined by $w^R = a_n ... a_2 a_1$. The concatenation of words u and v is written as uv. If $u, w \in \Sigma^*$ such that $w = vut$ for some $v, t \in \Sigma^*$, then u is a *subword* of w. If $A, B \subseteq \Sigma^*$, then the concatenation of A and B, denoted AB, is defined to be $\{xy | x \in A, y \in B\}$. For $A \subseteq \Sigma^*$, the set A^R is defined by $A^R = \{w^R | w \in A\}$.

Let Σ be a finite alphabet. A *string-rewriting system* T on Σ is a subset of $\Sigma^* \times \Sigma^*$, and each element (u, v) of T is a *rewriting rule*. A string-rewriting system T induces a congruence on Σ^*. The *congruence generated by* T is the reflexive transitive closure $\leftrightarrow_T^*$ of the relation defined as follows: if $(u, v) \in T$ or $(v, u) \in T$, then for every $x, y \in \Sigma^*$, $xuy \leftrightarrow_T xvy$. If $x, y \in \Sigma^*$, $x \leftrightarrow_T^* y$ then x and y are congruent modulo T, denoted $x \equiv y \,(mod\ T)$. The *congruence class* of $z \in \Sigma^*$ with respect to T is a set $[z]_T = \{w \in \Sigma^* \mid w \leftrightarrow_T^* z\}$. This notation is extended to sets $A \subseteq \Sigma^*$ as $[A]_T = \{w \in \Sigma^* | \text{ there exits } x \in A \text{ such that } x \leftrightarrow_T^* w\}$. Therefore $[A]_T = \cup \{[x] | x \in A\}$. The set $\{[w]_T | w \in \Sigma^*\}$ of congruence classes forms a monoid $\mathcal{M}_T$ under the operation $[x]_T \cdot [y]_T = [xy]_T$ with identity $[\lambda]_T$. For $x \in \Sigma^*$, the set $\Delta_T^*(x)$ of all descendants of x is defined by $\Delta_T^*(x) = \{w \in \Sigma^* \mid x \to_T^* w\}$. This notation is extended to sets $A \subseteq \Sigma^*$ as $\Delta_T^*(A) = \{w \in \Sigma^* | \text{ there exits } x \in A \text{ such that } x \to_T^* w\}$. For $x \in \Sigma^*$, the set $\langle x \rangle_T^*$ of all ancestors of x is defined by $\langle x \rangle_T^* = \{w \in \Sigma^* | w \to_T^* x\}$. This notation is extended to sets $A \subseteq \Sigma^*$ as $\langle A \rangle_T^* = \{w \in \Sigma^* | \text{ there exits } x \in A \text{ such that } w \to_T^* x\}$.

For a string-rewriting system T, write $x \to_T y$ if $x \leftrightarrow_T y$ and $|x| > |y|$; write $\to_T^*$ for the reflexive transitive closure of $\to_T$; write $\to_T^+$ for the irreflexive transitive closure of $\to_T$. A string x is *irreducible* if there is no y such that $x \to_T y$ and $IRR(T)$ denotes the set of all irreducible with respect to T strings. For any finite T the set $IRR(T)$ is a regular set and a finite state acceptor recognizing $IRR(T)$ can be effectively constructed from T [2]. A string-rewriting system T is called *monadic* if $(u, v) \in T$, $|u| > |v|$ implies $|v| \leq 1$. A string-rewriting system T is called *Church-Rosser* if $x \leftrightarrow_T^* y$ implies that, for there exists z, $x \to_T^* z$ and $y \to_T^* z$. The property of being a Church-Rosser string-rewriting system can be tested in polynomial time for finite systems [8].

If a string-rewriting system T is Church-Rosser, then each congruence class has a unique irreducible element and its word problem can be solved in linear time [5].

A nonempty set C such that $C \subseteq \Sigma^*$ is a *code* over Σ if for all $n, m \geq 1$ and words $x_{i_1}, x_{i_2}, ..., x_{i_n}, x_{j_1}, x_{j_2} ..., x_{j_m}$ from C, the condition

$$x_{i_1} x_{i_2} ... x_{i_n} = x_{j_1} x_{j_2} ... x_{j_m}$$

implies $x_{i_1} = x_{j_1}$. The last equality means that $m = n$ and $x_{i_k} = x_{j_k}$, $k = 1, ..., n$. Thus if C is a code, then any sequence from C^* can be uniquely presented as concatenation of words of C, i.e. uniquely decoded. From the definition follows that the empty word $\lambda \notin C$, and any nonempty subset of C is also a code [4]. The property of being a code can be effectively tested for a finite set C [4, 11].

3 Church-Rosser Codes

A nonempty set $C_T, C_T \subseteq \Sigma^+$ is a code, defined by T, if for all $x_1, ..., x_n$, $y_1, ..., y_m \in C_T$ the condition

$$x_1 \cdot ... \cdot x_n \equiv y_1 \cdot ... \cdot y_m \ (mod \ T)$$

implies $x_i \leftrightarrow^*_T y_i$, $i = 1, ..., n$ and $n = m$. C_T is called a *Church-Rosser code* if T has Church-Rosser the property. Informally it means that any word w of C_T^* can be uniquely represented as a product of congruence classes modulo T of words of C_T. From the above definition follows that if C_T is a code defined by T then $\lambda \notin \Delta^*_T (C_T) \cap IRR(T)$.

Here we are interested in the following problem:

Instance: A Church-Rosser string rewriting system T on Σ, and a finite subset C of Σ^*.
Problem: Is C a Church-Rosser code defined by T ?

The following lemma simplifies further considerations.

Lemma 1. *Let $C = \{w_1, ..., w_n\}$ be a Church-Rosser code defined by T. Then any set $C' = \{w'_1, ..., w'_n\}$ such that $w'_i \in [w_i]_T$ be also a Church-Rosser code defined by T.*

Proof: If C' is not a Church-Rosser code then there exist words $s = x_1 \cdot ... \cdot x_n$ and $t = y_1 \cdot ... \cdot y_m$, $x_i, y_j \in C'$ such that $s \equiv t \ (mod \ T)$ but there exists k, $1 \leq k \leq \max(n, m)$ such that $x_k \not\leftrightarrow^*_T y_k$. Let k_0 be the smallest value of k such that $x_{k_0} \not\leftrightarrow^*_T y_{k_0}$. For any x_i and y_j there are words w^i and w^j from C such that $x_i \leftrightarrow^*_T w^i$ and $y_j \leftrightarrow^*_T w^j$. Therefore s and t can be transformed into $s' = x'_1 \cdot ... \cdot x'_n$ and $t' = y'_1 \cdot ... \cdot y'_m$ by replacing every x_i and y_j with corresponding w^i and w^j. By construction, $s' \equiv t' \ (mod \ T)$. From $x_{k_0} \not\leftrightarrow^*_T y_{k_0}$ follows that $x'_{k_0} \not\leftrightarrow^*_T y'_{k_0}$. Contradiction with the fact that C is a Church-Rosser code defined by T. $\square$

The next theorem presents necessary and sufficient conditions for set to be a Church-Rosser code.

Theorem 2. *Let T be a Church-Rosser string-rewriting system over Σ. A nonempty set $C \subseteq \Sigma^+$ is a Church-Rosser code defined by T if and only if the following conditions hold:*

(i) set $C_{IRR(T)}$ is a code;

(ii) for any $w_1, w_2 \in C^*_{IRR(T)}$, $w_1 \neq w_2$ *exists no* u *such that* $w_1 \rightarrow^*_T u$ *and*
$w_2 \rightarrow^*_T u$,

where $C_{IRR(T)} = \Delta^*_T(C) \cap IRR(T)$

Proof: Let $C_T = \{w_1, ..., w_n\}$ be a Church-Rosser code defined by T. By Lemma 1 the set $C_{IRR(T)} = \{w'_1, ..., w'_n\}$ such that $w_i \rightarrow^*_T w'_i \in IRR(T)$, $i = 1, ..., n$ is also a Church-Rosser code defined by T. Therefore for all $x_1, ..., x_n$, $y_1, ..., y_m \in C_{IRR(T)}$ from $x_1 \cdot ... \cdot x_n \equiv y_1 \cdot ... \cdot y_m \, (mod \; T)$ follows $x_i \leftrightarrow^*_T y_i$. Since $x_i, y_i \in IRR(T)$, $x_i \leftrightarrow^*_T y_i$ means $x_i = y_i$. Thus $C_{IRR(T)}$ is a code.

Suppose that *(ii)* does not hold. Then there exist $w_1, w_2 \in C^*_{IRR(T)}$, $w_1 \neq w_2$ and $u \in IRR(T)$ such that $w_1 \rightarrow^*_T u$ and $w_2 \rightarrow^*_T u$, i.e. $w_1 \leftrightarrow^+_T w_2$. $w_1 \neq w_2$ implies that there exists k, $1 \leq k \leq \max(m, n)$ such that $x_k \neq y_k$, i.e. $x_k \not\leftrightarrow^*_T y_k$ since $x_k, y_k \in IRR(T)$. It means that $C_{IRR(T)}$ is not a Church-Rosser code defined by T.

Suppose that conditions *(i)* and *(ii)* are satisfied. We will show that C is a Church-Rosser code defined by T. According to Lemma 1 C is a Church-Rosser code defined by T if and only if $C_{IRR(T)}$ is a Church-Rosser code defined by T.

Let $w_1, w_2 \in C^+_{IRR(T)}$ be such that $w_1 \leftrightarrow^*_T w_2$.

If $w_1, w_2 \in IRR(T)$, then $w_1 = w_2$. If $w_1 \in IRR(T)$ and $w_2 \notin IRR(T)$, then $w_2 \rightarrow^+_T w_1$. Contradiction with *(ii)*. The case $w_1 \notin IRR(T)$ and $w_2 \in IRR(T)$ is analogous to the previous one.

If both $w_1 \notin IRR(T)$ and $w_2 \notin IRR(T)$ then, since T is a Church-Rosser system, must exist $u \in IRR(T)$ such that $w_1 \rightarrow^*_T u$ and $w_2 \rightarrow^*_T u$. According to *(ii)* it means that $w_1 = w_2$. $\square$

It is not always easy to verify that a given set of words is a Church-Rosser code with respect to T. The following theorem gives easy-testable sufficient conditions for a finite set to be a Church-Rosser code defined by a finite string-rewriting system.

Theorem 3. *Let* T *be a Church-Rosser string-rewriting system over* Σ, C *be a nonempty subset of* Σ^+, *and* $C_{IRR(T)} = \Delta^*_T(C) \cap IRR(T)$. *If*

(i) set $C_{IRR(T)}$ *is a code, and*
(ii) $C^*_{IRR(T)} \subseteq IRR(T)$,

then C *is a Church-Rosser code defined by* T.

The proof is similar to the previous one.

The following theorem shows that the necessary and sufficient conditions of being a Church-Rosser code defined by T are decidable when T is a finite monadic Church-Rosser string-rewriting system.

Theorem 4. *Let* T *be a finite Church-Rosser string-rewriting system over* Σ. *There exists an algorithm to test whether* C *is a Church-Rosser code defined by* T, *where* C *is a finite set such that* $C \subset \Sigma^+$.

Proof: We shall show that conditions from Theorem 2 can be tested algorithmically.

For any finite set C and a Church-Rosser system T a set $C_{IRR(T)} = \Delta_T^*(C) \cap IRR(T)$ can be constructed in linear time [5]. Algorithms to check whether a finite set $C_{IRR(T)}$ is a code can be found in [4, 11].

Condition (ii) from Theorem 2 can be reformulated as following: for any $w_1, w_2 \in C_{IRR(T)}^*$ from $w_1 \neq w_2$ follows $w_1 \not\leftrightarrow_T^* w_2$. We shall show that the last condition is decidable. Denote $L = \{u\#u^R | u \in IRR(T), \# \notin \Sigma\}$ and $L_1 = \{w_1\#w_2^R | \exists u : w_i \to_T^* u, i = 1, 2, \# \notin \Sigma\}$. Since $L_1 = \langle\{u\#u^R | u \in IRR(T)\}\rangle_T^*$, then L_1 is context-free and from T and a context-free grammar for L_1 one can construct a context-free grammar that generates L_1 [5]. For any finite set $C_{IRR(T)}$, R_C is defined by a regular set $R_C = C_{IRR(T)}^* \# \left(C_{IRR(T)}^*\right)^R$. Then $L_2 = L_1 \cap R_C$ is a context-free language such that if $w_1\#w_2^R \in L_2$ then $w_1 \leftrightarrow_T^* w_2$ and $w_1, w_2 \in C_{IRR(T)}^*$. Therefore condition (ii) from Theorem 2 holds if and only if $L_2 \subseteq D$, where $D = \{u\#u^R | u \in \Sigma^*, \# \notin \Sigma\}$. From the language L_2 can be constructed a context-free language $L_2' = \{w_1\#\overline{w}_2 | w_1\#w_2^R \in L_2\}$, where $\overline{\Sigma}$ is a new alphabet, disjoint from Σ, that is chosen with a bijection $a \mapsto \overline{a}$ between Σ and $\overline{\Sigma}$, and for $w \in \Sigma^*, w = a_1 a_2 ... a_n$, the word $\overline{w}$ is defined by $\overline{w} = \overline{a}_n ... \overline{a}_2 \overline{a}_1$. Then the property $L_2 \subseteq D$ holds if and only if $L_2' \subseteq [\lambda]_{T'}$, where $T' = \{(a\overline{a}, \lambda), (\overline{a}a, \lambda) | a \in \Sigma, \overline{a} \in \overline{\Sigma}\} \cup \{(\#, \lambda)\}$. Since T' is a Church-Rosser system and $\mathcal{M}_{T'}$ is a group then $[\lambda]_{T'}$ is a group language for the group with decidable word problem [3]. According to Lemma 2 from [1], it is decidable whether a context-free language L is inclosed into a group language of a group with decidable word problem. Thus the property $L_2 \subseteq D$ is decidable and can be tested algorithmically. $\square$

The last result holds also for regular monadic Church-Rosser string-rewriting systems [6].

Let us consider how the encoded messages can be decoded or factorized into words of C, where C is a Church-Rosser code defined by T.

In order to decode $w \in C^*$, we have to find $w' \in C_{IRR(T)}^* \cap [w]_T$ and then decode w'. Since C is a Church-Rosser code defined by T then $\left|C_{IRR(T)}^* \cap [w]_T\right| = 1$ for any $w \in C^*$. For a finite system T, $[w]_T$ can be even non context-free and it is undecidable whether $[w]_T$ is some restricted type of context-free language, e.g. context-free, deterministic context-free, linear context-free, regular [9]. From other side, $|w| \geq |w'|$. Therefore one can find w' by analyzing a finite number of strings w'' such that $|w| \geq |w''|$ and $w \leftrightarrow_T^* w''$.

If T is a monadic Church-Rosser system then $[w]_T$ is a deterministic context-free language and $w' \in C_{IRR(T)}^* \cap [w]_T$ can be found.

Theorem 3 can be used to construct an efficient decoding algorithm.

Suppose that C conforms with conditions of Theorem 3. Then $C_{IRR(T)}^* \subseteq IRR(T)$ and $w' \in C_{IRR(T)}^* \cap [w]_T$ is an irreducible form of w, i.e. $w \to_T^* w' \in IRR(T)$. For any string w and a finite system T, the normal form $w' \in IRR(T)$, $w \to_T^* w'$ can be found in linear time [5]. Thus decoding time for Church-Rosser code C is equal to $\mathcal{O}\left(T(n) + n\right)$, where $T(n)$ denotes decoding time for the code $C_{IRR(T)}$.

In the following examples we apply the above theorems to test whether finite sets are Church-Rosser codes.

Example 1. $T = \{(aab, bb)\}$ is a Church-Rosser system. $C = \{ab, bb\}$ is a code such that $C = C_{IRR(T)}$ and $\left(C_{IRR(T)}\right)^* \subset IRR(T)$. Thus C is a Church-Rosser code defined by T. $\square$

Example 2. $T = \{(aba, a), (bab, b)\}$ is a Church-Rosser system. $C = \{ab, aab\}$ is a code such that $C = C_{IRR(T)}$ and $\left(C_{IRR(T)}\right)^* \not\subset IRR(T)$. There exist $w_1, w_2 \in C^*$ such that $w_1 = aabab \rightarrow_T aab = w_2$. Therefore C is not a Church-Rosser code defined by T. $\square$

4 Conclusion

We have proposed a new approach to define codes with respect to string-rewriting systems with Church-Rosser property. In such a way, we can define infinitely many codes that are equivalent in the sense that they have the same normal form code and messages encoded in these codes can be decoded uniformly. The proposed notion of code has been used to construct public-key cryptosystems proposed in [10].

References

1. A.V. Anisimov, *Finite–automata semigroup mappings*, Cybernetics, **5** (1981) 1–7.
2. J. Berstel, *Congruences plus que parfaites et langages algébrique*, Séminaire d'Informatique Théorique, Institut de Programmation (1976-77) 123-147.
3. J. Berstel and L. Boasson, *Context-free languages*, in: J. van Leeuwen, ed., Handbook of Theoretical Computer Science, Vol. **B**, Elsevier Science Publishers B.V., 1990, 59-102.
4. J. Berstel and D. Perrin, *Theory of Codes*, Academic Press, 1985.
5. R. V. Book, *Confluent and other types of Thue systems*, Journal of ACM, **29** (1982) 171-183.
6. R. V. Book, *Thue systems as rewriting systems*, J. Symb. Comp. **3** (1987) 39-68.
7. R. V. Book and F. Otto, *String-Rewriting Systems*, Springer: New-York, 1993.
8. D. Kapur, M. Krishnamoorthy, R. McNaughton and R. Narendran, *An $O\left(|T|^3\right)$ algorithm for testing the Church-Rosser property of Thue systems*, Theor. Comp. Sci. **35** (1985) 109-114.
9. P. Narendran, C. O'Dunlaing and H. Rolletschek, *Complexity of certain decision problems about congruential languages*, J. Comp. Syst. Sci. **30**, 343-358.
10. V. A. Oleshchuk, *On public-key cryptosystem based on Church-Rosser string-rewriting systems*, Computing and Combinatorics: First Annual International Conference (COCOON'95). Proceedings. LNCS **959** (1995) 264-269.
11. A. Sardinas and G. Patterson, *A necessary and sufficient condition for the unique decomposition of coded messages*, I.R.E. Int. Conv. Rec. **8** (1953) 104-108.

A New Algorithm for Finding Minimum-Weight Words in Large Linear Codes

Anne Canteaut *

INRIA Projet Codes
Domaine de Voluceau
78153 Le Chesnay Cedex, FRANCE
email: Anne.Canteaut@inria.fr

Abstract. An algorithm for finding small-weight words in large linear codes is developed and a precise analysis of its complexity is given. It is in particular able to decode random [512,256,57]-linear binary codes in 9 hours on a DEC alpha computer. We improve with it the previously best known attacks on some public-key cryptosystems and identification schemes based on error-correcting codes: for example we reduce the work factor involved in breaking McEliece's cryptosystem, since our algorithm requires 2^{64} elementary operations that is 128 times less than Lee-Brickell's attack.

1 Presentation of the Algorithm

Let C be an $[n, k]$-linear code over $GF(2)$. We present a probabilistic algorithm for finding a codeword of weight w, where w is small. This algorithm was elaborated with Florent Chabaud [2].

1.1 A Probabilistic Method

Since the probability that a randomly chosen codeword has weight w is very small, we need to only consider words verifying a given property so that their weight will be *a priori* small. All algorithms for finding short codewords use the same method [5, 6, 10]: they only take in account codewords which are particular linear combinations of a small number of rows of a systematic generator matrix. The heuristic proposed by Stern was shown to give the best results [3].

Let $N = \{1, \cdots, n\}$. For any subset I of N, $G = (V, W)_I$ denotes the decomposition of matrix G onto I, that means $V = (G_i)_{i \in I}$ and $W = (G_j)_{j \in N \setminus I}$, where G_i is the ith column of matrix G.

Definition 1. Let I be a k-element subset of N. I is an information window for code C iff $G = (Id_k, Z)_I$ is a systematic generator matrix for the code. The complementary set, $J = N \setminus I$, is called a redundancy window.

---

* Also with Ecole Nationale Supérieure de Techniques Avancées, laboratoire LEI, 32 boulevard Victor, 75015 Paris, FRANCE.

¿From now on we index the rows of Z with I since $G = (Id_k, Z)_I$ is a generator matrix for the code and we denote by Z^i the i-th row of matrix Z.

The idea suggested by Stern is to randomly choose at each iteration an information window I which is split into two parts I_1 and I_2 of same size, and a subset L of J of size ℓ. We only examine codewords c verifying the following property, where p and ℓ are fixed parameters for the algorithm:

$$\text{wt}(c_{|I_1}) = \text{wt}(c_{|I_2}) = p \text{ and } \text{wt}(c_{|L}) = 0 \tag{1}$$

until we find such a particular codeword whose restriction on $J \setminus L$ has weight $w - 2p$.

1.2 An Iterative Procedure

Stern's algorithm therefore explores a set of randomly selected information windows by performing at each iteration a Gaussian elimination on a generator matrix. In order to avoid this time-consuming procedure, we here propose to choose at each step the new information window by modifying only one element of the previous one. This method is analogous to the one used in the simplex method as suggested by Omura [9] and van Tilburg [12].

Definition 2. Two information windows I and I' are close iff:

$$\exists \lambda \in I, \exists \mu \in N \setminus I, \text{ such that } I' = (I \setminus \{\lambda\}) \cup \{\mu\}$$

Proposition 3. Let I be an information window such that $G = (Id_k, Z)_I$ is a generator matrix for C. Let be $\lambda \in I$, $\mu \in J$ and $I' = (I \setminus \{\lambda\}) \cup \{\mu\}$. I' is an information window iff $z_{\lambda, \mu} = 1$

The redundant part Z of the systematic generator matrix can then be updated by a simple pivoting operation:

Proposition 4. Let I and I' be two close information windows such that $I' = (I \setminus \{\lambda\}) \cup \{\mu\}$. Let $(Id_k, Z)_I$ and $(Id_k, Z')_{I'}$ be the corresponding systematic generator matrices. Then Z' is obtained from Z by:

- $\forall j \in J'$, $z'_{\mu, j} = z_{\lambda, j}$
- $\forall i \in I' \setminus \{\mu\}$,
  - $\forall j \in J' \setminus \{\lambda\}$, $z'_{i, j} = z_{i, j} + z_{i, \mu} z_{\lambda, j}$
  - $z'_{i, \lambda} = z_{i, \mu}$

1.3 Description of the Iterative Algorithm

The use of this iterative procedure leads then to the following algorithm:

Initialization:
    Randomly choose an information window I and apply a Gaussian elimination in order to obtain a systematic generator matrix $(Id_k, Z)_I$.

Until a codeword of weight w will be found:

1. randomly split I in two subsets I_1 and I_2 where $|I_1| = \lfloor k/2 \rfloor$ and $|I_2| = \lceil k/2 \rceil$. (the rows of Z are then split in two parts Z_1 and Z_2); select an ℓ-element subset L of J.
2. for each linear combination Λ_1 of p rows of matrix Z_1, compute $\Lambda_{1|L}$. For each linear combination Λ_2 of p rows of matrix Z_2, compute $\Lambda_{2|L}$.
3. if $\Lambda_{1|L} = \Lambda_{2|L}$, check whether $\mathrm{wt}((\Lambda_1 + \Lambda_2)_{|J\setminus L}) = w - 2p$.
4. randomly choose $\lambda \in I$ and $\mu \in J$. Replace I with $(I\setminus\{\lambda\})\cup\{\mu\}$ by updating matrix Z according to the preceding proposition.

2 Theoretical Running Time

We here give a computable expression for the work factor of this algorithm, *i.e.* the average number of elementary operations it requires.

2.1 Average Number of Operations by Iteration

The average number of elementary operations performed at each iteration is:

$$\Omega_{p,\ell} = 2p\ell\binom{k/2}{p} + 2p(n - k - \ell)\frac{\binom{k/2}{p}^2}{2^\ell} + \frac{k(n-k)}{2} + K(p\binom{k/2}{p} + 2^\ell) \quad (2)$$

The first 3 terms correspond respectively to steps 2, 3 and 4 of the algorithm and the last one corresponds to the cost of the dynamic memory allocation (K is the size of a computer word).

2.2 Expected Number of Iterations

Since the successive information windows are not independent, the algorithm must be modeled by a discrete-time stochastic process.

Let c be the codeword of weight w to recover and $supp(c)$ its support. Let I be the information window and I_1, I_2 and L the other selections corresponding to the i-th iteration.

The i-th iteration can then be associated with the random variable X_i whose state space is $\mathcal{E} = \{0, \dots, 2p - 1\} \cup \{(2p)_S, (2p)_F\} \cup \{2p + 1, \dots, w\}$ where

$$X_i = u \quad \text{iff } |I \cap supp(c)| = u, \ \forall u \in \{0, \dots, 2p - 1\} \cup \{2p + 1, \dots, w\}$$
$$X_i = (2p)_F \quad \text{iff } |I \cap supp(c)| = 2p \text{ and } (|I_1 \cap supp(c)| \neq p$$
$$\text{or } |I_2 \cap supp(c)| \neq p \text{ or } |L \cap supp(c)| \neq 0)$$
$$X_i = (2p)_S \quad \text{iff } |I_1 \cap supp(c)| = |I_2 \cap supp(c)| = p \text{ and } |L \cap supp(c)| = 0$$

The success space is then $\mathcal{S} = \{(2p)_S\}$ and the failure space is $\mathcal{F} = \{0, \dots, (2p)_F, \dots, w\}$.

Proposition 5. *The stochastic process $\{X_i\}_{i \in \mathbb{N}}$ associated with the algorithm is an homogeneous Markov chain.*

The corresponding transition matrix P can be easily computed [2]. Since this Markov chain is a transient chain, the following theorem can be applied for computing the expected number of iterations performed by the algorithm.

Theorem 6. *The expectation $\bar{N}$ of the number of iterations required until X_n reaches a success state is given by:*

$$\bar{N} = \sum_{i \in \mathcal{F}} \pi_0(i) \sum_{j \in \mathcal{F}} R_{i,j}$$

where R is the fundamental matrix i.e. $R = \sum_{m=0}^{\infty} Q^m = (Id - Q)^{-1}$

Assuming that the number of codewords of weight w is $\mathcal{A}_w$, the overall work factor required by the algorithm is then:

$$W_{p,l} = \frac{\Omega_{p,\ell}\bar{N}}{\mathcal{A}_w} \tag{3}$$

where $\bar{N}$ is given by theorem 6 and $\Omega_{p,\ell}$ by equation 2.

3 Applications

3.1 Decoding Random Linear Codes

We here study the work factor required for decoding an $[n, k, d]$-random linear binary code where d is given by the Gilbert-Varshamov's bound.

Using expression 3 we show that, for a fixed rate $r = k/n$, $\log_2(W)$ linearly depends on n only when parameters p and ℓ are optimized and that the work factor can be written in the form $W_{opt} = 2^{na(r)+b}$ (see figure 1).

Furthermore figure 2 shows that $a(r)$ is closed to the entropy function $H_2(r)$ multiplied by a fixed coefficient, where $H_2(x) = -x \log_2(x) - (1-x) \log_2(1-x)$.

Then we obtain the following result:

Proposition 7. *The theoretical work factor required for decoding a random $[n, k]$-binary code can be approximated by the following formula:*

$$W_{opt} = 2^{anH_2(k/n)+b} \text{ where } a = 5.511 \ 10^{-2} \text{ and } b = 12$$

The same study was made concerning the work factor required for recovering a minimum-weight codeword in a random linear binary code ($w = d$) and we similarly obtain:

Proposition 8. *The theoretical work factor required for finding a minimum-weight word in a random $[n, k]$-binary code can be approximated by the following formula:*

$$W_{opt} = 2^{cnH_2(\frac{k}{n}+r_0)+d} \text{ where } c = 0.12, \ d = 10 \text{ and } r_0 = 3.125 \ 10^{-2}$$

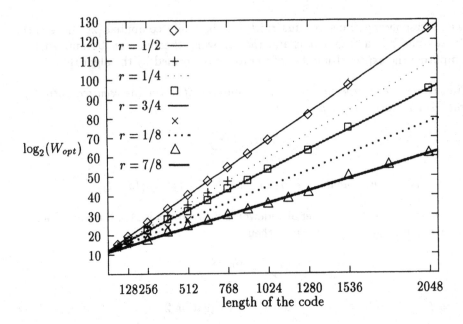

Fig. 1. Evolution of the theoretical work factor with optimized parameters for decoding random $[n, nr]$-binary codes

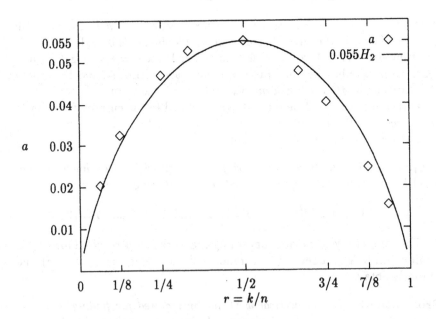

Fig. 2. Evolution of coefficient a vs. $r = k/n$

3.2 Cryptanalysis of Some Cryptosystems Based on Error-Correcting Codes

This algorithm can also be applied to attack McEliece's [7] and Niederreiter's [8] cryptosystems and some identification schemes based on syndrome decoding [4, 11] since they rely on the following NP-complete problem: given an $[n, k]$-linear binary code about nothing is known but one of its generator matrices G and given an n-bit vector y, we have to find a word of weight w in the coset of y, where w is equal to the error-correcting capability of the code for the cryptosystems, or slightly below its minimal distance for the identification schemes.

Considering the $[n, k + 1]$-code whose generator matrix is $\left(\dfrac{G}{y}\right)$, an attack on these systems consists in searching a word of weight w in this new code. We then obtain the following work factors.

cryptosystem	original McEliece	McEliece	Stern	McEliece
code	[1024,256]	[1024,614]	[512,256]	[512,260]
w	56	41	56	28
optimal parameters	$p = 2$ $\ell = 18$	$p = 2$ $\ell = 18$	$p = 2$ $\ell = 15$	$p = 1$ $\ell = 9$
work factor	$2^{64.2}$	$2^{66.0}$	$2^{69.8}$	$2^{40.9}$

Table 1. Work factors required for breaking some cryptosystems based on error-correcting codes

We otherwise notice that the parameters for McEliece's cryptosystem which maximize the work factor of this new attack are $n = 1024$, $k = 614$ and $w = 41$; the corresponding work factor is 2^{66}.

3.3 Experimental Results

We have made a great number of simulations for a small problem: decoding a $[256, 128, 29]$-random linear code whose minimal distance is obtained by Gilbert-Varshamov's bound. They confirm the validity of the previous theory concerning both optimized parameters and number of iterations (see table 2).

Decoding a $[256, 128, 29]$-random linear code requires 2 seconds on a DEC alpha station at 175 MHz. Thus decoding a $[512,256,57]$-one or breaking McEliece's cryptosystem with a $[512, 260]$ Goppa code requires around 9 hours on our computer.

Acknowledgments: I wish to thank Florent Chabaud with whom this algorithm was elaborated, and Hervé Chabanne who initiated this work.

parameters	theoretical work factor $\log_2(W)$	theoretical average iteration number	experimental average iteration number	deviation %	average CPU time (s)	corrected CPU time (s)
$p = 1, \ell = 5$	27.37	3961	4072	+2.80	2.76	2.68
$p = 1, \ell = 6$	26.80	4045	3985	-1.48	2.11	2.14
$p = 1, \ell = 7$	26.51	4139	4190	+1.23	2.07	2.04
$p = 1, \ell = 8$	26.56	4244	4338	+2.21	2.13	2.09
$p = 1, \ell = 9$	26.95	4362	4417	+1.26	2.90	2.87
$p = 2, \ell = 10$	29.80	432	433	+0.35	13.84	13.79
$p = 2, \ell = 11$	29.04	442	470	+6.51	13.00	12.21
$p = 2, \ell = 12$	28.51	454	446	-1.76	12.13	12.35
$p = 2, \ell = 13$	28.37	466	487	+4.51	17.70	16.91
$p = 2, \ell = 14$	28.68	480	508	+5.83	29.40	27.72

Table 2. Decoding random $[256, 128, 29]$-binary codes

References

1. A. Canteaut and H. Chabanne. A further improvement of the work factor in an attempt at breaking McEliece's cryptosystem. In P. Charpin, editor, *EUROCODE 94*, pages 163–167. INRIA, 1994.

2. A. Canteaut and F.Chabaud. Improvements of the attacks on cryptosystems based on error-correcting codes. Rapport interne du Département Mathématiques et Informatique LIENS-95-21, Ecole Normale Supérieure, Paris, July 1995.

3. F. Chabaud. On the security of some cryptosystems based on error-correcting codes. In A. De Santis, editor, *Advances in Cryptology – EUROCRYPT '94*, number 950 in Lecture Notes in Computer Science, pages 131–139. Springer-Verlag, 1995.

4. M. Girault. A (non-practical) three-pass identification protocol using coding theory. In J. Seberry and J. Pieprzyk, editors, *Advances in Cryptology – AUSCRYPT '90*, number 453 in Lecture Notes in Computer Science, pages 265–272. Springer-Verlag, 1991.

5. P.J. Lee and E.F. Brickell. An observation on the security of McEliece's public-key cryptosystem. In C.G. Günther, editor, *Advances in Cryptology – EUROCRYPT '88*, number 330 in Lecture Notes in Computer Science, pages 275–280. Springer-Verlag, 1988.

6. J.S. Leon. A probabilistic algorithm for computing minimum weights of large error-correcting codes. *IEEE Trans. Inform. Theory*, IT-34(5):1354–1359, September 1988.

7. R.J. McEliece. A public-key cryptosystem based on algebraic coding theory. *DSN progress report 42-44*, pages 114–116, 1978.

8. H. Niederreiter. Knapsack-type cryptosystems and algebraic coding theory. *Problems of Control and Information Theory*, 15(2):159–166, 1986.

9. J.K. Omura. Iterative decoding of linear codes by a modulo-2 linear program. *Discrete Math*, 3:193–208, 1972.

10. J. Stern. A method for finding codewords of small weight. In G. Cohen and J. Wolfmann, editors, *Coding Theory and Applications*, number 388 in Lecture Notes in Computer Science, pages 106–113. Springer-Verlag, 1989.

11. J. Stern. A new identification scheme based on syndrome decoding. In D.R. Stinson, editor, *Advances in Cryptology – CRYPTO '93*, number 773 in Lecture Notes in Computer Science. Springer-Verlag, 1994.

12. J. van Tilburg. On the McEliece public-key cryptosystem. In S. Goldwasser, editor, *Advances in Cryptology – CRYPTO '88*, number 403 in Lecture Notes in Computer Science, pages 119–131. Springer-Verlag, 1990.

Coding and Cryptography for Speech and Vision

E.V.Stansfield [1] and M.Walker [2]

[1] Racal Research Ltd, Worton Drive, Reading RG2 0SB
[2] Vodafone Ltd, The Courtyard, 2-4 London Road, Newbury RG14 1JX

Abstract. There are many areas of industry in which coding and cryptography play a key role, and in this paper applications to speech, video and data communications are described. In each case, the purpose of the coding and encryption processes is to improve the reliability, privacy and integrity of information transmission over different media, including telephone lines, mobile radio and satellites. Coding for error correction is often preceded by source coding and data compression techniques, and it is therefore inevitable that the applied error protection coding and cryptographic techniques depend on the nature of the signals to be transmitted. This is illustrated in our paper by considering the following application areas: compression and security of speech transmissions; transmission of still video pictures in a matter of a few seconds over voice band radio; and security for mobile telecommunications systems.

Dedication

This paper is dedicated to the memory of Dr Robert E Peile who died in January 1995. Robert had strong interests in co-designed modulation, coding and equalisation, cryptographic design and analysis, the theory of Boolean functions, the design and implementation of algebraic and convolutional decoding algorithms, and in many other applications of mathematics to communications and engineering. He published many papers, held a number of patents, and co-authored the book "Basic Concepts in Information Theory and Coding : The adventures of Secret Agent 00111" with Solomon Golomb and Robert Scholtz. The authors are privileged to have worked with Robert at Racal Research Ltd for many years, and his premature death is a great loss.

1 Introduction

The purpose of this paper is to present an overview of some commercial applications of coding and cryptography, and to provide an insight into the techniques currently used to create efficient and cost-effective secure systems.

We discuss a number of signal processing techniques which enable audio and video signals to be transmitted efficiently over digital communications channels. In particular, we describe a number of coding methods which are common to both speech and vision, and are mathematically very similar. Significant differences exist in their application, however, due to the fact that human ears and eyes are sensitive to different kinds of signal distortion. The impact of these subjective effects on the implementation of speech and vision coding techniques is highlighted in sections 2 and 3 respectively.

It is evident that transmitted information, whether it be derived from speech, visual images or written text, needs in many circumstances to be protected against eavesdropping. Access to the services provided by network operators to enable telecommunications must be protected so that charges for using the services can be properly levied against those that use them. The telecommunications services themselves must be protected against abuse which may deprive the operator of his revenue or undermine the legitimate prosecution of law enforcement. These issues are the subject of section 4, where developments in telecommunications security are outlined in the context of mobile communications.

2 Speech Coding

In this first section, we consider the coding of speech waveforms to permit their secure transmission in an efficient manner. Speech is the most natural means of human communication, and the acoustic wave which facilitates it is a highly redundant analogue signal. Conventional analogue scrambling techniques to guard against eavesdropping on speech communications [3,11,24] do not provide a high level of security. On the other hand, digitised speech can be secured to as high a level as is required by means of stream cipher techniques. This is one reason for the increasing popularity of digital speech in communications systems. However, there is a fundamental trade-off between data rate, speech quality and implementation complexity, and in addition there are issues which relate to the sensitivity of the coding algorithms to noise on the input speech signal and errors in the transmission path. Considerable research effort has been expended over the last twenty years on attempts to find coding algorithms which are capable of maintaining good speech quality at ever decreasing data rates within the bounds of reasonable complexity.

In essence, there are two basic methods for digitising speech, namely waveform coding and source coding. In both methods the signal is first uniformly sampled at a rate commensurate with the Nyquist limit, and each sample is then quantised to one of a number of levels so that its value can be represented by a fixed length digital word. However, once the speech signal sample values have been obtained, the way in which they are subsequently processed differs significantly between the two types of coding.

2.1 Waveform Coding

Waveform coding includes techniques such as Pulse Code Modulation (PCM) and Delta Modulation (DM) which encode the speech waveform on a sample by sample basis.

In PCM [4], sampling is generally undertaken at 8,000 samples per second, which is sufficient for speech signal frequencies up to 4,000 Hertz. The quantisation of the *amplitude* of each sample is taken to the nearest of one of $256 = 2^8$ levels, so that each sample value requires an 8-bit codeword. The transmitted data rate for PCM is thus 64,000 bit/s. This technique is widely used by terrestrial telephone network operators throughout the world for long distance trunk circuits. It has the advantages of low complexity and high reliability, and is capable of reproducing high quality speech over very long distances even if a number of such circuits in tandem are required to connect a call. That is why one can telephone Sidney in Australia from Cirencester and it will sound as if he is in the house next door!

Differential PCM [13] exploits the fact that there is generally a strong correlation between consecutive samples of a speech signal, and a reduction in data rate is possible by encoding the *difference* between the amplitudes of consecutive PCM samples. For the same perceived speech quality as PCM at a sampling frequency of 8000 Hertz, DPCM permits a reduction of about 1 bit per sample. The correlation between consecutive samples increases with sampling frequency, and with higher sampling rates fewer bits are needed to encode each difference without compromising the perceived speech quality. In the limit of 1 bit per sample the scheme is known as Delta Modulation (DM) [25]. A fundamental property of differential coding is that it is more tolerant of transmission errors than non-differential coding, which makes it attractive for applications where the transmission path includes radio links.

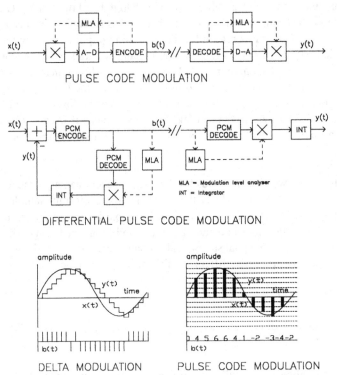

Fig.1 : Waveform coding

One problem with the design of a telephone network is the range of levels at which different people talk. In order to cope with this, the dynamic range of PCM has to be at least 30 dB, which requires more bits per sample to maintain a sufficient signal to noise ratio than would otherwise be needed. However, within the duration of a single telephone call, the speaking level of individuals tends not to vary very widely, and it is possible to reduce the number of quantisation bits required if the gain of the coder is made adaptive to the incoming signal level. This leads to schemes referred to as APCM, ADPCM and ADM where the 'A' stands for 'adaptive'. In the case of differential coding schemes, one common level adaption technique is to vary the gain of the feedback loop which compares the next input sample with a locally reconstructed previous output sample. This effectively changes the maximum slope which can be coded without overload, and in the case of ADM the method is more generally known as Continuously Variable Slope Delta Modulation (CVSD).

Other ways to reduce the data rate of waveform coders involve the prediction of future sample values based on mathematical modelling of the speech waveform. The feedback loop in DPCM and DM is effectively a zero-order prediction, wherein the current sample is used as a simple prediction of the next. By extrapolating the current signal slope in addition to the sample value, a first order adaptive predictive coder (APC) [16] is obtained. Higher orders of extrapolation based on a model of the speech signal are nowadays in widespread use, but the resulting coders are more reminiscent of analysis-synthesis vocoders than waveform coders, even though they retain the same basic structure.

The techniques outlined above have given rise to the implementation of speech coding techniques with data rates down to about 24,000 bits per second which provide high perceived quality, and as low as 16,000 bits per second with some loss in quality. Recently developed analysis-synthesis source coding techniques, on the other hand, are now able to provide high quality speech at data rates as low as 4,800 bits per second, and good quality at rates down to 1,600 bits per second. However, it should be noted that one of the penalties of source coding techniques is that they are less tolerant of acoustic noise at the microphone than waveform coding techniques, and the lower the data rate the more pronounced the differences tend to become.

2.2 Source Coding

In order to achieve lower data rates than waveform coding, source coding techniques exploit the fact that the signal to be coded is speech. The idea is to construct mathematical models of the human speech production and hearing mechanisms, and to mimic these in the encoding and decoding processes. The transmitted signal simply contains the parameters of the model. The methods are usually described as analysis-synthesis techniques, but at data rates of less than 4,800 bits per second they are also invariably known as voice-coders, or just simply vocoders.

There are in essence two basic methods of speech signal source coding, namely those which operate in the time domain and those which operate in the frequency domain. Time domain techniques are based directly on a model of the human vocal tract, and include the many variants of linear predictive coding (LPC). Frequency domain techniques, on the other hand, are based on models of the human organs of both speech production and perception, and include all the variants of sub-band coding (SBC).

A number of features of human speech are important for source coding. In particular, there are two basic types of speech sound, voiced and unvoiced. Voiced sounds are exemplified by vowels, and are quasi-periodic with the periodicity stemming from the vibration which occurs when air from the lungs is forced through the tightened vocal folds. The resulting puffs of air excite the nasal and vocal cavities to produce an acoustic speech waveform at the nose and lips. The fundamental frequency of the puffs of air is related to the perceived pitch of the speech. Unvoiced sounds are exemplified by fricatives, and have no perceived pitch. They arise as a result of air turbulence at a point of constriction in the vocal tract which excites the cavities. For both voiced and unvoiced sounds, the speech waveform can be modelled as the output of a vocal tract filter driven by a white periodic signal for voiced sounds or white noise for unvoiced sounds. This basic model of speech production is used in virtually all source coding techniques.

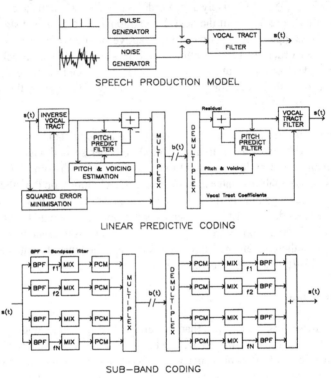

Fig.2 : Source coding

The features of the human hearing mechanism exploited in source coding techniques are that the ear performs a crude frequency analysis whilst remaining sensitive to the times at which changes occur within the speech signal. Furthermore, experiments have shown that for a given signal bandwidth the sensitivity of the ear decreases with increasing frequency. These properties are readily exploited to reduce the data rate requirements in frequency domain coders, but are not so easy to exploit in time domain coders. Other properties of hearing which are particularly useful in the design of pitch excited vocoders are that the ear is relatively insensitive to changes in phase, and also to small changes in signal level.

One important feature of any spoken language is the existence of a number of distinct sounds, called phonemes, from which a complete phonetic description of the speech can be constructed. Moreover, since speech is produced by moving the articulators, mechanical inertia imposes an upper limit on the rate at which the phonemes can be uttered [10]. In English there are in principle about 42 such phonemes, and a fast talker might manage to utter at most 10 per second. The information data rate of the written equivalent of the speech is thus only about $10\times\log_2 42 \approx 54$ bits per second. This implies that there is plenty of scope for source coding techniques to achieve significant reductions in data rate over and above those achieved to date through the application of differential and adaptive techniques to waveform coding.

Time Domain Source Coding

By far the most common technique for coding speech in the time domain is linear predictive coding (LPC) [17,22]. The speech signal is modelled as the output of a quasi-stationary all-pole filter excited by either white noise or a white periodic waveform :

$$s_n = e_n + \sum_{p=1}^{P} a_n(p)\, s_{n-p}$$

where s is the speech signal, e is the excitation signal, $a(p)$ is the pth filter coefficient, and subscript n denotes time index. The vector of P coefficients $A_n = \{a_n(p) : p=1,...,P\}$ defines the response of a vocal tract of order P at time $t = nT$, where T is the sampling period. This can be seen by taking z-transforms, where $z^{-1} = e^{-i\omega T}$ is the unit delay operator and ω is angular frequency, to obtain :

$$S_n(z)\, A_n(z) = S_n(z)\left[1 - \sum_{p=1}^{P} a_n(p)\, z^{-p} \right] = E_n(z)$$

and hence $S_n(z) = E_n(z)/A_n(z)$. The vocal tract filter response $A_n(z)^{-1}$ at time $t = nT$ is a function of time because the vocal tract changes shape continuously as the articulators move from producing one phoneme to the next. However, for time intervals of up to about 25 milliseconds, encompassing 200 samples at an 8000 Hertz rate, the vocal tract can be considered constant because mechanical inertia prevents the articulators from moving quickly. This is the reason why the vocal tract was previously described as being *quasi-stationary*. It means that if the filter response $A_n(z)$ is estimated at intervals not exceeding 25 ms, and a suitable mechanism is found to represent the excitation $E_n(z)$, then it is possible to construct high quality speech from the parameters.

The most common technique for estimating the vector A_n of vocal tract filter predictor coefficients from a vector of N speech samples $S_n = \{s_{n-i} : i=0,...,N-1\}$ is based on a squared error minimisation procedure [19]. The excitation sample e_n can be thought of as the error between the speech sample s_n and its linear predicted value based on the previous P speech samples. The white excitation sequence vector $E_n = \{e_{n-i} : i=0,...,N-1\}$ is thus equivalent to a vector of errors which can be expressed as

$$\begin{pmatrix} e_n \\ e_{n-1} \\ \cdots \\ e_{n-N+1} \end{pmatrix} = \begin{pmatrix} s_n \\ s_{n-1} \\ \cdots \\ s_{n-N+1} \end{pmatrix} - \begin{pmatrix} s_{n-1} & s_{n-2} & \cdots & s_{n-P} \\ s_{n-2} & s_{n-3} & \cdots & s_{n-P-1} \\ \cdots & \cdots & \cdots & \cdots \\ s_{n-N} & s_{n-N-1} & \cdots & s_{n-N-P} \end{pmatrix} \begin{pmatrix} a(1) \\ a(2) \\ \cdots \\ a(P) \end{pmatrix}$$

that is $E = S - CA$. The total squared error for the frame is $\epsilon = E^T E = (S - CA)^T(S - CA)$. Differentiating with respect to the vector A and setting the result to zero gives the conditions for which the total squared error ϵ is minimised, namely

$$\frac{\partial \epsilon}{\partial A} = -2C^T(S - CA) = 0 \quad \leftrightarrow \quad A = (C^TC)^{-1}C^TS$$

Observe that C^TC is the autocorrelation matrix of the speech samples, and that C^TS is a vector of speech autocorrelation coefficients. The least squares solution for the predictor coefficients A is thus readily computed from the autocorrelation of the speech samples. Efficient algorithms exist to perform the required matrix inversion [24].

Before the vector A of predictor coefficients can be transmitted in digital form, it must first be quantised and encoded. Scalar quantisation of the components of A is possible, but is not the most efficient in terms of the number of bits required. This is largely because the coefficients do not have a simple theoretical bound, and also because the sensitivity of the reproduced speech to small changes in them is a complex non-linear function of the coefficient values themselves. A better scheme is to convert the set of predictor coefficients $A = \{a(p) : p=1,...,P\}$ to an equivalent set of reflection coefficients $K = \{k_m : m=1,...,P\}$ in accordance with [19]

$$k_m = a_m^{(m)} \; ; \quad a_p^{(m-1)} = \frac{a_p^{(m)} + k_m a_{m-p}^{(m)}}{1 - k_m^2} \quad p=1,...,m-1 \; ; \quad m=P,...,1$$

The reflection coefficients have the useful property that they are guaranteed to lie in the bounded interval $(-1, 1)$, and are thus more easily quantised, but their sensitivity is still a non-linear function of their value. This problem is reduced by further transforming the coefficients to log-area ratios (LARs) in accordance with

$$LAR_m = Log \frac{1 + k_m}{1 - k_m} \quad m=1,...,P$$

and then quantising the LARs with a linear quantiser. This scheme is employed in the NATO Stanag 4198 vocoder for the transmission of secure voice at 2,400 bit per second [24], and has now been widely used for over a decade.

More recently, an alternative scheme for quantising the predictor coefficients which has gained in poularity employs so-called line spectral frequencies (LSF) [7]. These stem directly from the model of the vocal tract in terms of predictor coefficients. If the model order is increased by one, two different transfer functions $P(z)$ and $Q(z)$ can be defined according to whether the extra section has a reflection coefficient of plus or minus unity. These are related to the vocal tract transfer function $A(z)$ by

$$P(z) = A(z) - z^{-N-1}A(z^{-1}) \qquad \text{and} \qquad Q(z) = A(z) + z^{-N-1}A(z^{-1})$$

It can be shown that the polynomials $P(z)$ and $Q(z)$ have zeros which alternate on the unit circle. As a consequence, $P(z)$ and $Q(z)$ are defined uniquely by the frequencies of their zeros, and since $A(z) = \frac{1}{2}[P(z) + Q(z)]$ this means that the LSFs also define the predictor coefficient vector A. The main advantage of quantising LSFs instead of LARs lies in the fact that they are strongly related to the frequencies and bandwidths of the speech formants - vocal tract resonances - which are subjectively very important. The quantisation of LSFs can thus be optimised in a relatively simple way to provide the best perceived speech quality.

All of the quantisation methods mentioned so far are scalar in the sense that each component of the coefficient vector is quantised independently of the other components. However, this is by no means the most economical method in terms of the number of bits required to achieve a given fidelity in the reproduced speech. In vector quantisation, each set of coefficients is regarded as a vector in P-dimensional space. To encode the vector, a codebook is searched to determine which of a pre-determined set of possible vectors is nearest, and the associated index then defines the quantised vector [16]. This scheme requires the least number of bits for a given fidelity of reproduced speech, because the codebook can be populated with vectors which have been obtained from samples of real speech waveforms. This means that no bits are wasted in allowing for the coding of vectors which do not occur in real speech. As a consequence, vector quantisation and coding enables the transmission of digital speech with the highest perceived quality for a given data rate.

It remains to find a means to estimate, encode and transmit the excitation samples E_n associated with each frame of speech samples S_n. As with most things in speech coding, there are two ways to approach this problem, depending on whether the excitation waveform is to be parameterised or encoded as a waveform. In pitch excited vocoders, the excitation is transmitted as a set of parameters, and in voice excited vocoders it is transmitted as a coded waveform.

At the simplest level, the excitation has a white spectrum and is either random noise for unvoiced sounds or periodic pulses for voiced sounds. By examining the speech waveform, it is possible to determine which type of sound is being coded, and to transmit a single voiced-unvoiced bit to the receiver. This must also be accompanied by an indication of the pitch period, so that a suitable pulse waveform with the correct periodicity can be synthesised. The transmitted speech parameters in each frame of a pitch excited LPC vocoder thus consists of the predictor coefficients in some form, along with a voiced-unvoiced bit, the pitch period and an root mean square (rms) gain factor to set the overall signal level. Speech is reproduced at the receiver simply by operating the speech production model and updating the coefficients for every new frame of received parameters. This scheme is employed in the NATO Stanag 4198 vocoder mentioned earlier, and provides speech quality which is satisfactory for military applications [24]. However, it is by no means perfect, and is not considered acceptable for use in public telephone networks. Voice excited vocoders operating at higher data rates are needed for these applications.

For voice excited LPC vocoders, the first step is to estimate the excitation signal $E(z)$ by filtering the speech signal $S(z)$ through a filter whose response is the inverse of the vocal tract response. Since the inverse filter response is just $A(z)$, and the vector A can be estimated in the manner described earlier, the process of obtaining an estimate of the excitation signal $E(z) = S(z) A(z)$ is straightforward. Note that this signal is frequently referred to as the *residual*, because it is indeed the residue of the speech signal after the effects of the vocal tract have been removed. Once the residual has been obtained, there are numerous methods available for its quantisation and encoding. Each gives rise to a different type of speech coder with its own characteristics. However, space does not permit a detailed survey, and only a brief outline of the basic principles can be given here. The reader is referred to the references for more information.

Multi-pulse methods [16] improve upon the pitch excited technique by assigning a number of pulses to represent the excitation within each pitch period. The process of determining the magnitude and position of each pulse is based on a perceptually weighted comparison between the reproduced speech and the original. Multi-pulse LPC (MP-LPC) is capable of producing high quality speech at a data rate of 9,600 bits per second, and is used in the current Aerosatcom system to provide airline passengers on trans-Atlantic flights with a telephone service. An earlier form of multi-pulse LPC called regular pulse excited LPC (RPE-LPC), in which the pulse positions are assigned within a pre-defined grid, is used in the digital GSM cellular mobile telephone network [16]. The speech coder in this system operates at the somewhat higher data rate of 13,000 bits per second, and also provides high quality speech.

More recently, the residual encoding technique which shows the most promise for low data rates with high speech quality is code-excited LPC (CELP). This method is a form of vector quantisation in which each input vector is a sub-frame of the residual signal. The codebook can be populated with either segments of real speech residuals, or just random sequences [16]. The index of the most appropriate vector in the codebook is chosen by comparing the synthesised speech to the original with a perceptual error weighting. The technique has been known for a long time, but it is only in the last few years that cost-effective implementation has become possible, due to the large amount of processing needed to determine the nearest vector in the codebook. The US DoD have produced a Federal Standard CELP vocoder algorithm [26] which operates at a data rate of 4,800 bits per second, and provides speech quality which approaches what is acceptable for use in a public telephone network.

Much of recent speech coding research is aimed at finding ways to reduce the computation required for codebook searches. Variations on the basic CELP are the splitting of codebooks in a hierarchical tree structure, and the use of fixed and adpative codebooks to reduce the search time. Of particular merit is the use of both an adaptive codebook to facilitate the economic encoding of periodic voiced speech, and a fixed gain-shape codebook to encode the basic waveshape. This dual codebook structure has proved to be a very effective means for efficiently encoding the residual. Another currently evolving technique which shows promise is Prototype Waveform Interpolation (PWI). The idea here is to obtain waveforms at the analyser which represent single periods of the residual in voiced speech [15], and to reproduce speech at the synthesiser by a process of pitch synchronous interpolation between the individual periods.

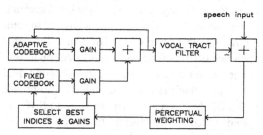

Fig.3 : Code excited LPC

The most recent significant development in CELP known to the authors is the time envelope vocoder [1]. This is reported to provide good quality speech at a data rate of only 1,600 bits per second. The improvement is achieved by adjusting the gain parameter of a CELP in such a way as to ensure that the time domain enevelope of the synthetic speech is matched to that of the original. The scheme is effective because it provides a more accurate reproduction of the transitions between phonemes when compared with other methods.

Frequency Domain Source Coding

Frequency domain source coding techniques achieve their advantage over waveform coding techniques by exploiting the properties of the human ear. The relevant features of hearing are that the sensitivity of the ear to changes in frequency is a decreasing function of frequency, and that if two frequencies are close together, then the louder one will tend to mask the quieter one, to an extent which depends on their separation in both frequency and amplitude.

The basic idea in all frequency domain source coding techniques is to have a parallel bank of filters which covers the entire speech spectrum. At the encoder, the speech signal is applied to the input of all filters in the bank, and the components at the output of each filter are encoded and transmitted separately. Speech is reproduced at the receiver by decoding the signals in each band, applying them as inputs to individual filters in a similar bank, and then combining the filter outputs. By making the filter bandwidths increase with frequency in an appropriate way, the individual signal components contained in each filter will, on average, contribute equally to the overall speech quality, and thereby improve coding efficiency. Further improvements can be achieved by making the number of bits allocated to the transmission of the signal in each frequency band dynamically adaptive to the relative importance of its contribution to overall speech quality, within the constraint of a pre-determined data rate. Compared with waveform coding, the method achieves compression because it concentrates the transmitted data to just those regions of the speech spectrum where it makes the greatest contribution to perceived speech quality.

The different types of frequency domain speech coder are largely distinguished by the way in which the individual signals in each frequency band are encoded. Conceptually, the simplest method is to frequency shift each signal to baseband and then apply a waveform coding technique such as ADPCM. This kind of structure with either eight or sixteen bands is known as sub-band coding (SBC), and is capable of providing high quality speech at a data rate of 16,000 bits per second. At the lower data rates of 8,000 and even 4,800 bits per second it is still able to provide good quality speech. As an aside, it is interesting to note that the so-called digital audio broadcast (DAB) system currently on trial in the UK [23] has a similar structure to SBC, and uses around one thousand frequency bands to reduce the adverse effects of differential time delays caused by multiple propagation paths in the radio links.

There are a number of variations on the basic theme of SBC, depending on the way in which the frequency domain signals are obtained. An early method was to apply the discrete Fourier transform (DFT) to the samples of speech for a single frame, and then average the spectral magnitudes over the bandwidths of the equivalent sub-bands. Dynamic bit allocation was then applied to the spectral magnitudes in each band in such a way as to maximise the perceptual benefit of each transmitted bit. Recovery of the speech at the receiver was achieved by reconstructing the speech spectrum with an appropriate phase, and then performing an inverse DFT. This scheme was called adaptive transform coding (ATC), and provided good quality speech at a data rate of 16,000 bits per second. The main problem with it was that it produced audible artifacts at the boundaries between frames which could sometimes be subjectively disturbing. Methods to improve the quality of ATC by the use of windows and overlapping frames certainly reduced the artifacts, but at the expense of an increase in the overall data rate.

A limiting case of the SBC is to measure the amplitude of the speech signal in each filter channel, and then transmit just these amplitudes along with the pitch period and voiced-unvoiced information in an identical way to that described previously for a pitch excited LPC vocoder. Synthetic speech can then be produced at the receiver by injecting a common artificial periodic or noise like excitation signal into all the filters in the synthesiser bank, with the gain of each one adjusted to correspond to the signal level measured in the analyser. Such a scheme is known as a channel vocoder, and historically it was the very first type of speech coder to be designed. It exploits the fact that the ear is relatively insensitive to phase, it uses a crude frequency analysis which matches the processing performed by the ear, and it uses a periodic or noise like excitation to ensure that the relative timing of changes in the speech waveform are properly reproduced. At a data rate of 2,400 bits per second the speech quality is adequate for military applications, and has been widely used for this purpose over many years. However, the quality is not good enough for use in a public telephone network.

A development of frequency domain speech coding which has achieved prominence in the last year or two is known as multi-band excitation (MBE). It is a successful attempt to combine the data rate reduction advantage of pitch excited vocoders with the quality advantage of voice excited vocoders. Careful examination of the spectrum of speech revealed that for many speech sounds it is not always clear if it is voiced or unvoiced, and moreover the voicing attributes can vary with frequency. In MBE these features are modelled by exciting each frequency band with an artificial excitation in which the type of excitation varies from band to band in accordance with measurements made at the analyser. An improved version of MBE known as IMBE [16] which operates at data rate of 4150 bits per second has become a standard for Inmarsat-M satellite communications terminals, and an advanced version called AMBE is being developed for use with future Aerosatcom applications.

2.3 Error Correction

In the previous two sections a wide variety of different schemes have been described for the transmission and reception of speech in digital form. Real communications links introduce errors at different bit error rates (BERs), and the errors can occur either randomly or in bursts. Just like any other kind of digital communications, speech is not immune from the effects of errors, and it is frequently necessary to take measures to counter their adverse effects. However, there are a number of distinct advantages to transmitting speech compared with data communications, and these will now be discussed.

In waveform coding techniques, the transmitted bits have a different significance depending on which position they occupy within the digital word representing each sample. Gray coding mitigates the worst effects of this, but in cases of high error rates additional forward error correction (FEC) will be needed. If differential encoding is employed, the overall sensitivity to errors is reduced. However, except in the case of delta modulation, the transmitted data will still contain some bits which are more sensitive to errors than other bits. Clearly, the most efficient use of FEC is to protect the most important bits with the strongest code, and protect the less important bits either with a weaker code or no protection at all, depending on their relative sensitivity.

The sensible application of FEC in the way suggested above applies equally to both waveform and source coding techniques. With source coded speech, however, there exists a variety of methods in addition to FEC which can be employed to help combat the adverse effects of errors. One particularly useful property of speech which can be exploited in conjunction with source coding is the fact that although speech is a continuous process of transitions from one phoneme to another, there are often periods of up to 100 milliseconds, or even longer, during which the characteristics change very little, for example during vowel sounds. With source coding techniques this means that the model parameters will often change very little from one frame to the next. If any particular frame is identified as containing one or more errors, and the surrounding frames do not contain errors, then interpolation can be applied to remove the worst effects of the errors. Since no extra data needs to be transmitted to implement this kind of error correction, it could be described as a "zero redundancy forward error correction" scheme.

Another useful property of speech is that there exist periods of silence between words and phrases which contain no useful information. If the extra system delay can be tolerated, these pauses can be used to send the additional information required by FEC schemes to correct errors, and thereby reduce the overall transmitted data rate.

Two ideas which have in the past been suggested for use in speech coding systems are variable bit rate coding and embedded coding. The idea of varying the data rate was first conceived as an attempt to overcome peak hour congestion problems on digital trunk circuits by allocating a lower data rate to each user during the busy hour, and thereby allow more users on to the system. Embedded coding is a variant of this in which the parameters of a low data rate speech coder are identified as an integral but distinct part of those for a higher data rate speech coder. When congestion occurs, only the parameters of the low data rate speech coder are transmitted in order to reduce the overall data rate requirement. Both of these ideas could be used to advantage in combatting the effects of transmission errors, since the extra data available to the higher rate coder could be used to provide FEC for the lower rate coder if the channel error rate increased to the extent that it could not be tolerated by the higher quality coder operating at the higher data rate.

Finally, it should be noted that the sensitivity of vocoded data to transmission errors will increase as the data rate is reduced, because the transmitted data is itself less redundant. This suggests that there must be an optimum transmitted data rate for speech which depends on the prevailing error rate, but the authors are not aware of any work to establish what this optimum might be.

3 Image Coding

In recent years there has been a growing demand for the transmission of moving pictures over communications channels normally dedicated to speech transmission. Applications include remote surveillance, videophone, and tele-conferencing. Research into the design of image processing algorithms to achieve these objectives has proceeded in parallel with the development of speech coding algorithms, but in the past their implementation has been hindered by the need for very considerable processing. However, the advent of high speed digital signal processing devices has made feasible the implementation in real-time of not only very complex speech coding algorithms but also complex image processing algorithms. This has prompted a significant increase in research activity in both fields during the last few years. As with speech coding, image coding also has some fundamental trade-offs to consider, namely frame update rate, transmission data rate, picture quality, algorithm complexity and sensitivity of the coded signals to transmission errors.

Space limitations prevent an in-depth discussion of all the available algorithms, so we will concentrate on just three of the more popular techniques used to date to provide cost effective low bandwidth still image transmission. The JPEG technique [5] is based on the application of a two dimensional discrete cosine transform (2-DCT), and the lapped orthogonal transform (LOT) technique [18] overcomes some of the unwanted artifacts produced by the JPEG algorithm. Finally, we describe a fractal coding technique [2,12] which has proven to be excellent for coding pictures which contain a lot of texture. An overview will be given of the basic ideas behind these three techniques, highlighting their important features and inter-relationships.

A still image, such as a black and white photograph, is fundamentally a time invariant function of two variables. A colour image is just a superposition of three such functions, one for each colour. In the interests of brevity we will restrict our attention to simple black and white images, but the concepts are equally applicable to colour images.

In order to transmit an image function electonically, it is first transformed into a one dimensional signal which varies with time. This can be achieved by scanning the image, in the same way as a facsimile machine. When the transmission is digital, the scanned image is sampled at regular intervals and the grey level coded to one of 256 levels, giving 8 bits per sample. For a moderate sized picture with a resolution of 512×480 picture elements (called pixels, or sometimes pels), the image will require 8×512×480 = 1,966,080 data bits to be transmitted. With a telephone line modem data rate of 9,600 bits per second, the overall transmission time will be about 3.5 minutes, which may be only just acceptable. If the telephone line quality is poor, the transmitted data rate will be reduced to 2,400 bits per second, and the picture will take nearly 14 minutes to transmit, which is almost certainly not acceptable. This simple example, analogous to PCM in speech coding, serves to illustrate why picture compression is needed. Even with moderate compression factors of 25:1, the unacceptable 14 minute transmission time with a poor quality telephone circuit becomes an acceptable half a minute.

In order to compress a picture, use is made of the fact that the image is two dimensional, and that there are usually strong correlations between one pixel and its neighbours in all directions. A two dimensional version of the differential PCM encoding scheme can be used to advantage, but unlike with speech, there are no general mathematical models which can be used to represent a picture. As a consequence, in order to achieve a compression algorithm which can be applied to all pictures, only the properties of the human visual receptors can be exploited. The cones on the retina of the human eye in effect undertake a two-dimensional sampling, but due to continuous random eye movement the sampling points of an image are not fixed. This makes picture compression just as difficult as speech compression, if not more so.

3.1 Discrete Cosine Transform Methods

A common feature of pictures which lends itself to compression is that there are frequently large areas where the intensity and texture do not change very much. This suggests that for those areas it should be possible to encode a small representative area and simply repeat it as necessary. In order to identify such areas a two-dimensional transform to estimate the spatial frequencies present in the picture has proved to be most useful. In particular, for images with a high adjacent pixel correlation, the output of the two dimensional discrete cosine transform (2-DCT) is very close to the output of the theoretically optimum Karhunen-Loeve transform (KLT) which identifies the eigenvalues and eigenvectors of the image [9]. Based on this idea, and its strong relationship with sampling by the retina of the eye, two techniques which have gained prominence in the compression and digital transmission of still pictures are a standard produced by a joint photographic experts group (JPEG), and a technique called lapped orthogonal transforms (LOT) which overcomes some of the artifacts of JPEG caused by the data blocking which is applied prior to taking the 2-DCT.

For an image represented by the two dimensional function $g(x,y)$, and sampled to give values denoted by $g_{mn} = g(m\Delta_x, n\Delta_y) : m,n=0,...,N-1$, where Δ_x and Δ_y are the sampling distances in the x and y coordinate directions respectively, the $N \times N$ point 2-DCT $G_{pq} : p,q=0,...,N-1$ of the image is given by :

$$G_{pq} = \frac{2}{N} \alpha_p \alpha_q \sum_{m=0}^{N-1} \sum_{n=0}^{N-1} g_{mn} \cos\left[\left(\frac{2m+1}{2N}\right)p\pi\right] \cos\left[\left(\frac{2n+1}{2N}\right)q\pi\right]$$

where $\alpha_0 = 1/\sqrt{2}$ and $\alpha_p = 1$ for all $p \neq 0$. The DCT has the property that, to within a scale factor, the kernel is the same for both the forward and the inverse transforms :

$$g_{mn} = \frac{2}{N} \sum_{p=0}^{N-1} \sum_{q=0}^{N-1} \alpha_p \alpha_q G_{pq} \cos\left[\left(\frac{2m+1}{2N}\right)p\pi\right] \cos\left[\left(\frac{2n+1}{2N}\right)q\pi\right]$$

From a computational point of view, and this is the second main reason for its popularity, the $N \times N$ point 2-DCT has the property of being *separable*, which means that it can be decomposed into a sequence of N one dimensional DCTs each of dimension N. This can be seen more clearly when the 2-DCT equation is written in the form

$$G_{pq} = \sqrt{\frac{2}{N}} \alpha_p \sum_{m=0}^{N-1} \gamma_{mq} \cos\left[\left(\frac{2m+1}{2N}\right)p\pi\right]$$

where

$$\gamma_{mq} = \sqrt{\frac{2}{N}} \alpha_q \sum_{n=0}^{N-1} g_{mn} \cos\left[\left(\frac{2n+1}{2N}\right)q\pi\right]$$

The method employed in JPEG for coding a large image is to divide the picture into non-overlapping blocks of 8×8 pixels, and to apply the 2-DCT to each block separately [5]. Since the spatial frequency values G_{pq} for each block are an equivalent representation of the image, there is no loss of information at this stage. The reason why the 2-DCT permits compression lies in the fact that, for highly correlated pixel values within a block, only low spatial frequency components will exhibit any significant magnitude. As a consequence, there will be little loss of fidelity when the high spatial frequency components are omitted, thereby allowing a reduction in the total number of bits needed to represent the image.

With the redundancy of an image reduced by the 2-DCT, the next step is to quantise the spectral components in such a way that the resulting quantisation is minimised visually. This is achieved in JPEG by linearly quantising each spectral frequency component, using a fixed step size which varies with spatial frequency. The one exception is the dc component G_{00}. Since this will tend to vary only slightly from one block to the next, only the *difference* between the values of G_{00} in consecutive blocks is encoded (apart from the first block). The process applied to the dc component is thus analogous to the use of DPCM in waveform coding of speech. For the remaining components, it is analogous to ATC with a fixed bit allocation for each spectral component.

Once the quantisation process has been completed, it remains to encode the quantised spatial spectrum for transmission. The key feature of JPEG encoding is that the spatial frequency components of each block are first rearranged in decreasing order of importance, and then run-length coding is applied. Since the frequency components G_{pq} tend to decrease in magnitude as the indices p and q move away from zero, the spectral values are simply read sequentially from the matrix in a zig-zag pattern starting at the dc position $p=q=0$. At the end of this process, for which there is no analogy in speech coding, an economical digital representation of the picture is available. For certain kinds of picture, even more compression may be possible, and a Huffman coder can be used to reduce the entropy still further.

Subjectively, the JPEG algorithm can produce high quality images with compression ratios of 24:1 or more. In some parts of a picture, however, the edges of the blocks used for the 2-DCT may be visible. The reason for the appearance of these is the same reason why ATC applied to speech exhibits artifacts at frame boundaries. The problem can be overcome in speech coding by the use of sub-band filters whose time domain responses extend beyond frame boundaries. An equivalent process in picture coding is the use of lapped orthogonal transforms (LOT). The idea is to have a transform size which exceeds the block size, so that spatial frequency components in a given block arise not only from pixels in that block but also from those in adjacent blocks. However, by ensuring that the basis functions used in the LOT decay toward zero at the edges, and by ensuring that they are orthogonal in the regions where they overlap, then in the reconstructed picture block edge effects are completely eliminated [18]. In principle, the overlapping ensures that the transitions across block boundaries are smooth, and the orthogonality ensures that there will be no spillover of information from one block into an adjacent block. The use of LOT thus maintains the integrity of the image for all types of picture material without any block artifacts. Quantisation and coding for the LOT technique is the same as that described for JPEG.

In any communications system there will inevitably be occasions when the transmission path introduces errors, and in picture coding the results of errors are often very visible. The use of FEC will mitigate them, but it requires an overhead which may not always be obtainable. However, just as the properties of speech can be exploited to help overcome the worst effects of errors without incurring any extra overhead, the nature of pictures can be similarly exploited. Run-length coding was used in JPEG to take advantage of the fact that there are often large areas of a picture where the intensity does not change very much. However, since run-length coding is not particularly robust against errors due to error propagation effects, JPEG encoded pictures without the protection offered by FEC will noticeably degrade when transmission errors occur. An alternative to run-length coding which promises better protection without the extra overhead demanded by FEC is error resilient positional coding (ERPC). This recently developed scheme [8] operates over large blocks of sparse data by encoding the positions of the non-zero spectral values in such a way that the probability of error propagation is inversely proportional to the importance of the spectral sample.

3.2 Fractal Methods

Much has been written about fractals in recent years, and they have generated a lot of interest in both the engineering and mathematics communities. Much of the interest has stemmed from their novelty value of being able to produce complex computer images from very simple equations, with very little real application. However, fractals can be used as the basis of a method to compress images, especially those in which there is a lot of texture such as in rural scenes containing trees and grass etc[2,12].

Fractals are based on "self-similarity". The idea is that each part of the picture can be obtained from another part through a linear transformation. Moreover, if the transformations are contractive, then they are in themselves sufficient to define the whole picture. To see how this works, consider an *affine* transform A defined by

$$A(x, y, g) = \begin{pmatrix} a & b & 0 \\ c & d & 0 \\ 0 & 0 & e \end{pmatrix} \begin{pmatrix} x \\ y \\ g \end{pmatrix} + \begin{pmatrix} u \\ v \\ w \end{pmatrix}$$

In fractal picture coding, g denotes the grey scale at a point of the image with position coordinates (x, y). The variables $\{a, b, c, d, e, u, v, w\}$ are constants which define a particular transform. The transform operates on grey scale images by taking each point (x, y, g) and mapping it to the point $A(x, y, g)$. To see how this can be used to encode an image, we first divide it into non-overlapping squares which we call ranges. For each range, we search the whole image for an area larger than the range square such that under some fixed affine transform A the magnitude g of the grey level at every point (x, y) in the larger area maps closely to the corresponding points and grey levels defined by $A(x, y, g)$ in the chosen range square. When such an affine transform has been found for each range, the set of transforms is a complete representation of the image. This can be shown by starting off with a different image and repeatedly applying the set of affine transforms to it, overwriting the points of the original image until no further changes occur. The final image must then be identical to the original image from which the affine tranforms were obtained.

The process works only if the affine transform is contractive, which is to say that under the transformation areas and intensities are reduced. This is guaranteed if $e < 1$ and $|ad - bc| < 1$. The difficult part is finding suitable transformations. One method is to define a limited set of possible contractive affine transforms and select that which gives the best result for each range. Specifically, in one particular application, the ranges were set to be blocks of size 4×4 pixels, and all affine maps were defined to have a contraction factor of 2. Eight such maps were defined by four possible rotations and/or a reflection, along with a limited set of possible grey level transformations. Under these restrictions, the domains for each of the possible transformations is a block of 8×8 pixels. Each possible domain block can be pre-processed by averaging over blocks of 2×2 pixels to contract them into equivalent range blocks. A systematic search can then be made for each range block to find the most suitable domain block and affine transform which gives a match. The position of the domain block and the number of the affine transform for each range block then defines a complete fractal coding of the picture.

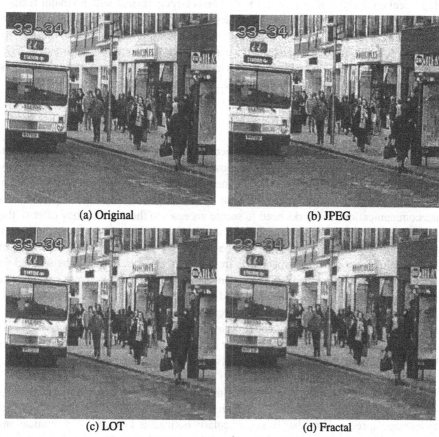

(a) Original (b) JPEG

(c) LOT (d) Fractal

Fig.4 : Comparison of image coding techniques

As already indicated, fractal coding can produce good results for pictures which contain a lot of texture, with compression factors in excess of 24:1. If the pictures contain a lot of detail, the quality of the reproduced images is less good. And in the particular case of a picture of a field of grass containing a single sign post, the image quality of the reproduced sign post will be decidedly poor. However, it is possible to mitigate this deficiency by artificially appending to the side of the image a suitable signpost during the encoding process, and then simply leaving out the appendage from the reproduced image. Such ideas are the subject of current research efforts.

Figure 4 compares reconstructed images for three different types of coding with a compression factor of about 32:1. At this level of compression, the block effects in JPEG are visible, and have been removed by the LOT technique at the expense of some loss of edge clarity. The fractal technique, on the other hand, exhibits a slight fuzzy quality which is not present in either JPEG or LOT coded images.

The subjective effect of transmission errors on fractal coded images is very much dependent on the type of picture. If the picture is largely texture with no major features then the errors will tend to give a blotchy appearance, but if the picture contains a lot of detail then the errors will be more apparent. Of course, FEC can be used to mitigate the worst effects, but this will again be at the expense of having to transmit more data. With fractal coded pictures it is also not clear as to which bits are more important than others, and hence the judicial application of FEC is a subject of further research.

4 Transmission Security

Whether it be for the transmission of speech or images, advances in the use of digital technology have been parallelled by a growing demand for communications security, particularly in the area of public telecommunications. Security for telecommunications can be viewed from three different perspectives. First, the operators of telecommunications networks need to secure accesses to the services they offer so that they can be properly charged to those who use them, and naturally the subscriber wants to be confident that he is only going to be charged for the services that he consumes. Second, there is a growing expectation from users that services should be secured for their benefit, for example to prevent conversations from being eavesdropped or to maintain privacy of their location. The third, and to date the least well addressed, aspect of telecommunications security is to protect against abuse of the services offered. This abuse may take the form of exploitation by criminals of advanced service features in order to defraud the network operators, or security features provided for the benefit of honest users may be exploited by criminals to avoid detection by the police.

The use of digital techniques means that telecommunications may benefit from being secured using quite sophisticated cryptographic techniques. The growing use of such technology in recent years has been particularly noticeable in mobile communications. This is primarily because mobile communications are perceived to be vulnerable in all three aspects of telecommunications security. These security problems arise in the form of mobile telephone cloning to impersonate genuine subscribers, by using scanners to intercept voice traffic on the radio path, and through the relative lack of visibility which the network has about the services being consumed by a subscriber who has roamed onto another network. It is therefore appropriate to consider the three different aspects of telecommunications security and the emerging techniques being used to address them in the context of mobile communications.

4.1 Security in GSM

The starting point for any discussion of security in mobile communications is without any doubt GSM, the *Global System for Mobile Communications* [20]. It has set the baseline standard for subsequent systems, like DECT, TETRA and UMTS, by providing a set of security features which are fully integrated into the telecommunications system, transparent to the user and which can be managed by the network operator. The GSM security features are subscriber authentication, subscriber identity confidentiality and confidentiality of user traffic, and it is useful for our subsequent discussions briefly to review the first and last of these.

Subscriber authentication uses a challenge-response protocol with a symmetric, or conventional, cryptographic algorithm. When a subscriber attempts to access the network to register or make a call, the network operator sends a challenge, which we denote RAND1, over the air-interface to the subscriber's mobile terminal. This challenge, which is 128 bits long, is generated by the subscriber's home network, i.e., the network where he has his subscription. For example, if a Vodafone subscriber tries to make a call whilst visiting the SFR network in France, RAND1 is generated in the UK and transmitted across international signalling lines to France. Within the mobile terminal a response, which we denote RES1, is computed as a function $RES1 = A(RAND1, K)$ of the challenge and an authentication key K using the authentication algorithm A, known as A3 in GSM. The key K is specific to the subscriber and is 128 bits in length. The algorithm A is particular to the home network operator. The response RES1 is 32 bits long, and is signalled over the air-interface back to the network for checking. This consists of the visited network, SFR in the example above, comparing the value it receives from the mobile terminal with a value for RES1 which it has already obtained from the home network. The home network is able to perform the computation in advance of the terminal because it generates RAND1 and knows K for the particular subscriber.

Confidentiality of user traffic, in particular speech, is achieved by encrypting transmissions on the radio path between the mobile terminal and the cellular base station that is serving it. The encryption takes place after digital speech encoding and error correcting coding as part of the process of building blocks of traffic and control data to transmit in the TDMA frame slots. The algorithm used is a stream cipher, known in GSM as A5, which generates bursts of 114 bits of key stream which are modulo 2 added to the data bits for transmission in a single TDMA frame. Like nearly all good key stream generators the cipher algorithm has two inputs, a cipher key Kc and a message key. The cipher key is 64 bits long, and is generated as part of the authentication process, $Kc = B(RAND1, K)$ where B is a key generation algorithm, known as A8. The message key is derived from the TDMA frame number, so changes with every 114 bits of transmitted data. The mechanism ensures that a fresh cipher key may be generated for each call, and that synchronisation of the cipher is not lost during handover, or at other times when frames of data may be lost. The key generation algorithm is particular to the network operator, and Kc is generated by the home network and passed to the visited network along with the authentication parameters RAND1 and RES1. Usually the cipher key generation and authentication algorithms are combined to provide an A/B algorithm - A38 in GSM terminology.

The authentication and cipher key computations in the mobile terminal are executed within a fundamental part of the GSM system, the subscriber interface module, or SIM. This is a smart card which contains all the subscriber related data needed for network access, including the authentication key K. The mobile telephone only becomes personalised and able to make calls (except perhaps emergency calls) when the SIM is inserted. The SIM is designed as a security device, and as such provides secure storage for K and execution of the algorithms that use it. It is the means by which the network operator manages K and other subscriber related data, it provides for automatic generation of the cipher keys Kc, and it ensures that the GSM security features are completely transparent to the subscriber.

4.2 A Security Architecture for Mobile Communications

GSM style security features have been adopted and enhanced in systems like DECT, the *Digital European Cordless Telecommunications* system [27], and TETRA, the *Trans European Trunked Radio* system [6], with the result that a security architecture for public mobile communications is gradually emerging, at least for systems under development in Europe. A somewhat simplified view of the architecture is illustrated in figure 5. As well as the GSM security features, it provides alternatives for managing cryptographic parameters, including keys, when roaming, support for network authentication, and additional options for generating cipher keys.

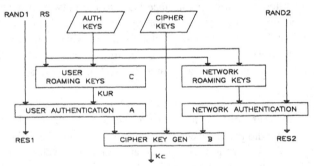

Fig.5 : Architecture for mobile communications security

When a GSM subscriber roams to another network, it is necessary for the visited network to request RAND1, RES1 pairs from the home network in order to authenticate call or registration attempts from the subscriber. Each such pair may only be used once, so there can be a considerable signalling overhead. An alternative is that a roaming key should be generated by the home network for use by the particular visited network for the particular subscriber. This key is generated as a function KUR = C(RS, K), where RS is a random value, K is the subscriber's authentication key and C is the user roaming key generation algorithm. The value RS is sent to the visited network along with the user roaming key KUR. It may be used for just one visit by the particular subscriber, it may be used for several subscribers for several visits, depending on the policies of the operators involved. When the user attempts to access the visited network, it is now the visited network which generates the challenge RAND1 and computes and checks the response RES1 = A(RAND1, KUR). The subscriber, or rather the subscriber's SIM (or whatever the smart card is called for the particular system), is able to compute KUR because the visited network sends it RS, and of course it knows K and can execute C. For the scheme to work, it is necessary that the visited network knows the authentication algorithm A, so this ceases to be network specific and in effect needs to become a standard available to all network operators. However, the roaming key generation algorithm may be particular to the home network.

Network authentication is achieved by mirroring the process used for subscriber authentication, including the provision of network roaming keys for the subscriber to use with one particular visit to one particular network if desired. The process is illustrated by the RAND2, RES2 parameters in figure 5. Cipher key generation is enhanced by allowing the key to be computed by the visited network, rather than signalled from the home network, and by making it an outcome of both subscriber and network authentication.

The security architecture has been designed so that none of the various cryptographic algorithms used for authentication and key generation needs to be reversible. This feature was introduced deliberately to ease export restrictions on telecommunications products which incorporate these algorithms. Another characteristic of the architecture is that it only addresses security for an access network. For example, encryption is confined to the local loop between the terminal and the base station, with an enhancement in the case of TETRA, whereby end-to-end encryption may be provided between pairs of terminals communicating through the same base station. The main reason for this is that today's mobile telecommunications networks have been designed as access networks to fixed systems, with the majority of mobile originated calls terminating in telephones connected to the public switched telecommunications network (PSTN) or analogue cellular systems. This means that the speech signals undergo a variety of tandem encodings and decodings between the originating and terminating telephones. This makes end-to-end digital encryption impossible, and even simple end-to-end analogue scrambling may be prevented by the use of the source coding techniques discussed in section 2.2.

4.3 A Glance into the Near Future

One of the characteristics of the security architecture is that all of the cryptographic algorithms are symmetric, and public key or asymmetric algorithms play no part at all. With the gradual emergence of third generation mobile systems, like UMTS, the *Universal Mobile Telecommunications System,* and FPLMTS, the *Future Public Land Mobile Telecommunications System,* we may well see architectures based on public key techniques. Indeed, work currently underway in the European Telecommunications Standards Institute (ETSI) and the International Telecommunications Union (ITU) allows for asymmetric as well as symmetric solutions for addressing security in third generation systems. However, it is still far from clear to what extent public key techniques offer benefits for mobile communications.

One step on the road to the third generation of mobile systems will be the appearance of mobile satellite systems. Over the next five years or so we will see the opening of mobile satellite services, like the Globalstar, Iridium and Inmarsat Project 21 (I-CO) systems, and users and operators of such services will expect levels of security at least as good as those on terrestrial mobile networks like GSM. Will the security architecture be flexible enough to accommodate such systems? For some systems where the satellites operate as transponders, often called bent pipe architectures, it appears that the answer is yes, with some minor enhancements, because such systems are still fundamentally radio access systems to fixed networks - mobile local loops. However, for systems where the satellites play a more active role and process the signals in a way analogous to a terrestrial base station or even a switch, then the situation is not so clear. Can a `local loop' security architecture address the security requirements for such systems? This is an area where further research is needed.

In the long term it is likely that public telecommunications networks will evolve so as to make it possible to provide users with end-to-end digital speech encryption. If users demand such a service and, as is likely to be the case, they insist that it should be provided to them transparently, then network operators who offer the service will have to address the following problem. How do they reconcile the need to provide users with the means to exchange cryptographic keys in a secure, and preferably transparent way, whilst at the same time making certain that they do not prevent legal interception of communications under warrant? One unsuccessful attempt to address the problem was the US government's Clipper initiative [21], a proposal that appears to have floundered because it failed to satisfy a number of important requirements. Other suggestions, which make use of trusted third parties and public key techniques, may be found in [14]. There is no doubt that this is an important problem which deserves proper attention.

5 Summary

We have described the principles which underly techniques for the efficient coding of speech and image signals. The success of a particular method depends on the fidelity of the reconstructed signals as perceived by a human being. It has been shown how the subjective nature of perception can be incorporated into the encoding process to reduce signal redundancy and achieve a high level of compression. The properties of human hearing and vision can also be exploited to help combat the adverse effects of transmission errors, which eases the requirements for forward error control. The digitally encoded signals are amenable to being secured against eavesdropping and fraudulent intervention, and a number of suitable cryptographic techniques for this purpose have been discussed.

6 References

1. Atkinson I. A., Kondoz A.M., Evans B.G.: 1.6 kbit/s LP vocoder using time envelope. IEE Electronics Letters **31** No.7 (30 March 1995) pp.517-518
2. Beaumont J.M.: Advances in block based fractal coding of still images. IEE Colloquium "Application of fractal techniques in image processing", Digest **1990/171** (3 December 1990)
3. Beker H.J., Piper F.: Secure speech communications. Academic Press (1985)
4. Cattermole K.W.: Principles of PCM. Iliffe (1969)
5. C-Cube Microsystems: CL550 JPEG Image Compression master Processor. Preliminary data Book (Novemer 1990)
6. Chater-Lea, D.: Security in PMR systems. IEE Colloquium "Security in Networks" Digest **1995/024** (February 1995)
7. Cheetham B.M., Wong W.T.K.: Adaptive LSP filter for sub-band coding of speech. IEE Colloquium "Digital signal processing" Digest **1987/14** (26 January 1987)

8. Cheng N.-T., Kingsbury N.G.: The ERPC - an efficient error resilient technique for encoding positional information or sparse data. IEEE Trans **COM-40** No.1 (January 1992) pp.140-148
9. Clarke R.J.: Relation between Karhunen-Loeve and cosine transforms. IEE Proc-F **128** No.6 (November 1981) pp.356-360
10. Flanagan J.L.: Speech Analysis, synthesis and perception. Springer-Verlag (1972)
11. Huang F., Stansfield E.V.: Time sample speech scrambler which does not require synchronisation. IEEE Trans **COM-41** No.11 (November 1993) pp.1715-1722
12. Jacquin A.E.: A novel fractal block coding technique for digital images. IEEE ICASSP (1990) pp.2225-8
13. Jayant N.S., Noll P.: Digital coding of waveforms. Prentice Hall (1984)
14. Jefferies, N., Mitchell, C. and Walker, M.: Proposed architecture for trusted third party services. Proc. Cryptography : policy and algorithms conference, Brisbane (July 1995)
15. Kleijn W.B.: Continuous representations in linear prediction. IEEE ICASSP (1991) pp.201-4
16. Kondoz A.M.: Digital speech. Wiley (1994)
17. Makhoul J.: Linear prediction : a tutorial review. Proc.IEEE **63** No.4 (1975) pp.561-580
18. Malvar H.S.: Signal Processing with lapped transforms. Artech House (1992)
19. Markel J.D., Gray A.H.: Linear prediction of speech. Springer-Verlag (1975)
20. Mouly, M. and Pautet, Marie-Bernadette: The GSM system for mobile communications. (1992)
21. NIST, FIPS Publication 185: Escrowed Encryption Standard. (February 1994)
22. Rabiner L.R., Schafer R.W.: Digital processing of speech signals. Prentice Hall (1978)
23. Shelswell P.: The COFDM modulation system : the heart of digital audio broadcast. IEE ECEJ (June 1995) pp.127-136
24. Stansfield E.V., Harmer D., Kerrigan M.F.: Speech processing techniques for HF radio security. IEE Proc-I **136** No.1 (February 1989) pp.25-46
25. Steele R.: Delta modulation systems. Pentech (1975)
26. Suen A.-N., Wang J.-F., Yao T.-C.: Dynamic partial search scheme for stochastic codebook of FS1016 CELP coder. IEE Proc.Vis.Image Signal Processing **142** No.1 (February 1995) pp.52-58
27. Walker, M.: Security in mobile and cordless telecommunications. IEEE Computer Society (1992)

7 Acknowledgement

The authors would like to acknowledge the assistance of their colleagues during the writing of this paper, and the managements of their respective companies for permission to publish it.

Some Constructions of Generalised Concatenated Codes Based on Unit Memory Codes

Victor Zyablov[1], Sergo Shavgulidze[2] and Jorn Justesen[3]

[1] Institute for Problems of Information Transmission, Russia Academy of Sciences
[2] Department of Digital Communication Theory, Georgian Technical University
[3] Institute of Telecommunication, Technical University of Denmark

Abstract. Various schemes of concatenated coding on the basis of unit memory codes are considered. Constructive lower bounds for the free distance and also for extended row, column and segment distances are obtained. For fixed rates of concatenated codes the reasonable range of rates of inner codes are given. Characteristics of interleaved codes are investigated. It is shown, that the generalised concatenated codes ensure better distance bounds than the ordinary concatenated codes.

1   Introduction

The idea of combining two or more stages of coding in error correction system was introduced to circumvent the difficulties of constructing good codes and efficient decoding methods [1, 2, 3, 4]. Concatenated codes with inner convolutional code and outer Reed-Solomon (RS) code appeared to provide the best performance for implementable systems, and this combination remains popular for channels with a significant error probability [5]. However, the analysis of the distance properties is more complicated, and the convolutional code must be characterized by several sequences of minimum distances [3, 6].

In this paper we extend the analysis of concatenated codes to codes with convolutional outer codes over large alphabets. Such convolutional codes have been recently constructed from RS block codes [7, 8, 9], and they may be decoded by extensions of the algebraic decoding procedures for RS codes. Thus it is interesting to study the improvements in performance and distance that may be obtained by combining these outer codes with inner block and convolutional codes.

We consider four schemes of concatenated coding (CC) on the basis of unit memory (UM) code (Fig.1).

The first scheme CC1 contains an inner block code.

Information sequence is written as a semi-infinite sequence of binary k_i by k_o matrices. Each column with length k_i is considered as a q-ary symbol of $GF(q)$, $q = 2^{k_i}$. As the result, the semi-infinite sequence of q-ary symbols is obtained. This sequence is encoded by an outer q-ary unit memory code with parameters (n_o, k_o, q). Then each symbol of the outer code is considered as a binary vector with length k_i over $GF(2)$ and it is encoded by an inner $(n_i, k_i, 2)$ block code.

Information sequence

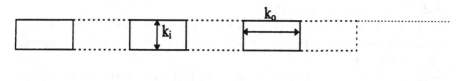

Code sequence of outer code

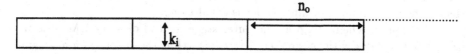

Code sequence of concatenated code 1

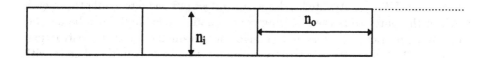

Code sequence of concatenated code 2

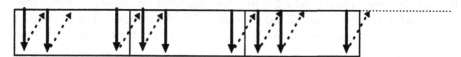

Code sequence of concatenated code 3

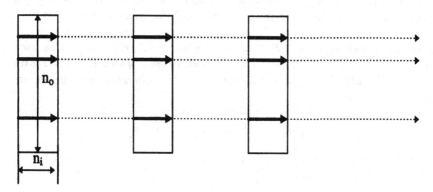

Fig. 1.

As the result of encoding we have the semi-infinite output sequence of a binary UM concatenated code with parameters $(n_c = n_i n_o,\; k_c = k_i k_o,\; 2)$. The rate of the code $R_c = (k_i k_o)/(n_i n_o) = R_i R_o$.

In the second scheme CC2, instead of an inner block code we use an inner $(n_i, k_i, 2)$ UM code and a sequence of symbols of the outer code is an information sequence of the inner code.

In the third scheme CC3 the symbols, that are located on the same positions in the blocks of the outer code sequence are combined into subsequences. Then these subsequences are encoded independently by the inner $(n_i, k_i, 2)$ UM codes. Hence, the total number of inner codes is equal to n_o.

In the fourth scheme CC4 at the outer stage instead of q-ary UM code we use q-ary RS codes with parameters (n_o, k_o, q). Thus the blocks of q-ary symbols are encoded independently with each other. Inner encoding is carried out analogically to CC3.

It is obvious, that as the result of encoding CC2, CC3 and CC4 give the semi-infinite sequence of binary concatenated code with the same parameters as CC1. It should be also noted, that these constructions have different memories. CC1 and CC4 are UM concatenated codes. In CC2 if we put one zero symbol after each block of the outer code the resulting concatenated code would also be an UM code. Asymptotically it does not change the parameters of this construction. As far as CC3 is concerned, strictly speaking, this is convolutional code with memory two.

In this paper we investigate the distance properties of the above given constructions. We obtain the constructive lower bounds for the free distance and also for extended row, column and segment distances. Then we investigate the characteristics of interleaved codes. Finally we show that the generalised concatenation improves the distance characteristics in comparison with the concatenated case.

2 Lower Bounds on Distances

Below we investigate the lower bounds on distances for concatenated codes given in Fig.1 on the basis of corresponding lower bounds for inner and outer codes.

Theorem 1. *For every R, $0 < R < 1$, and every $\varepsilon > 0$, we can find a T_0 such that if $T \geq T_0$ for $N \to \infty$ there exist non-catastrophic q-ary UM codes with block length N, rate R and period T, for which the following inequalities are satisfied*

$$\frac{d^r(l)}{N} \geq (l+1)H_q^{-1}(1 - \frac{l}{l+1}R) - \varepsilon, \qquad l = 1, 2, \ldots, T,$$

$$\frac{d^c(l)}{N} \geq lH_q^{-1}(1 - R) - \varepsilon, \qquad l = 1, 2, \ldots, T,$$

$$\frac{d^s(l)}{N} \geq lH_q^{-1}(1 - \frac{l+1}{l}R) - \varepsilon, \qquad l = 1, 2, \ldots, T,$$

$$\frac{d^f}{N} \geq \min_{l=1,2,\ldots} \{d^r(l)\},$$

where $d^r(l)$, $d^c(l)$, $d^s(l)$ and d^f are extended row distance, extended column distance, extended segment distance and free distance of the UM code respectively.

The proof of Theorem 1 is given in [10].

Corollary 2. For every R, $0 < R < 1$, and every $\varepsilon > 0$, we can find a T_0 such that if $T \geq T_0$ for $N \to \infty$ there exist non-catastrophic UM codes with block length N, rate R, period T and alphabet $q \to \infty$, for which the following inequalities are satisfied

$$\frac{d^r(l)}{N} \geq (l+1)(1 - \frac{l}{l+1}R) - \varepsilon, \qquad l = 1, 2, \ldots, T,$$

$$\frac{d^c(l)}{N} \geq l(1 - R) - \varepsilon, \qquad l = 1, 2, \ldots, T,$$

$$\frac{d^s(l)}{N} \geq l(1 - \frac{l+1}{l}R) - \varepsilon, \qquad l = 1, 2, \ldots, T,$$

$$\frac{d^f}{N} \geq 2 - R - \varepsilon.$$

It should be noted that among the known UM codes over large alphabets [7, 8, 9], only codes from [9] have the optimal value of d^f and only at rates $R \leq 1/3$. For all these codes the extended distances $d^r(l)$, $d^c(l)$ and $d^s(l)$ increase slowly with l as it is guaranteed by Theorem 1.

Taking into consideration that CC1, CC2 and CC4 are UM concatenated codes, we normalize the distances by quantity $n_c = n_i n_o$.

For a given rate of concatenated code - R_c we denote the lower bounds of normalized distances as

$$D_c^r(l, R_c) \leq \frac{d_c^r(l)}{n_i n_o}, \quad D_c^c(l, R_c) \leq \frac{d_c^c(l)}{n_i n_o},$$

$$D_c^s(l, R_c) \leq \frac{d_c^s(l)}{n_i n_o}, \quad D_c^f(R_c) \leq \frac{d_c^f}{n_i n_o},$$

where $d_c^{(*)}(l)$ and d_c^f denote the extended distance and free distance of the concatenated code respectively.

As the inner codes in CC1 we chose block codes that meet Gilbert-Varshamov bound

$$D_i^{GV}(R_i) = \frac{d_i}{n_i} = H_2^{-1}(1 - R_i).$$

As an outer code we chose UM code that meets the lower distance bounds of Theorem 1

$$D_o^r(l, R_o) \leq \frac{d_o^r(l)}{n_o}, \quad D_o^c(l, R_o) \leq \frac{d_o^c(l)}{n_o}, \quad D_o^s(l, R_o) \leq \frac{d_o^s(l)}{n_o}.$$

Then taking into account that $R_o = R_c/R_i$, we have

Theorem 3. *For CC1 lower bounds on extended row, column and segment distances for every $\varepsilon > 0$ satisfy the equations*

$$D_c^r(l, R_c) \geq D_i^{GV}(R_i) D_o^r(l, \frac{R_c}{R_i}) - \varepsilon, \tag{1}$$

$$D_c^r(l, R_c) \geq D_i^{GV}(R_i) D_o^c(l, \frac{R_c}{R_i}) - \varepsilon,$$

$$D_c^s(l, R_c) \geq D_i^{GV}(R_i) D_o^s(l, \frac{R_c}{R_i}) - \varepsilon.$$

In CC2 we chose inner and outer UM codes which meet the lower distance bounds of Theorem 1.

Theorem 4. *For CC2 lower bounds on extended row, column and segment distances for every $\varepsilon > 0$ satisfy the equations*

$$D_c^r(l, R_c) \geq D_i^{GV}(R_i) D_o^r(l, \frac{R_c}{R_i}) - \varepsilon, \tag{2}$$

$$D_c^c(l, R_c) \geq D_i^{GV}(R_i) D_o^c(l, \frac{R_c}{R_i}) - \varepsilon,$$

$$D_c^r(l, R_c) \geq D_i^{GV}(R_i) D_o^s(l, \frac{R_c}{R_i}) - \varepsilon.$$

The proof of Theorem 4 is given in [10].
It can be easily proved, that

Theorem 5. *For CC4 lower bounds on extended row, column and segment distances for every $\varepsilon > 0$ satisfy the equations*

$$D_c^r(l, R_c) \geq D_i^r(l, R_i)(1 - \frac{R_c}{R_i} + \frac{1}{n_c/n_i}) - \varepsilon, \tag{3}$$

$$D_c^c(l, R_c) \geq D_i^c(l, R_i)(1 - \frac{R_c}{R_i} + \frac{1}{n_c/n_i}) - \varepsilon,$$

$$D_c^s(l, R_c) \geq D_i^s(l, R_i)(1 - \frac{R_c}{R_i} + \frac{1}{n_c/n_i}) - \varepsilon.$$

Taking into consideration that CC3 is convolutional code with memory two, we normalize the extended distances by quantity $n_c = n_i n_o$ (the length of output block of concatenated code encoder) and the free distance by quantity $2n_i n_o$ (constraint length).

For a given rate of concatenated code - R_c we denote the lower bounds of normalized distances as

$$D_c^r(l, R_c) \leq \frac{d_c^r(l)}{n_i n_o}, \quad D_c^c(l, R_c) \leq \frac{d_c^c(l)}{n_i n_o},$$

$$D_c^s(l, R_c) \leq \frac{d_c^s(l)}{n_i n_o}, \quad D_c^f(R_c) \leq \frac{d_c^f}{2n_i n_o},$$

where $d_c^{(*)}(l)$ and d_c^f denote the extended distance and free distance of the concatenated code respectively.

In CC3 we chose inner and outer UM codes which meet the lower distance bounds of Theorem 1. We denote

$$Y^r = \left\lfloor \frac{D_o^r(l, \frac{R_c}{R_i})}{D_o^c(1, \frac{R_c}{R_i})} \right\rfloor, \qquad Y^c = \left\lfloor \frac{D_o^c(l, \frac{R_c}{R_i})}{D_o^c(1, \frac{R_c}{R_i})} \right\rfloor \qquad Y^s = \left\lfloor \frac{D_o^s(l, \frac{R_c}{R_i})}{D_o^c(1, \frac{R_c}{R_i})} \right\rfloor,$$

$$X^r = \frac{D_o^r(l, \frac{R_c}{R_i})}{D_o^c(1, \frac{R_c}{R_i})} - Y^r + 1,$$

$$X^c = \frac{D_o^c(l, \frac{R_c}{R_i})}{D_o^c(1, \frac{R_c}{R_i})} - Y^c + 1,$$

$$X^s = \frac{D_o^s(l, \frac{R_c}{R_i})}{D_o^c(1, \frac{R_c}{R_i})} - Y^s + 1.$$

Theorem 6. *For CC3 lower bounds on extended row, column and segment distances for every $\varepsilon > 0$ satisfy the equations*

$$D_c^r(l, R_c) \geq \{D_i^r(Y^r, R_i)X^r + D_i^r(Y^r - 1, R_i)(1 - X^r)\}D_o^c(1, \frac{R_c}{R_i}) - \varepsilon, \quad (4)$$

$$D_c^c(l, R_c) \geq \{D_i^c(Y^c, R_i)X^c + D_i^c(Y^c - 1, R_i)(1 - X^c)\}D_o^c(1, \frac{R_c}{R_i}) - \varepsilon,$$

$$D_c^s(l, R_c) \geq \{D_i^s(Y^s, R_i)X^s + D_i^s(Y^s - 1, R_i)(1 - X^s)\}D_o^c(1, \frac{R_c}{R_i}) - \varepsilon.$$

The proof of Theorem 6 is given in [10].

Later on we consider the asymptotic case when $n_i \to \infty$, $n_o \to \infty$ and thus $n_c \to \infty$.

By means of optimum inner and outer codes it is easy to obtain the lower bounds on free distance.

Theorem 7. *Lower bound on free distance of CC1 and CC2 is defined as*

$$D_c^f(R_c) \geq \max_{R_i} D_i^{GV}(R_i)(2 - \frac{R_c}{R_i}). \tag{5}$$

Theorem 8. *Lower bound on free distance of CC4 is defined as*

$$D_c^f(R_c) \geq \max_{R_i} D_i^{TJ}(R_i)(1 - \frac{R_c}{R_i}). \tag{6}$$

where $D_i^{TJ}(R_i)$ is Thommesen-Justesen bound for binary UM codes given by Theorem 1.

Theorem 9. *Lower bound on free distance of CC3 is defined as*

$$D_c^f(R_c) \geq \max_{R_i} \left\{ \min_l D_c^r(l, R_c) \right\}. \tag{7}$$

where $D_c^r(l, R_c)$ is given by the first formula of (4).

For various rates of concatenated code R_c and different rates of inner code R_i we have calculated the performance of extended row distance as a function of number of blocks $- l$. Taking into account that the characteristics of CC1 and CC2 coincide with each other we consider only CC1 and CC4. Our analysis shows that for a fixed rate R_c it is possible to choose the reasonable range of values of inner code rates $- R_i$. For example, for CC1 when $R_c = 0.5$, the best value of free distance is obtained when $R_i = 0.501$ and the best growth of extended row distance (the best "slope") when $R_i = 0.675$ and the reasonable range of inner code rates lies between $0.501 - 0.675$ (Fig.2). For CC4 (Fig.3) the reasonable range of inner codes rates is situated between $0.675 - 0.824$. The best slope we again have when $R_i = 0.675$, but the best value of free distance is obtained when $R_i = 0.824$. Similar performances we have at other concatenated code rates. We notice some advantage of CC4 over CC1 (or CC2). On the one hand having the same best slope the CC4 gives the better value of free distance and extended row distances when l is small and on the other hand starting from almost the same value of best free distance CC4 yields the better growth of extended row distance with l. As far as CC3 is concerned, due to the memory two, it has the superior extended row distances, than all other constructions. For $R_c = 0.5$ its characteristics are given in Fig.4. It should be noted that for CC3 for all R_c the inner code rate that maximizes the free distance gives the growth of extended row distances which is very close to the best growth. For $R_c = 0.5$, the best value of free distance is obtained when $R_i = 0.654$ and the best slope is given when inner code rate $R_i = 0.675$.

The results of calculation show that the bounds (5) and (6) are practically the same, although for middle rates R_c the bound for CC4 is a bit better. For example, when $R_c = 0.5$ for CC1 and CC2 $D_c^f(R_c) = 0.1100$ and for CC4 $D_c^f(R_c) = 0.1112$. In Fig.5 the new bound $D_c^f(R_c)$ (5) and (6) together with the best known bounds for binary block codes - $D^{GV}(R)$ and UM codes - $D^{TJ}(R)$ are given. The new bound for CC3 (7) is inferior than the same bounds for CC1, CC2 and CC4 if we normalize the free distance by quantity $2n_i n_o$ (constraint length). But if we normalize the free distance by $n_c = n_i n_o$ (the length of output block of concatenated code encoder), then the bound for CC3 exceeds the bound for CC4 (Fig.6).

3 Interleaving and Generalised Concatenation

Taking into consideration that the free distance of UM code can not be achieved per one block, it is possible in CC2 and CC3 to use the interleaving and at the same time to preserve the lower bounds on distances which were obtained in the previous section. For the simplicity we investigate only free distance of interleaved concatenated codes.

In CC2 we use the some number of outer codes (see Fig.7). For a given rate of concatenated code - R_c we denote the lower bound of ratio as

$$D_c^f(l_i, R_c) \leq \frac{d_c^f}{n_i l_i n_o},$$

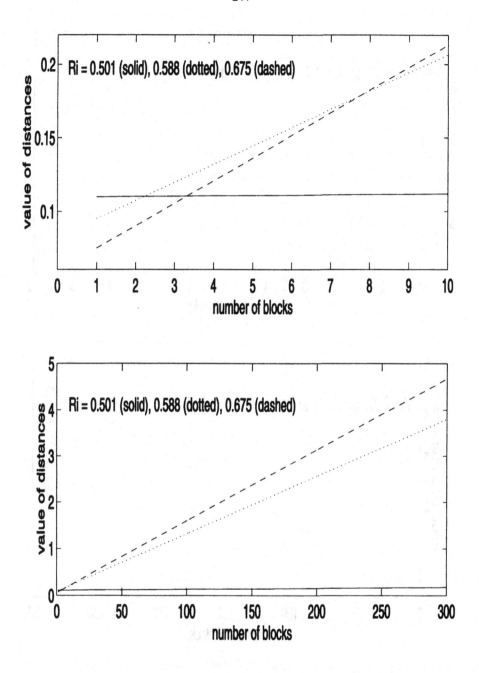

Fig. 2.

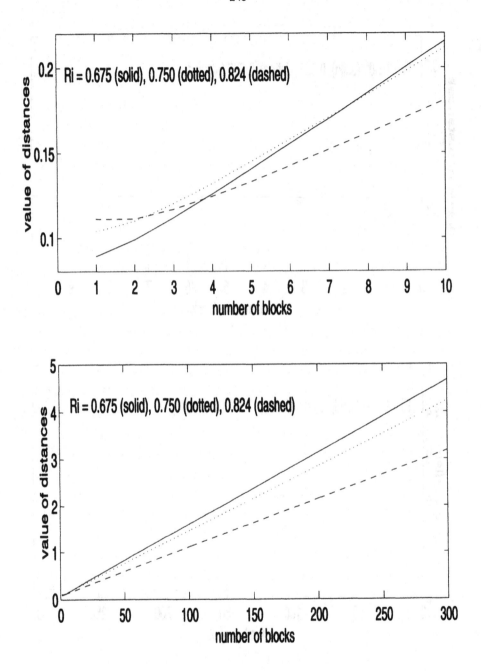

Fig. 3.

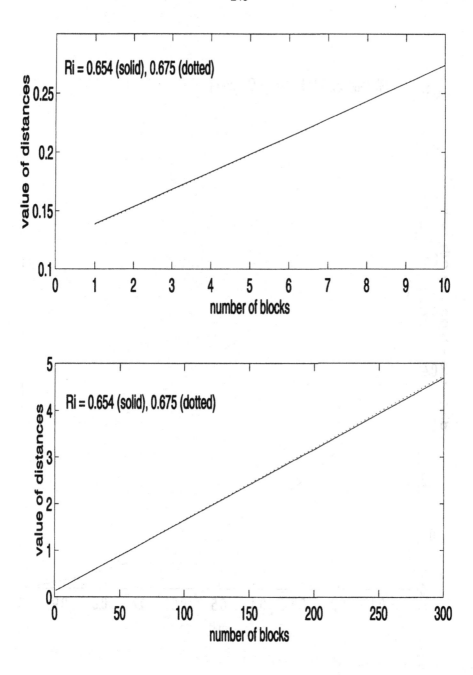

Fig. 4.

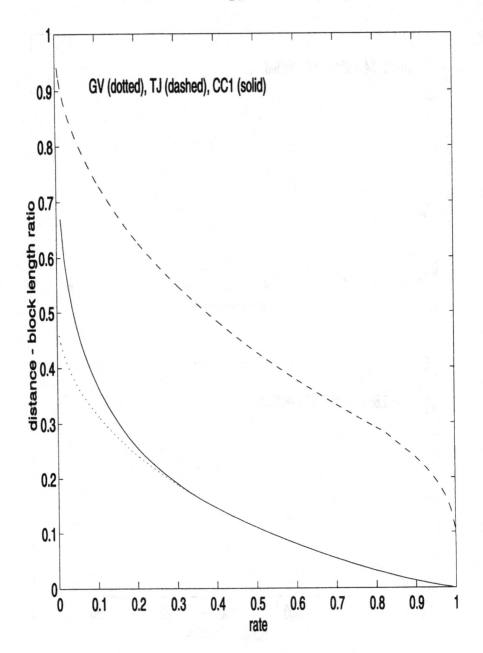

GV (dotted), TJ (dashed), CC1 (solid)

Fig. 5.

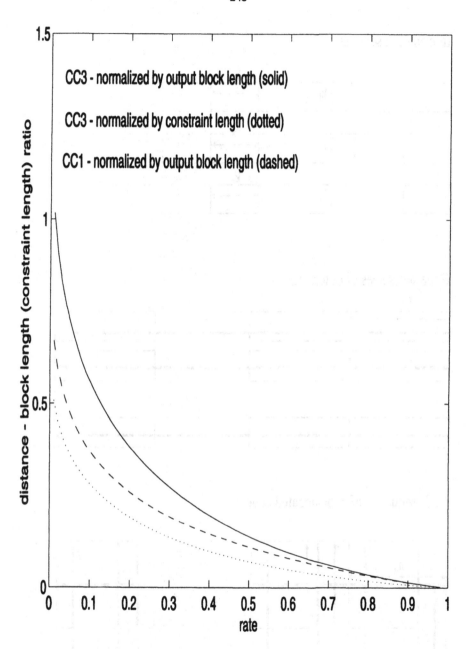

Fig. 6.

Information sequence

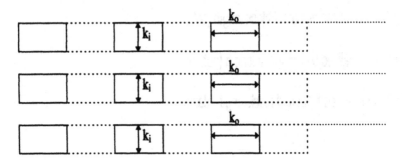

Code sequences of outer codes

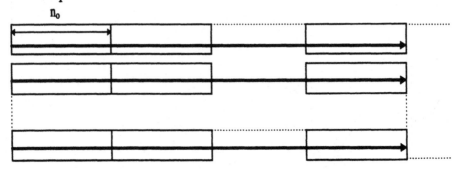

Code sequence of concatenated code 2

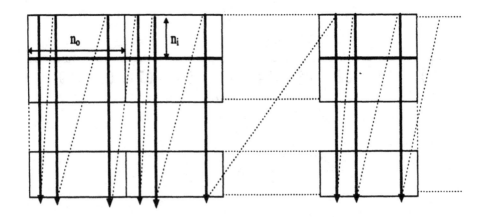

Fig. 7.

where d_c^f denotes the free distance of interleaved CC2, l_i is a number of blocks of inner code in one column in Fig.7, that is the number of outer codes. It is easily shown, that

Theorem 10. *For interleaved version of CC2 lower bound on free distance satisfies the equation*

$$D_c^f(l_i, R_c) \geq \min\{A, B\},$$

where

$$A = \max_{R_i}\left\{D_i^{GV}(R_i)(2 - \frac{R_c}{R_i})\right\},$$

$$B = (1/l_i)\max_{R_i}\left\{D_i^{TJ}(R_i)(2 - \frac{R_c}{R_i})\right\}.$$

In CC3 we use many inner codes (see Fig.8). For a given rate of concatenated code - R_c we denote the lower bound of ratio as

$$D_c^f(l_o, R_c) \leq \frac{d_c^f}{n_i n_o l_o},$$

where d_c^f denotes the free distance of interleaved CC3, l_o is a number of blocks of outer code in one row in Fig.8. Thus the number of inner codes equal to $n_o l_o$.
We denote

$$Y_a^r = \left\lfloor \frac{D_o^r(l, \frac{R_c}{R_i})}{D_o^c(l_o, \frac{R_c}{R_i})} \right\rfloor, \qquad X_a^r = \frac{D_o^r(l, \frac{R_c}{R_i})}{D_o^c(l_o, \frac{R_c}{R_i})} - Y_a^r + 1,$$

It is easily shown, that

Theorem 11. *For interleaved version of CC3 lower bound on free distance satisfies the equation*

$$D_c^f(l_o, R_c) \geq \min\{A, B\},$$

where

$$A = \max_{R_i}\left\{[D_i^r(Y_a^r, R_i)X_a^r + D_i^r(Y_a^r - 1, R_i)(1 - X_a^r)](1 - \frac{R_c}{R_i})\right\},$$

$$B = (1/l_o)\max_{R_i}\left\{D_i^{TJ}(R_i)(2 - \frac{R_c}{R_i})\right\}.$$

It is an interesting question when the free distance of concatenated code is a product of free distances of inner and outer codes.
For CC2 It follows from Theorem 10 that this happens at smallest integer $l_i = l_i'$ for which the following condition holds

$$l_i D_i^{GV}(R_i) \geq D_i^{TJ}(R_i).$$

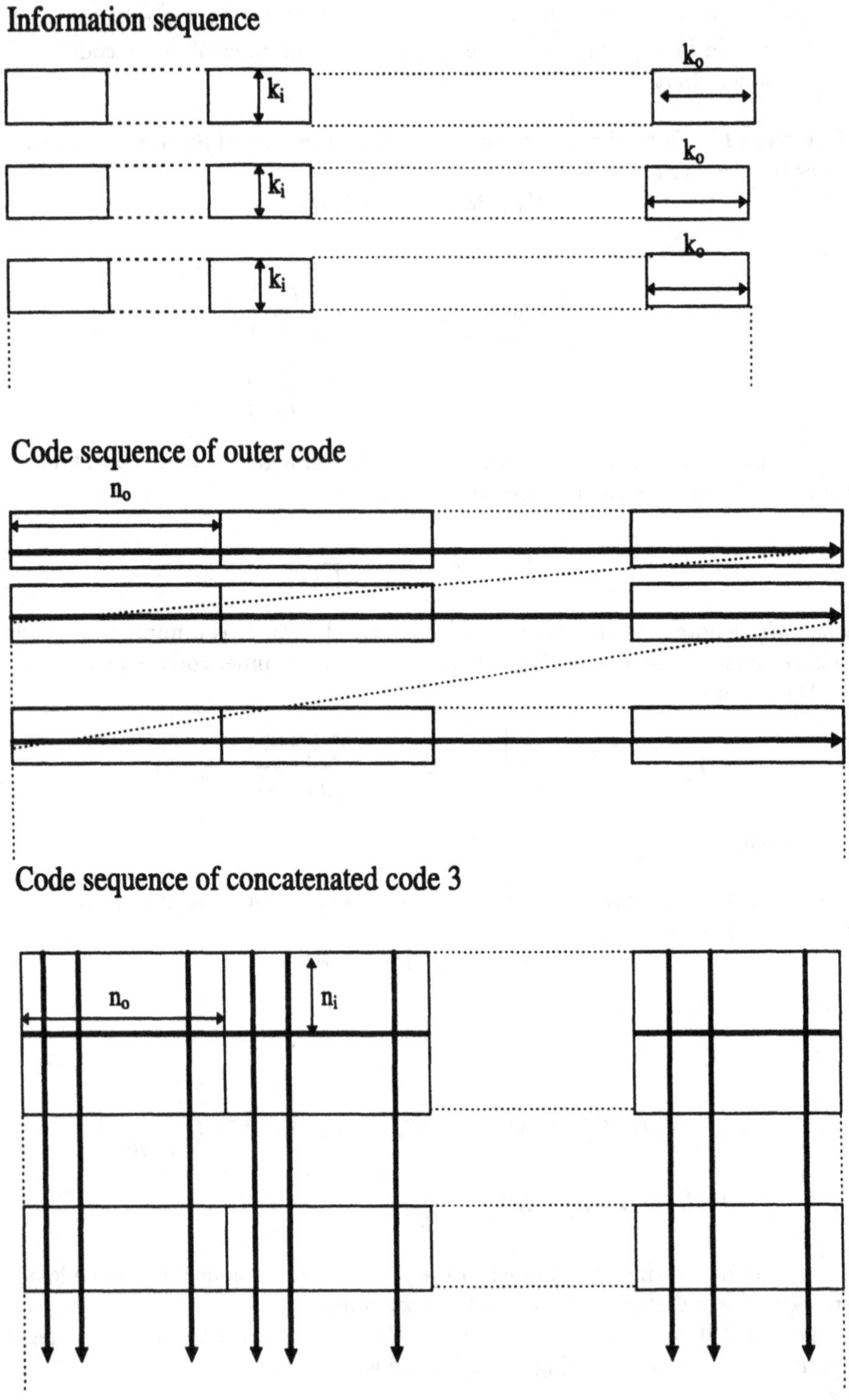

Information sequence

Code sequence of outer code

Code sequence of concatenated code 3

Fig. 8.

Thus, we can increase the number of outer codes and at the same time preserve the growth of weight of CC2 until $l_i < l'_i$.

Similarly for CC3 it follows from Theorem 11 that the same situation arises at smallest integer $l_o = l'_o$ for which the following condition holds

$$l_o(1 - \frac{R_c}{R_i}) \geq (2 - \frac{R_c}{R_i}).$$

Thus, we can increase the number of inner codes and at the same time preserve the growth of weight of CC3 until $l_o < l'_o$.

Hence, starting from the above given values of $l_i \geq l'_i$ (or $l_o \geq l'_o$) the free distance of CC2 (or CC3) equals

$$d_c^f = d_i^f d_o^f,$$

where d_i^f and d_o^f are free distances of inner and outer codes respectively.

In Fig.9 for concatenated code rate $R_c = 0.5$ we give the performance of $D_c^f(l_i, R_c)$ as a function of l_i (CC2) and $D_c^f(l_o, R_c)$ as a function of l_o (CC3).

Our final topic is the generalised concatenated coding (GCC). For the sake of simplicity we investigate the case without interleaving.

We consider the constructions with inner nested binary block (GCC1) or binary UM (GCC2, GCC3, GCC4) codes. The rate of j-th inner subcode $R_{i,j} = R_{i,1}(m-j+1)/m$, $j = 1, 2 \ldots m$, where $R_{i,1}$ is the mother code rate, and m is the order of GCC. Then the rate of GCC code $R_c = \frac{R_{i,1}}{m} \sum_{j=1}^{m} R_{o,j}$, where $R_{o,j}$ is the rate of j-th outer code. We use such inner codes, where mother code and all subcodes simultaneously meet the best known lower bounds for distances. The existence of such block codes is proved in [2], and the existence of binary UM codes with subcodes for which the bounds of Theorem 1 are satisfied is proved in [4]. We dwell our attention to the case when all information bits in GCC are equally protected. Then we have

Theorem 12. *Lower bound on free distance of GCC1 and GCC2 is defined as*

$$D_c^f(R_c) \geq \max_{R_{i,1}} \frac{2 - \frac{R_c}{R_{i,1}}}{\frac{1}{m} \sum_{j=1}^{m} 1/D_i^{GV}(R_{i,j})}. \tag{8}$$

Theorem 13. *Lower bound on free distance of GCC4 is defined as*

$$D_c^f(R_c) \geq \max_{R_{i,1}} \frac{1 - \frac{R_c}{R_{i,1}}}{\frac{1}{m} \sum_{j=1}^{m} 1/D_i^{TJ}(R_{i,j})}. \tag{9}$$

Theorem 14. *Lower bound on free distance of GCC3 is defined as*

$$D_c^f(R_c) \geq \max_{R_{i,1}} \left\{ \min_l D_c^r(l, R_c) \right\}, \tag{10}$$

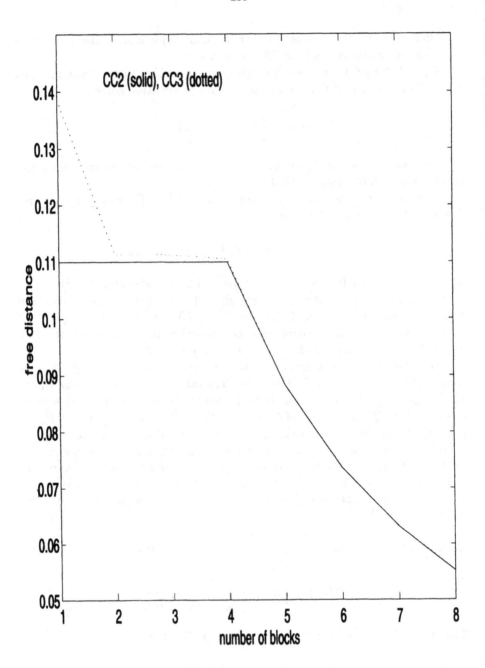

Fig. 9.

254

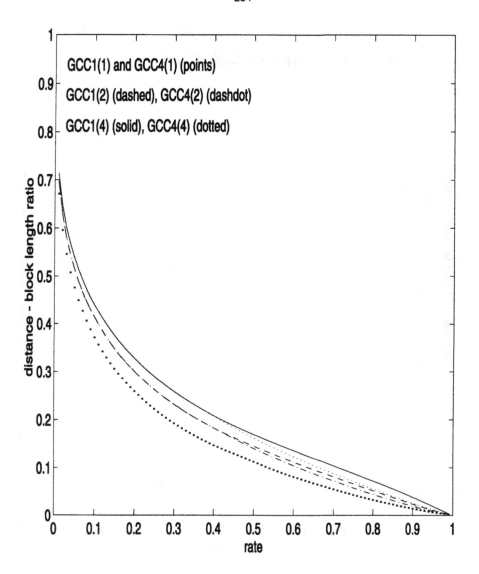

Fig. 10.

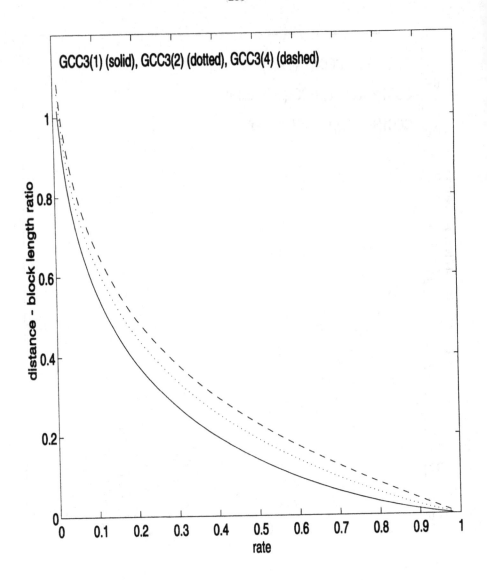

GCC3(1) (solid), GCC3(2) (dotted), GCC3(4) (dashed)

Fig. 11.

where

$$D_c^r(l, R_c) = \frac{D_o^c(1, \frac{R_c}{R_{i,1}})}{\frac{1}{m} \sum_{j=1}^m 1/(D_i^r(Y^r, R_{i,j})X^r + D_i^r(Y^r - 1, R_{i,j})(1 - X^r))},$$

$$Y^r = \left\lfloor \frac{D_o^r(l, \frac{R_c}{R_{i,1}})}{D_o^c(1, \frac{R_c}{R_{i,1}})} \right\rfloor, \qquad X^r = \frac{D_o^r(l, \frac{R_c}{R_{i,1}})}{D_o^c(1, \frac{R_c}{R_{i,1}})} - Y^r + 1.$$

In Fig.10 and Fig.11 we give free distance bounds (8), (9) and (10) for GCC1(m), GCC4(m) and GCC3(m) when the order of generalised concatenation $m = 1, 2$ and 4. The results of calculation show that the distance bounds become better when m increases, thus the generalised concatenation ameliorates the characteristics of codes. It should be also noted that starting from the middle rates GCC1(m) outperforms GCC4(m) when $m > 1$.

References

1. G.D.Forney, Jr., Concatenated Codes. Cambridge, MA: MIT, 1966.
2. E.L.Blokh and V.V.Zyablov, Linear Concatenated Codes. Moscow, Nauka, 1982.
3. J.Justesen, C.Thommesen and V.V.Zyablov, "Concatenated codes with inner convolutional code," IEEE Trans. Inform. Theory, vol. IT-34, pp.1217-1225, Septem. 1988.
4. V.V.Zyablov and S.A. Shavgulidze, Generalised Concatenated Constructions on a Base of Convolutional Codes. Moscow, Nauka, 1991.
5. Consultative Committee for Space Data Systems, "Recommendation for Space Data Systems Standard, Telemetry Channel Coding," CCSDS 101.0-B-2, Blue Book, Issue 2, Jan.1987.
6. C.Thommesen and J. Justesen, "Bounds on distances and error exponents of unit memory codes," IEEE Trans. Inform. Theory, vol. IT-29, pp.637-649, Septem. 1983.
7. J.Justesen, "Bounded distance decoding of unit memory codes,". IEEE Trans. Inform. Theory, vol. IT-25, pp. 240-243, Mar. 1993.
8. U.Sorger, "A new construction of partial unit memory codes based on Reed-Solomon codes," Probl. Peredach. Inform., vol. 30, pp. 16-20, 1994.
9. V.Zyablov and V.Sidorenko, "On periodic (partial) unit memory codes with maximum free distance," Lecture Notes in Computer Science, vol. 29, Springer-Verlag, Berlin, pp. 74-79, 1994.
10. J.Justesen, C.Thommesen, S.Shavgulidze and V.Zyablov, Bounds on Distances of Convolutional Concatenated Codes Based on Unit Memory Codes. Tech. Rep. Institute of Telecommunication IT - 158, Technical University of Denmark, Lyngby, 1995.

A Note on the Hash Function of Tillich and Zémor

Willi Geiselmann

Department of Computer Science, Royal Holloway, University of London, Egham, Surrey TW20 0EX, U.K.

Abstract. The hash function based on the group $SL_2(\mathbf{F}_{2^n})$ [4] is studied by embedding the generators of $SL_2(\mathbf{F}_{2^n})$ into finite fields. Using this embeddings, clashing sequences can be found by calculationg discrete logarithms in the field $\mathbf{F}_{2^n}$.

At Crypto 1994 Tillich and Zémor have proposed the following hash function [4], based on the group $SL_2(\mathbf{F}_{2^n})$:

With $\mathbf{F}_{2^n}$ we denote the field with 2^n elements. Let $A := \begin{pmatrix} \alpha & 1 \\ 1 & 0 \end{pmatrix}$ and $B := \begin{pmatrix} \alpha & \alpha + 1 \\ 1 & 1 \end{pmatrix}$ be elements of $SL_2(\mathbf{F}_{2^n})$, the group of 2×2-matrices with determinant 1 over $\mathbf{F}_{2^n}$. And $\alpha \in \mathbf{F}_{2^n}$ denotes the zero of a generating polynomial $p(x)$ of the field $\mathbf{F}_{2^n} \simeq \mathbf{F}_2[x]/p(x)$.

The hash value $h \in \mathbf{F}_{2^n}^{2 \times 2}$ of a stream $a_0 \ldots a_m$ of 0 and 1 is calculated as the product $M_0 \cdot \ldots \cdot M_m$ with $M_i := \begin{cases} A \text{ if } a_i = 0 \\ B \text{ if } a_i = 1. \end{cases}$

At ASIACRYPT '94 a first "attack" on this hash function was proposed [1]. It is based on a very bad choice of the defining polynomial $p(x)$ of $\mathbf{F}_{2^n}$ and works only if the order of A respectively B is very small. Not to choose such a bad representation of $\mathbf{F}_{2^n}$ is only implicitly mentioned in the suggestion of Tillich and Zémor.

Here we present a method that works for any choice of the representation of the field. It is based on an embedding of the matrix rings generated by A and B into finite fields, represented as subrings of $\mathbf{F}_{2^n}^{2 \times 2}$. With this technique finding clashing sequences is (roughly spoken) as hard as calculating discrete logarithms in the fields $\mathbf{F}_{2^n}$ or $\mathbf{F}_{2^{2n}}$. These clashing sequences have the strange structure of very long sequences of zeros and ones and thus can not really be looked at as an attack. Nevertheless this result gives me a bad feeling for the security of this hash function.

1 The Matrix Representation of a Finite Field

In this section some elementary properties of finite fields are summarized. More details can be found in the book of Lìdl and Niederreiter [3] and in [2].

Definition 1. *The* trace *over* $\mathbf{F}_q$ *of an element* $\beta \in \mathbf{F}_{q^n}$ *is given by*

$$tr(\beta) := \sum_{i=0}^{n-1} \beta^{q^i}.$$

The trace function is used to define the dual bases:

Definition 2. *Let* $\mathcal{A} = (\alpha_0, \alpha_1, \ldots, \alpha_{n-1})$ *be a basis of* $\mathbf{F}_{q^n}$ *over* $\mathbf{F}_q$. *Then the dual basis* $(\beta_0, \beta_1, \ldots, \beta_{n-1})$ *to* $\mathcal{A}$ *is defined by* $tr(\alpha_i \cdot \beta_j) = \delta_{i,j}$ *for* $0 \le i, j < n$.

It is known, that for every basis of a finie field there exists a unique dual basis. The concept of dual bases opens a very elegant way to define the matrix representation of a finite field. This way is better fitted to finite fields than the original one, known from representation theory of groups and ends up in the same representation. More details can be found in [2].

Definition 3. *Let* $\mathcal{A} = (\alpha_0, \alpha_1, \ldots, \alpha_{n-1})$ *be a basis of* $\mathbf{F}_{q^n}$ *over* $\mathbf{F}_q$ *and let* $(\beta_0, \beta_1, \ldots, \beta_{n-1})$ *be its dual basis. Then the* matrix representation $I_{\mathcal{A}}$ *with respect to* $\mathcal{A}$ *is given by*

$$I_{\mathcal{A}} : \mathbf{F}_{q^n} \to \mathbf{F}_q^{n \times n}$$
$$\gamma \mapsto \left(tr(\gamma \cdot \alpha_i \cdot \beta_j) \right)_{0 \le i,j \le n-1}.$$

For those who are not familiar with the matrix representation, there are two more intuitive interpretations:

- The matrix $I_{\mathcal{A}}(\gamma)$ is the linear mapping realizing the multiplication by γ in $\mathbf{F}_{q^n}$ with respect to the basis $\mathcal{A}$. With this interpretation, the entries of the matrix can be calculated in a different way, solving a system of linear equations over $\mathbf{F}_q$.
- The ring of the matrices $I_{\mathcal{A}}(\gamma)$ (for $\gamma \in \mathbf{F}_{q^n}$) with the matrix addition and multiplication is (a representation of) the field $\mathbf{F}_{q^n}$.

2 Embeddings

Now we come back to the hash function of Tillich and Zémor. Let

$$A := \begin{pmatrix} \alpha & 1 \\ 1 & 0 \end{pmatrix} \quad \text{and} \quad B := \begin{pmatrix} \alpha & \alpha+1 \\ 1 & 1 \end{pmatrix}$$

be the matrices mentioned above, with α a zero of a defining polynomial of $\mathbf{F}_{2^n}$ over $\mathbf{F}_2$. It is known from [4] that A and B generate $SL_2(\mathbf{F}_{2^n})$. The way to embed the subrings of $SL_2(\mathbf{F}_{2^n})$ generated by A and B in a field contained in $\mathbf{F}_{2^n}^{2 \times 2}$ is based on the following observation:

Remark. The matrix A is a zero of the polynomial $\hat{A}(x) := x^2 + \alpha x + x^0$ and B is a zero of the polynomial $\hat{B}(x) := x^2 + (\alpha+1) x + x^0$.

Following the idea that A (resp. B) and its powers are elements of some finite extension field F_A of $\mathbf{F}_{2^n}$ with $F_A \subset \mathbf{F}_{2^n}^{2 \times 2}$, one first has to find an element fulfilling the same equations as A, i.e. $x^2 + \alpha x + 1 = 0$. In the case "$\hat{A}$ irreducible" the matrix representation of $\mathbf{F}_{(2^n)^2}$ gives the desired embedding.

Without loss of generality we restrict ourselves to $\hat{A}$ for the rest of this section; for $\hat{B}$ a nearly identical calculation gives the similar result.

2.1 The Irreducible Case

Let $\hat{A} = x^2 + \alpha x + 1 \in \mathbf{F}_{2^n}[x]$ be irreducible in this section. Thus it holds: $\mathbf{F}_{(2^n)^2} \simeq \mathbf{F}_{2^n}[x]\big/ \hat{A}$ and the matrix representation $I_\mathcal{B}$ with the basis $\mathcal{B} := (x, 1)$ gives: $I_\mathcal{B}(x) = A$ and $I_\mathcal{B}(1) = I$, the identity matrix. This construction embeds $\mathbf{F}_{(2^n)^2} \simeq \mathbf{F}_{2^n} \cdot A + \mathbf{F}_{2^n} \cdot I$ into the ring of 2×2-matrices.

To examine $SL_2(\mathbf{F}_{2^n})$ we are mostly interested in matrices with determinant 1. There are $2^n + 1$ matrices in $F_1 := SL_2(\mathbf{F}_{2^n}) \cap (\mathbf{F}_{2^n} \cdot A + \mathbf{F}_{2^n} \cdot I)$. (Each $M \in F_1$ defines $(2^n - 1)$ different matrices $\sqrt{\gamma} \cdot M$ with determinant γ for all $\gamma \in \mathbf{F}_{2^n}^*$. Thus separating $(\mathbf{F}_{2^n} \cdot A + \mathbf{F}_{2^n} \cdot I)$ into sets of matrices with the same determinant results in $2^n - 1$ sets of $2^n + 1$ elements and the set of the 0-matrix.) In addition this observation gives an elementary proof for "the order of A divides $2^n + 1$".

This shows the following

Theorem 4. Let $A := \begin{pmatrix} \alpha & 1 \\ 1 & 0 \end{pmatrix} \in SL_2(\mathbf{F}_{2^n})$ and $\hat{A}(x) := x^2 + \alpha x + 1 \in \mathbf{F}_{2^n}$ be irreducible, then a matrix $M \in \mathbf{F}_{2^n}^{2 \times 2}$ is a power of A, iff the following conditions are fulfilled:

- $M = \lambda A + \mu I$ for some $\lambda, \mu \in \mathbf{F}_{2^n}$.
- $\det(M) = 1$ and
- $M^{\mathrm{ord}(A)} = I$.

Let us now look on the more strange case where $\hat{A}$ is reducible.

2.2 The Reducible Case

Let $x^2 + \alpha x + 1$ be reducible, i.e. there exists a zero $\beta \in \mathbf{F}_{2^n}$ of $\hat{A} \in \mathbf{F}_{2^n}[x]$ with

$$(x + \beta)\left(x + \frac{1}{\beta}\right) = x^2 + \alpha x + 1.$$

Remark. Let $I \in \mathbf{F}_{2^n}^{2 \times 2}$ denotes the identity matrix and A be defined as above, then the 2-dimensional $\mathbf{F}_{2^n}$-vector space

$$V := \mathbf{F}_{2^n} \cdot A + \mathbf{F}_{2^n} \cdot I$$

with the matrix multiplication is an abelian $\mathbf{F}_{2^n}$-algebra. (Note: $A^2 = \alpha \cdot A + I$.)

Lemma 5. *Let* $M_1 = \lambda_1 A + \mu_1 I$, $M_2 = \lambda_2 A + \mu_2 I \in V$ *with* $\lambda_1, \lambda_2, \mu_1, \mu_2 \in \mathbf{F}_{2^n}$ *and* $\beta \in \mathbf{F}_{2^n}$ *a zero of* $x^2 + \alpha\, x + 1$. *Then*

$$M_1 \simeq M_2 \;:\Longleftrightarrow\; \lambda_1 \cdot \beta + \mu_1 = \lambda_2 \cdot \beta + \mu_2$$

defines an equivalence relation on the algebra V.

Proof: Checking the conditions for an equivalence relation is trivial: the addition of the algebra is defined "component wise", thus it is directly well defined on the equivalence classes. It remains to check the multiplication to be well defined. The straight forward calculation gives the desired result:
Let $\lambda_{i,1} A + \mu_{i,1} I = M_{i,1} \simeq M_{i,2} = \lambda_{i,2} A + \mu_{i,2} I$ for $i = 1, 2$. Then the following equations hold:

$$\mu_{1,1} = \lambda_{1,1} \cdot \beta + \lambda_{1,2} \cdot \beta + \mu_{1,2} \tag{1}$$
$$\mu_{2,1} = \lambda_{2,1} \cdot \beta + \lambda_{2,2} \cdot \beta + \mu_{2,2} \tag{2}$$
$$0 = \beta^2 + \alpha \cdot \beta + 1 \tag{3}$$

Direct calculation gives for $i = 1, 2$:

$$
\begin{aligned}
M_{1,i} \cdot M_{2,i} &= \lambda_{1,i}\lambda_{2,i} \cdot (\alpha A + I) + (\lambda_{1,i}\mu_{2,i} + \lambda_{2,i}\mu_{1,i}) \cdot A + \mu_{1,i}\mu_{2,i} I \\
&= (\lambda_{1,i}\lambda_{2,i}\alpha + \lambda_{1,i}\mu_{2,i} + \lambda_{2,i}\mu_{1,i}) \cdot A + (\lambda_{1,i}\lambda_{2,i} + \mu_{1,i}\mu_{2,i}) \cdot I
\end{aligned}
$$

Thus we have:
$$M_{1,1} \cdot M_{2,1} \simeq M_{1,2} \cdot M_{2,2}$$

$$\Longleftrightarrow \quad (\lambda_{1,1}\lambda_{2,1}\alpha + \lambda_{1,1}\mu_{2,1} + \lambda_{2,1}\mu_{1,1}) \cdot \beta + \lambda_{1,1}\lambda_{2,1} + \mu_{1,1}\mu_{2,1}$$
$$= (\lambda_{1,2}\lambda_{2,2}\alpha + \lambda_{1,2}\mu_{2,2} + \lambda_{2,2}\mu_{1,2}) \cdot \beta + \lambda_{1,2}\lambda_{2,2} + \mu_{1,2}\mu_{2,2}$$

$$\overset{(1),(2)}{\Longleftrightarrow} \quad \lambda_{1,1}\lambda_{2,1}\alpha\beta + \lambda_{1,1} \cdot (\lambda_{2,1}\beta + \lambda_{2,2}\beta + \mu_{2,2}) \cdot \beta +$$
$$\lambda_{2,1} \cdot (\lambda_{1,1}\beta + \lambda_{1,2}\beta + \mu_{1,2}) \cdot \beta + \lambda_{1,1}\lambda_{2,1} +$$
$$(\lambda_{1,1}\beta + \lambda_{1,2}\beta + \mu_{1,2}) \cdot (\lambda_{2,1}\beta + \lambda_{2,2}\beta + \mu_{2,2})$$
$$= \lambda_{1,2}\lambda_{2,2}\alpha\beta + \lambda_{1,2}\mu_{2,2}\beta + \lambda_{2,2}\mu_{1,2}\beta + \lambda_{1,2}\lambda_{2,2} + \mu_{1,2}\mu_{2,2}$$

$$\Longleftrightarrow \quad (\beta^2 + \alpha\beta + 1) \cdot \lambda_{1,1}\lambda_{2,1} + 2 \cdot \lambda_{1,1}\lambda_{2,1}\beta^2 + 2 \cdot \lambda_{1,1}\lambda_{2,2}\beta^2 +$$
$$2 \cdot \lambda_{1,1}\mu_{2,2}\beta + 2 \cdot \lambda_{2,1}\lambda_{1,2}\beta^2 + 2 \cdot \lambda_{2,1}\mu_{1,2}\beta +$$
$$\lambda_{1,2}\mu_{2,2}\beta + \lambda_{2,2}\mu_{1,2}\beta + \mu_{1,2}\mu_{2,2}$$
$$= \lambda_{1,2}\mu_{2,2}\beta + \lambda_{2,2}\mu_{1,2}\beta + \mu_{1,2}\mu_{2,2} + (\beta^2 + \alpha\beta + 1) \cdot \lambda_{1,2}\lambda_{2,2}.$$

Equation (3) yields: The equivalence relation on the algebra V is well defined and thus finishes the proof. ∎

With these two steps, defining V and $V\!/\!\!\sim$, we have reached our aim to embed A into an extension field of $\mathbf{F}_{2^n}$. This is shown in the following

Lemma 6. *With the notations of above,*

$$\varphi: \quad V\!\!\Big/_{\!\sim} \quad \to \quad \mathbf{F}_{2^n}$$
$$\overline{\lambda A + \mu I} \mapsto \lambda \beta + \mu,$$

is an isomorphism of fields, where $\overline{\lambda A + \mu I}$ denotes the equivalence class of $\lambda A + \mu I$.
For each class $\overline{M_0} \neq \overline{0}$ of $V\!\!\big/_{\!\sim}$ and for each $\gamma \in \mathbf{F}_{2^n}$, there exists exactly one $M \in \overline{M_0}$ with $\det(M) = \gamma$, i.e. there is one $M \in \overline{M_0} \cap \mathrm{SL}_2(\mathbf{F}_{2^n})$.

Proof: By definition of $V\!\!\big/_{\!\sim}$ the mapping φ is an algebra homomorphism. For $\mu \in \mathbf{F}_{2^n}$ the class $\overline{\mu I}$ maps to μ, thus φ is surjective.
By definition of the equivalence classes, φ is injective, and each class contains exactly 2^n matices.
Let $\delta \in \mathbf{F}_{2^n}$ with $\varphi\left(\overline{M_0}\right) = \delta$, then each $M := \lambda A + \mu I \in \overline{M_0}$ fulfils $\mu = \lambda \beta + \delta$ and thus $\det(M) = (\lambda \alpha + \mu) \cdot \mu + \lambda^2 = \lambda^2(\beta^2 + \alpha \beta + 1) + \lambda \alpha \delta + \delta^2 = \lambda \alpha \delta + \delta^2$.
If $\delta \neq 0$, this linear equation for λ gives for each determinant γ exactly one solution. ∎

This Lemma gives us a method to describe the powers of A by simple equations. These are summarized in the following theorem.

Theorem 7. *Let $A := \begin{pmatrix} \alpha & 1 \\ 1 & 0 \end{pmatrix} \in \mathrm{SL}_2(\mathbf{F}_{2^n})$ and suppose the polynomial $\hat{A}(x) :=$ $x^2 + \alpha x + 1 \in \mathbf{F}_{2^n}$ factors into $\hat{A}(x) = (x + \beta)(x + \frac{1}{\beta})$. Then a matrix $M \in \mathbf{F}_{2^n}^{2 \times 2}$ is a power of A iff the following conditions are fulfilled:*

- $M = \lambda A + \mu I$ *for some* $\lambda, \mu \in \mathbf{F}_{2^n}$,
- $\det(M) = 1$ *and*
- $(\lambda \beta + \mu)^{\mathrm{ord}(A)} = 1$.

Proof: Remark 2.2 directly yields: $A^i \in \lambda A + \mu I$ for all $i \in \mathbf{N}$, $\lambda, \mu \in \mathbf{F}_{2^n}$. Condition 2 is obvious and the isomorphism of Lemma 6 gives condition 3.

A matrix M fulfilling the first two conditions is member of $V \cap \mathrm{SL}_2(\mathbf{F}_{2^n})$ (see Lemma 6 and Remark 2.2). The field elements fulfilling condition 3 are the powers of $\varphi(\overline{A})$. ∎

With these embeddings we have found a simple and very efficient way to describe the powers of A and B. Before we see how this result can be used to find clashing sequences, a very small example is introduced:

Example 1 Ground field: $\mathbf{F}_{2^3}$. Let the field $\mathbf{F}_{2^3}$ be represented as

$$\mathbf{F}_{2^3} \equiv \mathbf{F}_2[x]\Big/_{x^3 + x^2 + 1},$$

and $\alpha \in F_{2^3}$ be a zero of $x^3 + x^2 + 1 \in F_2[x]$. Then the matrix $A := \begin{pmatrix} \alpha & 1 \\ 1 & 0 \end{pmatrix}$ is a zero of the *reducible* polynomial $y^2 + \alpha y + 1 \in F_{2^3}[y]$. We have

$$F_{2^3}[y]\big/(y^2 + \alpha y + 1) \simeq$$

$$F_{2^3}[y]\big/(y + \alpha^2 + 1) \times F_{2^3}[y]\big/(y + \alpha^2 + \alpha + 1) \simeq F_{2^3} \times F_{2^3}.$$

this gives the isomorphism φ defined above with $\varphi(A) = \alpha^2 + 1$. The order of A divides 7 (note: $\varphi(A) \in F_{2^3}^*$) and thus is 7. The powers of A are all matrices in V with determinant 1 and form together with the 0-matrix a complete system of 8 representatives of $V\big/_\simeq$.

The matrix $B := \begin{pmatrix} \alpha & \alpha + 1 \\ 1 & 1 \end{pmatrix}$ is the matrix representation of the element y' due to the basis $(y', y' + 1)$ of the field $F_{2^3}[y']\big/(y'^2 + (\alpha + 1) y' + 1) \simeq F_{2^6}$. All the powers of B have determinant 1, they are a multiplicative subgroup of the group of the 9 matrices with determinant 1. Thus the order of B is 3 or 9. Because of $B^3 = \begin{pmatrix} \alpha^2 + \alpha & 1 \\ \alpha^2 & \alpha^2 + \alpha + 1 \end{pmatrix}$ the order of B is 9.

Now we come to the application of this embedding, to calculate a special type of clashing sequences of the hash function of Tillich and Zémor.

3 Clashing Sequences

With the results of the previous section it is a very easy thing to find clashing sequences for the SL_2 hash function:

- Choose for $C \in SL_2(F_{2^n})$ a small length $l (= 3, 4)$ and construct the equation
  $A^{i_1} B^{i_2} \ldots A^{i_l} = (\lambda_1 A + \mu_1 I) \cdot (\lambda_2 B + \mu_2 I) \cdot \ldots \cdot (\lambda_l A + \mu_l I) = C$;
- This gives 4 polynomial equations of degree l. Add the l equations for the determinants, i.e. $\det (\lambda_1 A + \mu_1 I) = 1$ and solve this system of polynomial equations over F_{2^n}. In general there exists solutions for $l = 3$, otherwise choose a different C or add a fourth "undetermined matrix" $(\lambda_4 B + \mu_4 I)$.
- Check the conditions with the orders of A and B. If they are not fulfilled, try again, otherwise calculate the discrete logarithms of the matrices calculated (regarded as field elements with Lemma 6 or the matrix representation).

The complexity of this algorithm depends on the orders of A and B. For small orders, the probability to succeed in the last step is small. The number of "undetermined matrices" should be increased and the equations with the small orders of A resp. B should be added to the system of equations. In any case we expect the discrete logarithms to dominate the complexity.

With the above algorithm a lot of such collisions can be found. Some of them work in general, independent of the underlying field. These can easily be checked with undetermined γ in the matrices:

$$A \cdot B^{-1} = B \cdot A^{-1}$$
$$A^2 \cdot B^{-3} = B^3 \cdot A^{-2}$$
$$A \cdot B \cdot A^{-4} \cdot B^2 = B^3 \cdot A$$
$$A \cdot B \cdot A^{-3} \cdot B^2 = B \cdot A^2$$
$$A^2 \cdot B \cdot A^{-2} \cdot B^4 = B^{-1} \cdot A^2.$$

I have implemented this method in AXIOM and for the field $\mathbf{F}_{2^{21}}$ with defining polynomial $p(x) = x^{21} + x^2 + 1$ (for this size the discrete logarithm procedure in AXIOM still works quite fast) I have found lots of sequences. As expected the sequences are all very long; one of the shorter ones is:

$$A^{79\,670} = B^3 \cdot A^7 \cdot B^{69\,216} \cdot A \cdot B^{88\,234},$$

where $\mathrm{ord}(A) = 2\,097\,151$ and $\mathrm{ord}(B) = 699\,051$.

References

1. C. Charnes, J. Pieprzyk; *Attacking the SL_2 Hashing Scheme*; Proceedings of ASIA-CRYPT '94, J. Pieprzyk (Ed.), LNCS, Springer, pp. 268–276.
2. W. Geiselmann, D. Gollmann; *Self-Dual Basis in $\mathbf{F}_{q^n}$*; Designs, Codes and Cryptography, Vol. 3, No. 4, pp. 333–345, 1993.
3. R. Lidl, H. Niederreiter; *Introduction to Finite Fields and Their Applications*; Cambridge University Press, 1986.
4. J-P. Tillich, G. Zémor; *Hashing with SL_2*; Proceedings of CRYPTO '94, Y. Desmet (Ed.), LNCS Vol 839, Springer, pp. 40–49, 1994.

Cryptanalysis of Harari's Identification Scheme

Pascal Véron*

G.E.C.T., Université de Toulon et du Var, B.P. 132,
83957 La Garde Cedex, FRANCE

Abstract. In this paper, it is shown that the first identification scheme based on a problem coming from coding theory, proposed in 1988 by S. Harari, is not secure.

1 Introduction

An identification scheme is a cryptographic protocol which enables party A (called the "prover" or Alice) to prove his identity polynomially many times to party B (called the "verifier" or Bob) without enabling B to misrepresent himself as A to someone else.

As it is often the case in cryptography, the first identification schemes were based on hard problems from number theory [5], [8], [12]. In the last few years, some new schemes, whose security depends on an NP-complete problem, have been proposed. The first identification scheme based on an NP-complete problem coming from error correcting codes (SD problem) have been proposed by S. Harari in 1988 [9]. The Syndrome Decoding problem can be stated as follows [1]:

Name   : SD
Input   : $H(k, n)$ a parity check matrix of a binary $[n, k]$ code, i a syndrome, p an integer.
Question : Is there a vector e of length n such that $He^t = i$ and $\omega(e) \leq p$?

Remark. e^t denotes the transpose of the vector e.

The purpose of this paper is to show that Harari's scheme is not secured. First, we give a deterministic method which allows a false prover to pass the protocol every second time. We give then new parameters so as to restore the security of the scheme, and we show that these latter cannot be reduced (so as to improve the practical performances of the scheme). Indeed, if the parameters are shortened, it is possible, for a dishonest verifier, to find prover's secret by using some probabilistic decoding algorithms.

---

* veron@marie.polytechnique.fr

2 Harari's Identification Scheme

This scheme is interactive. The principle is the following: the prover knows a secret quantity s which satisfies a public predicate. To identify himself, the prover must convince, with overwhelming probability, the verifier that he knows s without revealing any information about it.

Prover's Public : $H(1000, 2000)$ a parity check matrix of a binary code
Data   whose minimum weight, μ ($\in [50, 100]$), is odd,

Prover's secret : s a codeword of weight μ.

Remark. see [9] for how to construct a $[2000, 1000]$ code, whose minimum weight is, with high probability, μ.

The prover shows that he knows the secret s as follows:

1. A chooses an integer $\ell \in [100, 200]$,
2. A randomly computes ℓ binary vectors r_i, of length n and odd weights, and whose supports are disjoint,
3. A sends to B: $t_i = Hr_i{}^t$ and $w_i = \omega(r_i)$,
4. B randomly computes a binary vector e of length ℓ, whose weight w is odd and satisfies, $\frac{\ell}{3} \leq w \leq \frac{2\ell}{3}$, and sends it to A,
5. A randomly computes a permutation π of $\{1, \ldots, \ell\}$ and sends it to B,
6. A and B compute $t = e\pi$,
7. B randomly chooses $b \in \{t, \bar{t}\}$, where $\bar{t}$ is the binary complement of t,
8. A computes $r = \sum_{i=1}^{\ell} b_i r_i$, and sends to B the vector $y = r + s$,
9. B checks that:
   - First condition on the weights: $\omega(y) \neq \sum_{i=1}^{\ell} b_i w_i$ ($\omega(y)$ is even, $\sum_{i=1}^{\ell} b_i w_i$ is odd),
   - Condition on the syndrome : $Hy^t = \sum_{i=1}^{\ell} b_i t_i$,
   - Second condition on the weights :

$$\sum_{i=1}^{\ell} b_i w_i - \mu \leq \omega(y) \leq \sum_{i=1}^{\ell} b_i w_i + \mu$$

The practical performances of the scheme are shown in table 1.

Remark. For a smart card application, these three parameters are very important. In this context, the prover is a portable device with few memory and limited computing power. The Rom gives the size of the memory needed on the card. The prover's workfactor is the average number of binary operations computed by the prover during an identification process. The transmission rate is the number of bits exchanged between Alice and Bob.

Rom	1002000 bits
Prover's Workfactor	$2^{28.2}$
Transmission rate	153900 bits

Table 1. Harari's Scheme

3 Cryptanalysis

The attack is efficient if, at step 7, B chooses b equal to $e\pi$. We recall that e is a vector of odd weight $\frac{\ell}{3} + q$, $0 \leq q \leq \frac{\ell}{3}$. So as to simplify the description of this attack, we will assume that ℓ is divisible by 3, otherwise $\frac{\ell}{3}$ must be replaced by $\lceil \frac{\ell}{3} \rceil$. At step 5, the prover chooses a permutation π of $\{1, \ldots, \ell\}$. For more clearness, we will assume that the permutation chosen by the dishonest prover is such that

$$e\pi = (\underbrace{1 \cdots 1}_{\frac{\ell}{3}+q} 0 \cdots 0)$$

(It is easy to see that the method we are going to describe can be changed so as to work with any permutation π). It is possible, for a dishonest prover, to build three sequences $(w_i)_{1 \leq i \leq \ell}, (t_i)_{1 \leq i \leq \ell}, (x_j)_{0 \leq j \leq \frac{\ell}{3}}$ (w_i are odd integers, t_i are vectors of length k and x_j are vectors of length n) such that

$$\begin{array}{rcl} \omega(x_j) & = & \sum_{i=1}^{\frac{\ell}{3}+j} w_i \\ \omega(x_j) & \equiv & \frac{\ell}{3} + j \bmod 2 \\ Hx_j^t & = & \sum_{i=1}^{\frac{\ell}{3}+j} t_i + x \end{array} \qquad (3.1)$$

where $x = Hz^t$, z being a random vector of odd weight μ. Now, suppose that at step 8, the dishonest prover sends to B, $y = x_q + z$. This vector satisfies:

- $\omega(y) \neq \sum_{i=1}^{\frac{\ell}{3}+q} w_i = \sum_{i=1}^{\ell} b_i w_i$ (since $\frac{\ell}{3} + q$ is odd),
- $Hy^t = H(x_q^t) + x = \sum_{i=1}^{\frac{\ell}{3}+q} t_i = \sum_{i=1}^{\ell} b_i t_i$,
- $\omega(x_q) - \mu \leq \omega(y) \leq \omega(x_q) + \mu$, now $\omega(x_q) = \sum_{i=1}^{\frac{\ell}{3}+q} w_i = \sum_{i=1}^{\ell} b_i w_i$.

Thus, this vector satisfies the three conditions of the protocol. Hence, an intruder can misrepresent himself as Alice every second time (when $b = e\pi$) without knowing the secret s.

Here is how to construct the three sequences $(w_i)_{1 \leq i \leq \ell}, (t_i)_{1 \leq i \leq \ell}$ and $(x_j)_{0 \leq j \leq \frac{\ell}{3}}$ (let $\tilde{A}$ be a dishonest prover):

- $\tilde{A}$ randomly computes a word z of weight μ. Let $x = Hz^t$,
- $\tilde{A}$ randomly chooses ℓ odd integers $w_1, \ldots, w_\ell$, and computes a vector x_0 of weight $\sum_{i=1}^{\frac{\ell}{3}} w_i$. Let $t = H(x_0^t) + x$.
- $\tilde{A}$ computes $\frac{\ell}{3}$ vectors of length k, $t_1, \ldots, t_{\frac{\ell}{3}}$, such that $t = \sum_{i=1}^{\frac{\ell}{3}} t_i$,
- For j from 1 to $\frac{\ell}{3}$, $\tilde{A}$ computes:

- x_j a vector of weight $\sum_{i=1}^{\frac{\ell}{3}+j} w_i$,
- the vector $t_{\frac{\ell}{3}+j} = H(x_j^t) + \sum_{i=1}^{\frac{\ell}{3}+j-1} t_i + x$ (hence $H(x_j^t) = \sum_{i=1}^{\frac{\ell}{3}+j} t_i + x$),

• $\tilde{A}$ randomly chooses $t_{\frac{2\ell}{3}+1}, \ldots, t_\ell$, and sends to B, the quantities $w_1, \ldots, w_\ell$ and $t_1, \ldots, t_\ell$. It is easy to check that these sequences satisfy (3.1).

4 New Parameters

If the original protocol is repeated r times, then the probability of success of the previous attack is bounded by 2^{-r}. For $r = 20$, a dishonest prover has only one chance over one million to satisfy the protocol and the performances of the scheme are the following:

Rom	1002000 bits
Prover's Workfactor	$2^{32.5}$
Transmission rate	3078000 bits

Table 2. New Parameters

So as to reduce these parameters, it is enough to lower the size of the matrix H. The other schemes based on SD problem ([6], [14]) use a $(256, 512)$ parity check matrix. We are going to show that such a matrix cannot be used in Harari's scheme.

First, notice that if someone is able to find the vectors r_i from the pair (w_i, t_i) then he can find the secret s from the vector y sent by the prover. This is exactly the SD problem. To find r_i is equivalent to find a word c_i of weight $w_i + 1$ (whose last bit is 1) in the binary code whose parity check matrix is $(H \mid t_i)$.

Probabilistic algorithms defined by J.S. Leon [10] and J. Stern [13] are able, for some parameters, to solve this problem. Their efficiency depends essentially on the weight of the word to find. When Harari's scheme was proposed, there were very few results concerning the validity of these algorithms. Since the recent work of A. Canteaut and H. Chabanne [2], and F. Chabaud [3], their practical efficiency has been proved.

The success of the proposed attack is essentially due to the fact that the vectors r_i (computed at step 2) have disjoint supports and so they must verify

$$\sum_{i=1}^{\ell} \omega(r_i) \leq n \tag{4.1}$$

Thus the integers w_i are constrained.

Let $\tilde{B}$ be a dishonest verifier and let $n = 512$ and $k = 256$:

- Suppose first that the weight of the r_i satisfy $w_i \simeq \frac{n}{\ell}$. Notice that it is enough to choose $\ell = 60$ so as the number of vectors e that can be chosen by B be sufficiently high in order to prevent any imposture from a dishonest prover (by pre-computing for each possible query an answer which satisfies the protocol). Thus each vector r_i as a weight less or equal than 9. The search for a word of weight 10 in a $(513, 256)$ code needs about 2^{24} operations. Thus $\tilde{B}$ can compute all the vectors r_i in less than 2^{30} operations. It is clear that the efficiency of this attack grows with ℓ because of relation (4.1).

- To avoid this attack, taking into account the results given in [2], [3], the prover can try to choose some r_i with weight much more greater than 9 so that the knowledge of w_i and t_i be insufficient to find back r_i. But clearly this implies that all the other vectors will have a weight strictly less than 9. As an example, we will consider that the prover computes as much r_i as possible that cannot be found by Leon's or Stern's algorithm.

The search for a word of weight 45 in a $(513, 256)$ code needs about 2^{60} operations. Thus, for $\ell = 60$, the maximum number of vectors r_i (with disjoint supports) that can have weight 45 is 10, and in this case all the other vectors have weight no more greater than 2.

For any ℓ, this maximum number is given by $N = \lfloor \frac{512-\ell}{44} \rfloor$, and a simple computation shows that for ℓ being between 60 and 100, all the other vectors could not have their weight greater than 2. Thus, it is easy for $\tilde{B}$ to find back $\ell - N$ of the r_i in less than $(\ell - N)2^{20}$ operations (some r_i are columns from H, the others can be found by exhaustive search). Let $r_{i_1}, \ldots, r_{i_{\ell-N}}$ denote the vectors known by $\tilde{B}$. Now $\tilde{B}$ chooses at step 7 of the protocol $b = e\pi$. Notice that for any q $(0 \le q \le \frac{\ell}{3})$, $\frac{\ell}{3} + q < \ell - N$. Indeed, since $N \le \frac{512-\ell}{44}$, then $N < \frac{\ell}{3}$, as soon as $\ell > 33$. Thus the support of b can be included in $\{i_1, \ldots, i_{\ell-N}\}$. In this case, $\tilde{B}$ can find back s from the answer y. The probability of this event is (for any ℓ and $\omega(e) = \frac{\ell}{3} + q$):

$$P_{\ell,q} = \frac{\binom{\ell - N}{\frac{\ell}{3} + q}}{\binom{\ell}{\frac{\ell}{3} + q}}$$

When ℓ is fixed, $P_{\ell,q}$ is minimum when $q = \frac{\ell}{3}$. Thus the complexity C_ℓ of the attack is bounded by $\frac{\Omega_\ell}{P_{\ell,\frac{\ell}{3}}}$ where Ω_ℓ is the workfactor for finding back the $\ell - N$ vectors r_i. Table 3 shows the evolution of this value for $60 \le \ell \le 100$.

ℓ	$\log_2(C_\ell)$
60	44.5
70	44
80	42
90	41.9
100	41.9

Table 3. Workfactor of the attack

Thus, it is clear that a $(256, 512)$ matrix cannot be used in Harari's scheme. Similar results can be found for a $(512, 1024)$ matrix.

References

1. E.R. Berlekamp, R.J. Mc Eliece & H.C.A. Van Tilborg: On the inherent intractability of certain coding problems, IEEE Trans. Inform. Theory, (1978) 384–386
2. A. Canteaut & H. Chabanne: A further improvement of the work factor in an attempt at breaking Mc Eliece's cryptosystem, Proceedings of Eurocode'94, (1994) 163–167
3. F. Chabaud: On the Security Of Some Cryptosystems Based On Error-Correcting Codes, Pre-proceedings of Eurocrypt'94, (1994) 127–135
4. U. Feige, A. Fiat & A. Shamir: Zero-knowledge proofs of identity, Proc. 19th ACM Symp. Theory of Computing, (1987), 210–217
5. A. Fiat, A. Shamir: How To Prove Yourself: Practical Solutions to Identification and Signatures Problems, Advances in Cryptology, Crypto'86, Lecture Notes in Computer Science **263**, (1986) 186–194
6. M. Girault: A (non-practical) three-pass identification protocol using coding theory, Advances in Cryptology, Auscrypt'90, Lecture Notes in Computer Science **453**, (1990) 265–272
7. S. Goldwasser, S. Micali and C. Rackoff: The knowledge complexity of interactive proof systems, Proc. 17th ACM Symp. Theory of Computing, (1985), 291–304
8. L.C. Guillou and J.-J. Quisquater: A practical zero-knowledge protocol fitted to security microprocessor minimizing both transmission and memory, Advances in Cryptology, Eurocrypt'88, **330**, (1988) 123–128
9. S. Harari: A New Authentication Algorithm, Proceedings of Coding Theory and Applications, Lecture Notes in Computer Science **388**, (1988) 91–105
10. J.S. Leon: A probabilistic algorithm for computing minimum weights of large error-correcting codes, IEEE Trans. Inform. Theory, **IT-34(5)**: 1354–1359
11. F.J. MacWilliams & N.J.A. Sloane: The Theory of error-correcting codes, North-Holland, Amsterdam-New-York-Oxford, 1977
12. C.P. Schnorr: Efficient signature generation by smart cards, Journal of Cryptology, **4**, (1991) 161–174
13. J. Stern: A method for finding codewords of small weight, Coding Theory and Applications, Lecture Notes in Computer Science **434**, 173–180
14. J. Stern: A new identification scheme based on syndrome decoding, Crypto'93, Lecture Notes in Computer Science **773**, Springer-Verlag (1994) 13–21

Analysis of Sequence Segment Keying as a Method of CDMA Transmission

T.M. Quirke and M. Darnell

Communication & Signal Processing Research Consortium (Leeds), Department of Electronic & Electrical Engineering, University of Leeds, Leeds LS2 9JT

1 Introduction

The aim of the research described in this paper is to eliminate self-interference during the transmission of simultaneous signals over a common channel. Code-division multiple-access (CDMA) communication systems, also termed spread-spectrum multiple access (SSMA) systems, make use of a transmission bandwidth greatly in excess of the information bandwidth. A synchronous CDMA technique known as Sequence Segment Keying (SSK) that provides zero co-channel interference has been proposed and described by Darnell [1]. Previous work in this area has concentrated on the system aspects of SSK describing the application within the design of a practical radio communication system [2].

This paper provides a comprehensive analysis of this communication scheme by introducing a novel mathematical model of an SSK system. This has been initially tested by computer simulation and verified for an unlimited number of simultaneous users.

The conditions imposed on the pseudorandom (PR) bearer sequences for SSK limit those sequences which can be utilised. The lack of large sequence sets restricts the number of simultaneous users and to overcome this problem the use of independent varying chip periods for each PR bearer sequence is examined. This allows sequences that do not meet specified period criteria to be adjusted so that they are locked into the required sequence length ratios, thus providing a larger sequence set than was previously possible and hence increasing the flexibility and viability of SSK systems.

2 The SSK Concept

For an ideal CDMA system, a set of completely uncorrelated sequences with ideal impulsive autocorrelation functions should be used. In a conventional CDMA system, the crosscorrelation functions (CCFs) of the sequences are quasi-ideal rather than exactly zero; therefore, the superimposed CCF components will lead to co-channel interference which increases with the number of users. This progressive degradation will eventually limit the number of users. SSK provides a solution to this problem by reducing co-channel interference to zero at the expense of non-ideal autocorrelation, which are not a disadvantage in practice.

The SSK technique uses periodic uncorrelated sequences that have discrete spectral components which interleave with each other. This can be achieved by imposing certain conditions on the structure and transmission of the PR bearer sequences. A knowledge of the frequency spectrum of periodic sequences is necessary to determine the position of the discrete spectral components of the spreading code. If y(t) is a sequence with v symbols in one period and δ is the chip interval, its spectrum consists of a DC component and components at frequencies $\frac{n}{v\delta}$ (n integer ≥ 1) when the sequences has no structural symmetry.

Condition 1: All the even components, including the DC component, will be removed if the sequence is anti-symmetric; that is the second half of the sequence is the complement (term-by-term) of the first half. Such sequences are also known as inverse repeat (IR) sequences and all sequences in an uncorrelated set, except that with the shortest period, must have this property.

Condition 2: To ensure spectral interleaving and thus isolation, the periods of the sequence should be locked in the ratios $1:2:4:...:2^{n-1}$ for the most efficient operation

Condition 3: The received sequences must be time aligned with respect to each other to maintain the zero crosscorrelation property at the correlation detector.

Data Encoding: Data is encoded by dividing the sequence into integer chip segments which are inverted/not inverted depending on the data to be transmitted. To maintain the inverse-repeat format, any operation carried out on the first half of the sequence must be repeated on the second half. Figure 1 shows an example of this encoding procedure.

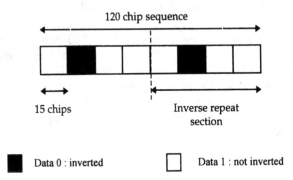

Fig. 1 Example of sequence encoding

By applying the above conditions, a suitable transmission format for a three user system can be derived, as shown in figure 2.

3 SSK Analysis

An analysis of an SSK system is now presented with the aid of an example in which all users employing an identical BPSK modulation format. The transmission format for a three-user transmission is shown in figure 3.

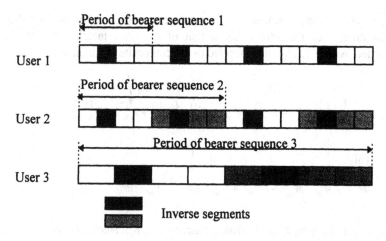

Fig. 2 3-user SSK transmission format with the same data word for all three users

The analysis model identifies one chip in each distinct section of the least common period; the following notation is adopted;

$$\sin\left(\omega_c t + \phi_m^i\right) = W_m^i \tag{1}$$

and

$$-\sin\left(\omega_c t + \phi_m^i\right) = W_{\overline{m}}^i \tag{2}$$

where i is the user index and ϕ_m indicates the modulated segment state ($m=1$, $\overline{m}=0$).

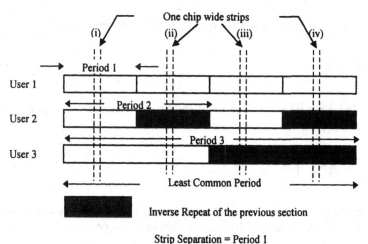

Fig. 3 Transmission format for three-user SSK.

If the receiver of the first user applies matched filtering only to the four strips under consideration, then the zero delay output of the filter matched to the PSK chip transmitted by user 1, U_1, when applied to the composite summed signal, C_s, is

$$R_{U_1 C_s}(0) = \frac{1}{\delta}\int_0^\delta \left\{ \left(W_m^{(1)} + W_m^{(2)} + W_m^{(3)} \right) W_m^{(1)} + \left(W_m^{(1)} + W_{\overline{m}}^{(2)} + W_m^{(3)} \right) W_m^{(1)} + \right.$$

with labels over the equation: (i) over first group; (ii) with IR$_2$ arrow pointing to $W_{\overline{m}}^{(2)}$ in second group.

$$\left. \left(W_m^{(1)} + W_m^{(2)} + W_{\overline{m}}^{(3)} \right) W_m^{(1)} + \left(W_m^{(1)} + W_{\overline{m}}^{(2)} + W_{\overline{m}}^{(3)} \right) W_m^{(1)} \right\} dt$$

with labels: (iii) with IR$_3$ arrow pointing to $W_{\overline{m}}^{(3)}$; (iv) with IR$_2$ arrow to $W_{\overline{m}}^{(2)}$ and IR$_3$ arrow to $W_{\overline{m}}^{(3)}$.

$$\tag{3}$$

which is essentially an autocorrelation function (ACF). Here, IR$_2$ and IR$_3$ indicate the inverse repetitions required by users 2 and 3 respectively, with the terms (i)-(iv) corresponding to the strips (i)-(iv) in figure 3. In a 3-user system the IR format allows four possible combinations of composite signal. The particular combination depends on the chip states of users 2 and 3, whilst the IR format ensures that all permutations yield identical matched filter outputs. The output of the user 1 complementary chip matched filter, $\overline{U}_1$, is given by:

$$R_{\overline{U}_1 C_s}(0) = \frac{1}{\delta}\int_0^\delta \left\{ \left(W_m^{(1)} + W_m^{(2)} + W_m^{(3)} \right) W_{\overline{m}}^{(1)} + \left(W_m^{(1)} + W_{\overline{m}}^{(2)} + W_m^{(3)} \right) W_{\overline{m}}^{(1)} + \right.$$

$$\left. \left(W_m^{(1)} + W_m^{(2)} + W_{\overline{m}}^{(3)} \right) W_{\overline{m}}^{(1)} + \left(W_m^{(1)} + W_{\overline{m}}^{(2)} + W_{\overline{m}}^{(3)} \right) W_{\overline{m}}^{(1)} \right\} dt \tag{4}$$

This represents the CCF. The summation of equation (3) reduces to

$$R_{U_1 C_s}(0) = \frac{1}{\delta}\int_0^\delta 4\left(W_m^{(i)} \right)^2 dt \tag{5}$$

where the user index can be generalised to i. Similarly the summation of equation(4) can be expressed as:

$$R_{\overline{U}_1 C_s}(0) = -\frac{1}{\delta}\int_0^\delta 4\left(W_m^{(i)} \right)^2 dt \tag{6}$$

The difference energy between the ACF and CCF can be used as a measure of cross-talk [3]. For an antipodal system with no co-channel interference this will be

$$E_d = 2\left[R_{U_1 C_s}(0) - R_{\overline{U}_1 C_s}(0) \right] = 4E_s' \tag{7}$$

were E_s' is the signal energy of the four chips representing user 1. Normalising equation (7) to a single chip by dividing by the number of strips gives

$$E_d = \frac{2\left[R_{U_1 C_s}(0) - R_{\bar{U}_1 C_s}(0)\right]}{4} = 4E_s \tag{8}$$

Where E_s is the energy of a single chip. This is expressed as

$$E_s = \frac{1}{\delta}\int_0^\delta \left(W_m^{(i)}\right)^2 dt \tag{9}$$

Substituting equations (5) and (6) into (8) will result an expression for signal energy identical to equation (9). Hence, for this example isolation has been verified.

The IR conditions ensure that every combination of received chips from the four strips is considered, yielding a binomial distribution. Each additional simultaneous user can be incorporated by extending the binomial distribution one stage further. By generalising the ACF equation (3) we obtain an expression for n users:

$$R_{U_1 C_s}(0) = \frac{1}{\delta}\int_0^\delta \sum_{k=0}^{k=n-1}\left(\frac{n-1}{(n-1-k)!k!}\right)\left(W_m^{(i)}\right)^2 dt \tag{10}$$

Adopting standard notation the CCF equation (4) is

$$R_{\bar{U}_1 C_s}(0) = -\frac{1}{\delta}\int_0^\delta \sum_{k=0}^{k=n-1}\binom{n-1}{k}\left(W_m^{(i)}\right)^2 dt \tag{11}$$

For perfect isolation, the relationship between signal energy and difference energy must remain constant, regardless of the value of n. For each additional user the period doubles (condition 2) the number of strips required to model the system must also double. Equations (10) and (11) can be substituted into (8), whilst including the requirement of condition 2. This gives a characteristic equation for the energy difference:

$$E_d = \frac{2\left[\frac{1}{\delta}\int_0^\delta \sum_{k=0}^{k=n-1}\binom{n-1}{k}\left(W_m^{(i)}\right)^2 dt - \left(-\frac{1}{\delta}\int_0^\delta \sum_{k=0}^{k=n-1}\binom{n-1}{k}\left(W_m^{(i)}\right)^2 dt\right)\right]}{2^{n-1}} = 4E_s \tag{12}$$

This can be simplified to

$$E_d = \frac{4\left[\frac{1}{\delta}\int_0^\delta \sum_{k=0}^{k=n-1}\binom{n-1}{k}\left(W_0^{(1)}\right)^2 dt\right]}{2^{n-1}} = 4E_s \tag{13}$$

The basic binomial expression has the following property

$$\frac{\sum_{k=0}^{k=n-1}\binom{n-1}{k}}{2^{n-1}} = 1 \tag{14}$$

Hence equation (13) is true for all real positive values of n [4]. The co-channel interference remains zero for an SSK-BPSK signalling system for any number of simultaneous users.

3.1 Simulation Results

The performance of an SSK system has been verified by computer simulation over an additive white Gaussian noise (AWGN) channel. Three simultaneous users were modelled using the following PR bearer codes [1]:

Sequence1: 15 Length binary m-sequence.
Sequence2: 30 chip binary IR sequence derived via modulo-2 addition of the second row of the Hadamard matrix (1010110...) and sequence 1.
Sequence3: 60 chip binary IR sequence constructed by modulo-2 addition of the third row of the Hadamard matrix (11001100...) and sequence 1.

Three data bits were encoded on each PR bearer sequence in the SSK format. The simplest search technique was adopted, i.e. a parallel search of all possible code words using a maximum likelihood algorithm for the final decision. Such a system requires a bank of 2^{Nb} matched filters where N_b is the number of data bits encoded onto the PR bearer sequence. Large N_b, even if sufficient processing power is available, is unlikely to be practical in real time.

SSK-PSK receiver architecture: The receiver assumes that synchronisation has been established and the signals are time aligned as in figure 2. For the sequences to be uncorrelated at the receiver, the correlation intervals must be equal to the period of the longest sequence; the IR properties cause a zero mathched filter response from the unmatched sequences.

Simulation results: The performance of the proposed CDMA scheme was evaluated by computer simulation. The theoretical performance in terms of BER for the segmented CDMA sequence is dependent on the variance of the noise which is inversely proportionally processing gain across the period of the longest sequence. Explicitly, the E_b/N_o is the product of the $SNR_o/2$ for single chip and the number of chips representing a data bit. Where E_b/N_o is the ratio of energy per bit to noise spectral density and SNR_o is the matched filter output signal-to-noise ratio. In figure 4, the results shown are a comparison between a three-user and single-user transmission.

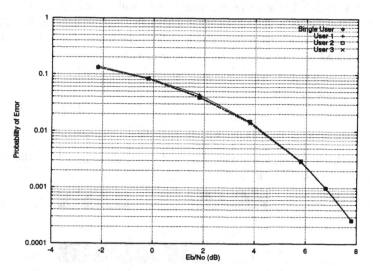

Fig. 4 Single and multi-user SSK-PSK performance over an AWGN channel

Although the number of users is small, the lengths of the sequences are very short compared with conventional CDMA and any co-channel interference would have a noticeable impact on the BER performance. The results, based on an accumulation of 5000 to 100 errors, depending on probability of error, support the theory and show no discernible difference in performance between single and multi-user operation. Thus it can be inferred that co-channel interference effects are negligible.

4 Analysis of Variable Chip Period SSK

The condition for the period ratios between the component bearer sequences is so strict that there maybe few sequences that can be used in a practical system. Specifically, the requirement that the sequence periods must have ratios $1...2^k$ also necessitates sequences of specific lengths having an IR structure. For many lengths, such sequences may not exist. It will be shown that the sequence length in chips can be relaxed whilst maintaining sequence period by using sequences with different chip intervals. This allows the synthesis of IR sequences without the strict sequence length ratios but with the required period ratios. If no degradation then exists, a major obstacle to the implementation of SSK will have been removed. A simplified scheme is illustrated in figure 5.

      The figure is labelled so that the duration of B>A and Z>X>U>Y and demonstrates the variation and distribution of fractionalised chips involved over sequence 1 chip period C.

If FSK modulation is adopted, the chips that are not correlated over the full chip period will lose their orthogonality. Now if we assume that the frequencies used are the same throughout and are separated with respect to the highest chip rate (in this case sequence 2) then all the chips will be orthogonal over the shortest chip period. In the case of sequence 2, the chips are split in half with respect to user 1 so MSK reception occurs so is still orthogonal. Although it is must be noted that the correlation function over fractional chips is a not a continuous linear function. The analysis presented here is for binary PSK-SSK modulation.

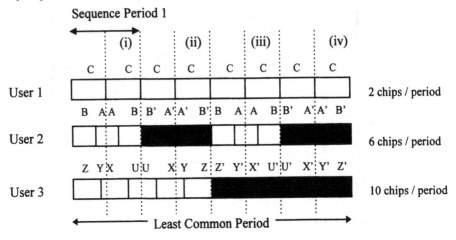

Fig. 5 Simplified version of variable-chip period SSK

The function ψ is defined by:

$$\Psi_{W_m^{(i)} W_m^{(k)}}(\delta) = \frac{1}{\delta}\int_0^\delta W_m^{(i)} W_m^{(k)} dt \tag{15}$$

so that

$$\Psi_{W_m^{(i)} W_m^{(k)}}(\delta) = -\Psi_{W_m^{(i)} W_{\overline{m}}^{(k)}}(\delta) \tag{16}$$

such that $k \in [1\text{-}n]$ users and where δ is the chip correlation period of the shortest chip. In the example given δ is the chip period of user 2. Using this notation, the ACF for user 1 is given by equation (17). Each line of the equation is the correlation of an individual strip.

$$R_{U_1C_s}(0) = \Psi_{W_m^{(1)} W_m^{(1)}}\left(\frac{3\delta}{2}\right) + \Psi_{W_m^{(1)} W_{\overline{m}}^{(2)}}\left(\frac{\delta}{2}\right) + \Psi_{W_m^{(1)} W_{\overline{m}}^{(2)}}(\delta) + \Psi_{W_m^{(1)} W_m^{(3)}}\left(\frac{9\delta}{10}\right) + \Psi_{W_m^{(1)} W_{\overline{m}}^{(3)}}\left(\frac{3\delta}{5}\right) \quad \text{(i)}$$

$$+ \Psi_{W_m^{(1)} W_m^{(1)}}\left(\frac{3\delta}{2}\right) + \Psi_{W_m^{(1)} W_m^{(2)}}\left(\frac{\delta}{2}\right) + \Psi_{W_m^{(1)} W_{\overline{m}}^{(2)}}(\delta) + \Psi_{W_m^{(1)} W_m^{(3)}}\left(\frac{3\delta}{10}\right) + \Psi_{W_m^{(1)} W_{\overline{m}}^{(3)}}\left(\frac{6\delta}{5}\right) \quad \text{(ii)}$$

$$+ \Psi_{W_m^{(1)} W_m^{(1)}}\left(\frac{3\delta}{2}\right) + \Psi_{W_m^{(1)} W_{\overline{m}}^{(2)}}\left(\frac{\delta}{2}\right) + \Psi_{W_m^{(1)} W_m^{(2)}}(\delta) + \Psi_{W_m^{(1)} W_{\overline{m}}^{(3)}}\left(\frac{9\delta}{10}\right) + \Psi_{W_m^{(1)} W_m^{(3)}}\left(\frac{3\delta}{5}\right) \quad \text{(iii)}$$

$$+\Psi_{W_m^{(1)}W_m^{(1)}}\left(\frac{3\delta}{2}\right)+\Psi_{W_m^{(1)}W_m^{(2)}}\left(\frac{\delta}{2}\right)+\Psi_{W_m^{(1)}W_{\overline{m}}^{(2)}}(\delta)+\Psi_{W_m^{(1)}W_{\overline{m}}^{(3)}}\left(\frac{3\delta}{10}\right)+\Psi_{W_m^{(1)}W_m^{(3)}}\left(\frac{6\delta}{5}\right) \qquad \text{(iv)}$$

$$(17)$$

Here, terms (i)-(iv) correspond to the strips (i)-(iv) in figure 5. Similarly, the CCF is given by equation (18).

$$R_{\overline{U}_1C_s}(0)=\Psi_{W_m^{(1)}W_{\overline{m}}^{(1)}}\left(\frac{3\delta}{2}\right)+\Psi_{W_m^{(1)}W_m^{(2)}}\left(\frac{\delta}{2}\right)+\Psi_{W_m^{(1)}W_{\overline{m}}^{(2)}}(\delta)+\Psi_{W_m^{(1)}W_m^{(3)}}\left(\frac{9\delta}{10}\right)+\Psi_{W_m^{(1)}W_{\overline{m}}^{(3)}}\left(\frac{3\delta}{5}\right)$$

$$+\Psi_{W_m^{(1)}W_{\overline{m}}^{(1)}}\left(\frac{3\delta}{2}\right)+\Psi_{W_m^{(1)}W_m^{(2)}}\left(\frac{\delta}{2}\right)+\Psi_{W_m^{(1)}W_m^{(2)}}(\delta)+\Psi_{W_m^{(1)}W_{\overline{m}}^{(3)}}\left(\frac{3\delta}{10}\right)+\Psi_{W_m^{(1)}W_m^{(3)}}\left(\frac{6\delta}{5}\right)$$

$$+\Psi_{W_m^{(1)}W_{\overline{m}}^{(1)}}\left(\frac{3\delta}{2}\right)+\Psi_{W_m^{(1)}W_m^{(2)}}\left(\frac{\delta}{2}\right)+\Psi_{W_m^{(1)}W_{\overline{m}}^{(2)}}(\delta)+\Psi_{W_m^{(1)}W_m^{(3)}}\left(\frac{9\delta}{10}\right)+\Psi_{W_m^{(1)}W_{\overline{m}}^{(3)}}\left(\frac{3\delta}{5}\right)$$

$$+\Psi_{W_m^{(1)}W_{\overline{m}}^{(1)}}\left(\frac{3\delta}{2}\right)+\Psi_{W_m^{(1)}W_m^{(2)}}\left(\frac{\delta}{2}\right)+\Psi_{W_m^{(1)}W_m^{(2)}}(\delta)+\Psi_{W_m^{(1)}W_m^{(3)}}\left(\frac{3\delta}{10}\right)+\Psi_{W_m^{(1)}W_{\overline{m}}^{(3)}}\left(\frac{6\delta}{5}\right) \qquad (18)$$

The difference energy E_d between the two correlators, for a three user system, is defined by equation (8). Hence E_d can be shown to be

$$E_d=16\Psi_{W_m^{(1)}W_m^{(1)}}\left(\frac{3\delta}{2}\right) \qquad (19)$$

and the signalling energy E_s' for four chips is

$$E_s'=4\Psi_{W_m^{(1)}W_m^{(1)}}\left(\frac{3\delta}{2}\right) \qquad (20)$$

Hence,

$$E_d=4E_s' \qquad (21)$$

Hence, isolation has been verified in this example. The correlation equations (17) and (18) now include additional correlation terms, compared with equations (3) and (4). These are dependant on the number of co-channel users and their chip periods. In the example above, the number of correlation elements compromising the correlation function has been increased by 8.

The number of correlation elements will increase further if the co-channel chip periods decrease below one half of the correlated chip period. A measure of this effect, γ, in terms of the number of differing chips, l, across one correlated chip periods can be expressed as

$$\gamma = \sum_{l=2}^{l=\infty}(l-1)\beta_l \tag{22}$$

where the term inside the brackets indicates the increase in the number of correlation elements and β_l is the number of sequences with l co-channel chips per correlated chip period. The IR structure ensures that every combination of co-channel chips in each strip will occur over the least common period. This is a binomial distribution dependent on the number of users and fractionalised chips. The number of correlation elements, N_{ce}, and hence the size of the distribution, can be expressed as

$$N_{ce} = \sum_{k=0}^{k=n-1}\binom{n-1}{k}+\gamma 2^{n-1} \tag{23}$$

However, the IR format will cause the majority of the correlation elements to cancel, as in equation (19). This leads to the characteristic expression for the ACF

$$R_{U_iC_x}(0)= \sum_{k=0}^{k=n-1}\binom{n-1}{k}\Psi_{W_m^{(i)}W_m^{(i)}}\left(\frac{L_l}{L_i}\delta\right) \tag{24}$$

where L_l is the length, in chips, of the sequence with the greatest number of chips over the correlation period and L_i is the length of the sequence being correlated. In the previous example $L_l =12$. Similarly the CCF can be expressed as

$$R_{\overline{U}_iC_x}(0)= -\sum_{k=0}^{k=n-1}\binom{n-1}{k}\Psi_{W_m^{(i)}W_m^{(i)}}\left(\frac{L_l}{L_i}\delta\right) \tag{25}$$

Due to the IR format of the sequences, E_d per chip will be

$$E_d = \frac{2\left[\sum_{k=0}^{k=n-1}\binom{n-1}{k}\Psi_{W_m^{(i)}W_m^{(i)}}\left(\frac{L_l}{L_i}\delta\right)-\left[-\sum_{k=0}^{k=n-1}\binom{n-1}{k}\Psi_{W_m^{(i)}W_m^{(i)}}\left(\frac{L_l}{L_i}\delta\right)\right]\right]}{2^{n-1}}=4E_s \tag{26}$$

This can be rearranged into equation (27).

$$E_d = \frac{4\left[\sum_{k=0}^{k=n-1}\binom{n-1}{k}\Psi_{W_m^{(i)}W_m^{(i)}}\left(\frac{L_l}{L_i}\delta\right)\right]}{2^{n-1}}=4\Psi_{W_m^{(i)}W_m^{(i)}}\left(\frac{L_l}{L_i}\delta\right) \tag{27}$$

which then can be reduced to

$$\frac{\sum_{k=0}^{k=n-1}\binom{(n-1)}{k}}{2^{n-1}}=1 \tag{28}$$

This is true for any positive real values of n [4], thus indicating that perfect isolation will be maintained when sequences have differing chip periods.

For each additional user, the least common period doubles; hence the processing gain also doubles for each additional user in an SSK system. The value for E_d for a repeated chip in an n user system is given by

$$E_d = 2^{n+1}\Psi_{W_m^{(i)}W_m^{(i)}}\left(\frac{L_l}{L_i}\delta\right) \tag{29}$$

This analysis can be repeated for FSK if the spectral separation between tones is determined by the smallest chip period.

4.1 Simulated Results for Variable Chip Period Encoding

Here, the sequences used are the same as in the previous simulation in this paper, except that sequence 3 is replaced with a 48 length 7 level m-sequence. An integer level transform [3] was applied to convert the sequence to antipodal format. To verify the above analysis, the CCF between the 48 length ssequence and the composite signal was evaluated over the least common period.

The chip period of sequence 3 was increased by 25% so that the correlation period is identical to that of the former 60 chip length sequence. The results of the correlation of the composite signal and user 3 signal alone are shown on figure 6; this

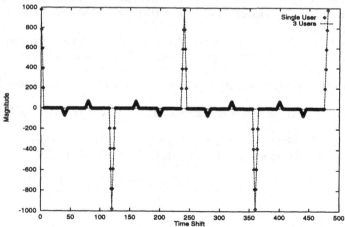

Fig. 6 ACF of 7 level 46 length sequence with a chip period 1.25 times greater than other sequence chip periods overlaid by CCF between 46 length sequence and 3-user composite signal.

profile is identical to the ACF of the user 3 sequence which indicates that there is no discrepancy between multi-user and single-user operation. This evidence verifies the previous theory. The peak-to-sidelobe ratio will be larger than that of the former sequence because an m-sequence has an ACF superior to a sequence generated using a Hadamard matrix.

5 Concluding Remarks

This paper has briefly introduced a mathematical analysis of SSK that can be applied to the various modulation options. The performance of synchronous SSK has been characterised.

The strip method of analysing SSK is a simple and effective way of determining performance and allows design choices to be made of the system parameters, e.g. sequence types, number of simultaneous users and data modulation rates.

The proposed variable chip period SSK has been analysed using the mathematical model and exhibits no degradation in performance. This enables sequences to be used that have superior ACF properties whilst still maintaining perfect isolation. Two advantages result from this; a greater range of possible sequences increasing system flexibility and an improved performance in noise due to higher peak-to-sidelobe ratios.

Computer simulation results support the theory. Future work in this area will include methods of systematically exploiting the inherent time (repetition) diversity of the SSK system to combat effects of fading in radio channels.

References

[1] Darnell, M. (1989). "The theory and generation of sets of uncorrelated digital sequence", in Cryptography and Coding (H.J. Beker and F.C. Piper eds), Oxford University Press.

[2] Darnell, M. and Miller, T.E. (1993). "An In-band CDMA System employing Sequence Segment Keying", in Cryptography and Coding III (M.J.Ganely ed), Oxford University Press

[3] Couch (1983). Digital and Analogue Communication Systems, Macmillan Publishing Co

[4] Fogeil M (1984). Handbook of Mathematical, Scientific and Engineering, Research and Education Association

[5] Miller, T.E. (1995). "Reliable Interference-Resistant Data Transition Techniques for Multi-User HF Communications". PhD Thesis, University of Hull, UK.

Constructions for Variable-Length Error-Correcting Codes

Victor Buttigieg[1] and Patrick G. Farrell[2]

[1] Dept. of Communications and Computer Engineering, University of Malta, Msida MSD06, Malta. Email vjbutt@unimt.mt

[2] The Manchester School of Engineering, Simon Building, The University of Manchester, Oxford Road, Manchester M13 9PL, UK. Email Farrell@manchester.ac.uk

Abstract. Two construction techniques for variable-length error-correcting (VLEC) codes are given. The first uses fixed-length linear codes and anticodes to build new VLEC codes, whereas the second uses a heuristic algorithm to perform a computer search for good VLEC codes. VLEC codes may be used for combined source and channel coding. It is shown that over an additive white Gaussian noise channel the codes so constructed can perform better than standard cascaded source and channel codes with similar parameters.

1 Introduction

Variable-length codes are normally used for source coding (e.g., the Huffman code [1]). When transmitted over noisy channels, variable-length codes are particularly susceptible, because in addition to the errors in the received symbol stream there is also loss of codeword synchronisation resulting in error propagation [2]. Thus it is necessary to protect source-encoded data with an error-correcting code. A natural question then arises: is it possible and effective to combine the source and error control coding functions to create variable-length error-correcting (VLEC) codes? The answer is yes; previous work by the authors [3, 4, 5] has demonstrated that combined source and error-control (channel) coding is both feasible and can provide better coding gain than the normal cascade of codes. Previous papers [3, 4, 5] have described the properties and decoding of such codes; the present paper will describe methods for constructing VLEC codes.

In Section 2 we define some important parameters of VLEC codes. Linearity is discussed in Section 3. Two construction techniques for VLEC codes are then presented in Section 4. The first uses fixed-length linear codes and anticodes [7] to build new VLEC codes, whereas the second uses a heuristic algorithm to perform a computer search for good VLEC codes. Problems associated with the construction of good VLEC codes are then discussed in Section 5.

2 Preliminaries

Let X be a code alphabet with cardinality q. A finite sequence $\mathbf{w} = x_1 x_2 \cdots x_l$ of code symbols is called a word over X of length $|\mathbf{w}| = l$, where $x_i \in X$, for all $i = 1, 2,$

$\cdots$, l. A set C of words is called a code. Let the code C have s codewords $\{\mathbf{c}_1, \mathbf{c}_2, \cdots, \mathbf{c}_s\}$ and let $l_i = |\mathbf{c}_i|$, $i = 1, 2, \cdots, s$. Without loss of generality, assume that $l_1 \le l_2 \le \cdots \le l_s$. Further, let σ denote the number of different codeword lengths in the code C and let these lengths be $L_1, L_2, \cdots, L_\sigma$, where $L_1 < L_2 < \cdots < L_\sigma$. C is a non-trivial VLEC code iff $\sigma > 1$. Let the number of codewords with length L_i be s_i, and the number of codewords with length less than L_i be $\tilde{s}_i$, i.e. $\tilde{s}_i = \sum_{j=1}^{i-1} s_j$. Note that $\tilde{s}_1 = 0$ and that $L_1 = l_{\tilde{s}_1+1} = l_1$, $L_2 = l_{\tilde{s}_2+1}, \cdots, L_\sigma = l_{\tilde{s}_\sigma+1} = l_s$ and $s = \sum_{i=1}^{\sigma} s_i$. We shall use $(s_1@L_1, b_1; s_2@L_2, b_2; \cdots; s_\sigma@L_\sigma, b_\sigma; d_{min}, c_{min})$ to denote such a code, where b_i is the minimum block distance for length L_i, d_{min} is the minimum diverging distance and c_{min} is the minimum converging distance. These are defined as follows:

Definition 1: The *minimum block distance for length L_i, b_i,* of a VLEC code C is defined as the minimum Hamming distance between all codewords with the same length L_i, i.e.

$$b_i = \min\{d_h(\mathbf{c}_j, \mathbf{c}_k): \mathbf{c}_j, \mathbf{c}_k \in C, j \ne k \text{ and } |\mathbf{c}_j| = |\mathbf{c}_k| = L_i\}, \tag{1}$$

where $d_h(\mathbf{a}, \mathbf{b})$ is the Hamming distance between the words $\mathbf{a}$ and $\mathbf{b}$.

There are σ different minimum block distances, one for each different codeword length. However, if for some length L_i there is only one codeword, i.e. $s_i = 1$, then in this case the minimum block distance for length L_i is undefined and this will be denoted by a '-'.

Definition 2: The *overall minimum block distance, b_{min},* of a VLEC code C is defined as the minimum value of b_i over all $i = 1, 2, \cdots, \sigma$.

Definition 3: The *diverging distance* between two codewords $\mathbf{c}_i = c_{i_1} c_{i_2} \cdots c_{i_{l_i}}$ and $\mathbf{c}_j = c_{j_1} c_{j_2} \cdots c_{j_{l_j}}$, $\mathcal{D}(\mathbf{c}_i, \mathbf{c}_j)$, where $\mathbf{c}_i, \mathbf{c}_j \in C$, $l_i = |\mathbf{c}_i|$ and $l_j = |\mathbf{c}_j|$, with $l_i > l_j$, is defined as

$$\mathcal{D}(\mathbf{c}_i, \mathbf{c}_j) = d_h(c_{i_1} c_{i_2} \cdots c_{i_{l_j}}, c_{j_1} c_{j_2} \cdots c_{j_{l_j}}). \tag{2}$$

Note that $\mathcal{D}(\mathbf{c}_i, \mathbf{c}_j) = \mathcal{D}(\mathbf{c}_j, \mathbf{c}_i)$.

The *minimum diverging distance* of the VLEC code C, d_{min}, is the minimum value of all the diverging distances between all possible pairs of unequal length codewords of C, i.e.

$$d_{min} = \min \{\mathcal{D}(\mathbf{c}_i, \mathbf{c}_j) : \mathbf{c}_i, \mathbf{c}_j \in C, |\mathbf{c}_i| \ne |\mathbf{c}_j|\} \tag{3}$$

Definition 4: The *converging distance* between two codewords $\mathbf{c}_i = c_{i_1} c_{i_2} \cdots c_{i_{l_i}}$ and $\mathbf{c}_j = c_{j_1} c_{j_2} \cdots c_{j_{l_j}}$, $\mathcal{C}(\mathbf{c}_i, \mathbf{c}_j)$, where $\mathbf{c}_i, \mathbf{c}_j \in C$, $l_i = |\mathbf{c}_i|$ and $l_j = |\mathbf{c}_j|$, with $l_i > l_j$, is defined as

$$\mathcal{C}(\mathbf{c}_i, \mathbf{c}_j) = d_h(c_{i_{l_i-l_j+1}} c_{i_{l_i-l_j+2}} \cdots c_{i_{l_i}}, c_{j_1} c_{j_2} \cdots c_{j_{l_j}}). \tag{4}$$

Again, note that $\mathcal{C}(\mathbf{c}_i, \mathbf{c}_j) = \mathcal{C}(\mathbf{c}_j, \mathbf{c}_i)$.

The *minimum converging distance* of the VLEC code C, c_{min}, is the minimum value of all the converging distances between all possible pairs of unequal length codewords of C, i.e.

$$c_{min} = \min \{\mathcal{C}(\mathbf{c}_i, \mathbf{c}_j) : \mathbf{c}_i, \mathbf{c}_j \in C, |\mathbf{c}_i| \ne |\mathbf{c}_j|\} \tag{5}$$

Definition 5: The set $F_N = \{\mathbf{f}_i : |\mathbf{f}_i| = N, \mathbf{f}_i = \mathbf{c}_{i_1} \mathbf{c}_{i_2} \cdots \mathbf{c}_{i_{\eta_i}}, \mathbf{c}_{i_j} \in C \ \forall \ j = 1, 2, \cdots, \eta_i\}$, is the *extended code of the VLEC code C of order N*.

Definition 6: The *free distance, d_{free},* of a VLEC code C is defined by

$$d_{\text{free}} = \min \{d_h(\mathbf{f}_i, \mathbf{f}_j) : \mathbf{f}_i, \mathbf{f}_j \in F_N, i \neq j, N = 1, 2, \cdots, \infty\} \tag{6}$$

where, F_N is the extended code of order N of C [1].

Theorem 1: The free distance of a VLEC code C is bounded by

$$d_{\text{free}} \geq \min(b_{\min}, d_{\min} + c_{\min}) \tag{7}$$

For the proof see [3]. It can be shown [6] that the free distance of VLEC codes is the single most important parameter affecting their performance on an additive white Gaussian noise (AWGN) channel.

3 Linearity

Standard error-correcting codes are usually classified to be either linear or non-linear. It turns out that a VLEC code can never be entirely linear. However, by suitably sub-dividing a code, linear sub-structures may be defined. Linearity is important because:

- it allows simpler encoding and decoding algorithms to be used, by exploiting the additional mathematical structure;
- it simplifies code construction, firstly by again utilising mathematical structures and secondly by reducing the possible domain from where the code is chosen.

Theorem 2: A non-trivial, uniquely decodable VLEC code is always non-linear.

Proof: Let $\mathbf{c}_i$ and $\mathbf{c}_j$ be two unequal length codewords of a non-trivial VLEC code C. Let $|\mathbf{c}_i|=l_i$ and $|\mathbf{c}_j|=l_j$. Then, for the code to be linear, $\mathbf{c}_i+\mathbf{c}_i=\mathbf{0}_{l_i}$ and $\mathbf{c}_j+\mathbf{c}_j=\mathbf{0}_{l_j}$ must both be codewords in C, where $\mathbf{0}_l$ represents the word with l all zero bits. However, if both $\mathbf{0}_{l_i}$ and $\mathbf{0}_{l_j}$ are codewords of C, then C is not uniquely decodable. This can be shown to be true by considering the codeword sequence consisting of l_j consecutive $\mathbf{0}_{l_i}$ which is indistinguishable from the codeword sequence consisting of l_i consecutive $\mathbf{0}_{l_j}$. Hence a uniquely decodable VLEC code can never be linear. ∎

Theorem 2 does not exclude us, however, from using linear sub-codes, or cosets of linear sub-codes, within the VLEC codes. It does state, however, that these sub-codes must be of fixed length. Thus, we can have two different linear structures imposed on VLEC codes, called respectively *vertical* and *horizontal linearity*.

Definition 7: Define the vertical sub-codes of a VLEC code C ($s_1@L_1$, b_1; $s_2@L_2$, b_2; $\cdots$; $s_\sigma@L_\sigma$, b_σ; $d_{\min}$, $c_{\min}$), with $C = \{\mathbf{c}_1, \mathbf{c}_2, \cdots, \mathbf{c}_s\}$, to be $V_1, V_2, \cdots, V_\sigma$ where $V_i = \{\mathbf{c}_{\bar{s}_i+1}^{L_{i-1}+1,L_i}, \mathbf{c}_{\bar{s}_i+2}^{L_{i-1}+1,L_i}, \cdots, \mathbf{c}_s^{L_{i-1}+1,L_i}\}$, $i = 1, 2, \cdots, \sigma$, L_0 is defined to be equal to 0 and $\mathbf{c}_j^{a,b} = c_{j_a} c_{j_{a+1}} \cdots c_{j_b}$ given that $\mathbf{c}_j = c_{j_1} c_{j_2} \cdots c_{j_{l_j}}$, $c_j \in C$. Then, a VLEC code C is said to be *vertically linear* iff its vertical sub-codes $V_1, V_2, \cdots, V_\sigma$ are all cosets of linear fixed-length codes.

Definition 8: Define the horizontal sub-codes of a VLEC code C ($s_1@L_1$, b_1; $s_2@L_2$, b_2; $\cdots$; $s_\sigma@L_\sigma$, b_σ; $d_{\min}$, $c_{\min}$), with $C = \{\mathbf{c}_1, \mathbf{c}_2, \cdots, \mathbf{c}_s\}$, to be $H_1, H_2, \cdots, H_\sigma$, where $H_i = \{\mathbf{c}_{\bar{s}_i+1}, \mathbf{c}_{\bar{s}_i+2}, \cdots, \mathbf{c}_{\bar{s}_i+s_i}\}$, $i = 1, 2, \cdots, \sigma$. Then, a VLEC code C is said to be

---

[1] Note that for certain values of N the set F_N may be empty.

horizontally linear iff its horizontal sub-codes H_1, H_2, $\cdots$, H_σ are all cosets of linear fixed-length codes.

For maximum likelihood decoding of VLEC codes using the modified Viterbi algorithm [3] only horizontal linearity is useful. This is also true if sequential decoding [5] of VLEC codes is employed [6]. Hence we are not interested in vertically linear VLEC codes.

4 Constructions

We now give two construction techniques for binary VLEC codes with free distance d_{free}. The first, the code-anticode construction, is suitable to construct horizontally linear VLEC codes, while the second, the heuristic construction algorithm, is suitable for both horizontally linear and non-linear VLEC codes. It is assumed that the required number of codewords is s.

Code - Anticode Construction

This construction splits a fixed-length code into cosets with specified minimum distances and then uses an anticode to remove certain columns of some of these cosets. This guarantees an overall minimum block distance and a minimum diverging distance, however the minimum converging distance necessary to give the required free distance may not always be satisfied.

1. Find a binary linear fixed-length code F with parameters (n, k, d) with minimum n such that $2^k \geq s$ and $d = d_{\text{free}}$, where n is the block length, 2^k is the number of codewords and d the minimum distance.

2. Let $d_{\min} = \lceil d_{\text{free}}/2 \rceil$.

3. Rearrange the columns of F such that the rightmost m columns form an anticode A with parameters (m, k, δ) with maximum m for $\delta = d_{\text{free}} - d_{\min}$, where m is the block length, 2^k is the number of codewords, and δ is the maximum distance for the anticode.

4. Re-order the codewords of F such that repeated codewords of A are consecutive.

5. Perform simple row operations and column permutations on the generator matrix for F such that the generator matrix of A contains the maximum number of consecutive 0's in the top most positions of each column, starting from the rightmost column.

6. Delete the rightmost m columns of the first s_1 codewords in F, where s_1 is the number of codewords with identical m rightmost columns.

7. Considering the remaining codewords in F, delete the rightmost $m-1$ columns of the next s_2 codewords in F, where s_2 is the number of codewords with identical $m-1$ rightmost columns.

8. Repeat Step 7 with the next $m-2$, $m-3$, ... rightmost columns until there are no further columns to delete.

Note that Step 5 will ensure that the maximum number of codewords with the same rightmost columns are generated, hence resulting in a VLEC code with a shorter average codeword length.

Theorem 3: The VLEC code constructed using the code-anticode construction from the linear fixed-length code with parameters (n, k, d) and the anticode with parameters (m, k, δ) is a horizontally linear VLEC code with overall minimum block distance at least equal to d and minimum diverging distance at least equal to $d - \delta$.

Proof: Step 1 in the code-anticode construction ensures that all codewords are at least distance d from each other. Now, by deleting the rightmost m columns in Step 6 of the first s_1 codewords in $\mathcal{F}$ results in two sub-codes, one of length $n - m$ with s_1 codewords (sub-code H_1) and one of length n with $s - s_1$ codewords (sub-code H_2'). Since the deleted columns from H_1 are identical (Step 6), then the codewords of H_1 form a linear code with minimum distance at least d. Also, since H_2' is a sub-code of the original code $\mathcal{F}$ (with the same length), then it also has minimum distance at least d. Hence the overall minimum block distance of the VLEC code $\{H_1 \cup H_2\}$ is at least d. In addition, since the m rightmost columns of $\mathcal{F}$ form an anticode with maximum distance δ, then by deleting the m rightmost columns from $\mathcal{F}$ will result in a code with minimum distance $d - \delta$. Hence, the minimum diverging distance between the codewords of H_1 and H_2' must at least be $d - \delta$. Exactly the same reasoning applies for all the other columns deleted. ∎

Unfortunately, as one may observe from Theorem 3, this construction algorithm does not guarantee a minimum converging distance. Since with this construction $d_{min} < b_{min}$, then using the bound given by expression (7) we require $c_{min} = b_{min} - d_{min}$ such that $d_{free} \geq b_{min}$. The required value of c_{min} for a given free distance may be obtained by adding a suitable modification vector to the code and/or performing column permutations in such a way as not to affect the values of b_{min} and d_{min}. In general there are 2^{L_σ} possible modification vectors each of length L_σ to test, since adding a modification vector of length L_σ to each codeword in C may affect the minimum converging distance but not the minimum block and diverging distances[2]. On the other hand, the total number of column permutations possible without affecting d_{min} and b_{min} is $\prod_{i=1}^{\sigma}(L_i - L_{i-1})!$, since only columns within a vertical sub-code may be permuted. However, even after performing these operations, the required converging distance still may not be reached. In this case the only alternative is to increase the design value of d_{min} (in Step 2) and repeat the construction. This in general will decrease the maximum value of m possible for $\mathcal{A}$, resulting in a longer average codeword length for C. Now, however, the required value of c_{min} is likewise decreased, making it easier to find a suitable modification vector or column permutations. Note that in most cases the total number of column permutations possible may be too large to test all, and in practice it becomes more feasible to find a modification vector instead. We will now illustrate the code-anticode construction with an example.

Example 1: Consider that we want to construct a horizontally linear VLEC code for the 26-symbol English source with free distance five. Therefore since $s = 26$, then we must choose $k = 5$. An optimum fixed-length linear code with $k = 5$ and minimum

---

[2] Note that when adding the modification vector to a shorter length codeword, only the corresponding number of bits from the leftmost position in the modification vector are considered.

distance five has block length $n = 13$. The generator matrix for this code is shown in Fig. 1.

From Step 2, the design minimum diverging distance is $\lceil \frac{5}{2} \rceil = 3$. Hence, we need to find an anticode with maximum distance 5–3=2. The maximum value for m possible in this case is 3. This is achieved by choosing two columns of $\mathcal{F}$ and their modulo-2 sum as the third column [8]. Three such columns in $\mathcal{F}$ are columns 2, 3 and 13, where the leftmost one is column 1. Hence, rearranging these columns to be in the rightmost positions, and performing simple row operations on the generator matrix of $\mathcal{F}$ so as to arrange its codewords to satisfy Steps 4 and 5, we get the generator matrix given in Fig. 2. The relative codebook is shown in Fig. 3.

By deleting the columns indicated by Steps 6-8 in the code-anticode construction, shown shaded in Fig. 3, we obtain an (8@10,5; 8@11,5; 16@12,5; 3,1) VLEC code, C_1. Since the minimum converging distance in this case is one, then the free distance may be less than five (using the bound given by expression (7)). In fact, for this code the free distance is four. To get the required minimum converging distance of two and hence a free distance of five we add the modification vector 1111100000000 to code C_1. Fig. 4 gives the resultant code, C_2, plus the required generator matrices and modification vectors for the constituent horizontally linear sub-codes.

Heuristic Construction Algorithm

The aim of this algorithm is to construct a VLEC code with specified overall minimum block, diverging and converging distances (and hence a minimum value for d_{free}) and with codeword lengths matched to the source statistics so as to give a minimum average codeword length for the specified free distance and the specified source.

We will now describe the basic construction concept. First, a minimum codeword length, L_1, must be specified. This must at least be greater than or equal to the minimum diverging distance required. Then, a fixed-length code with this length and with minimum distance equal to b_{min} with a maximum number of codewords must be constructed. For a horizontally linear VLEC code, all fixed-length codes constructed with this algorithm must be linear or a coset. Let the code so constructed be C and let the number of codewords in C be s_1. Next, all the possible L_1-tuples which are at distance d_{min} from the codewords of C are listed. Let this set of words be W. Obviously, if $d_{\text{min}} \geq b_{\text{min}}$, then W will be empty. However, the bound on d_{free} given by expression (7) suggests that it is best to choose $b_{\text{min}} = d_{\text{min}} + c_{\text{min}}$. Therefore we may take $d_{\text{min}} = \lceil d_{\text{free}}/2 \rceil$ and $b_{\text{min}} = d_{\text{free}}$. In which case, $d_{\text{min}} < b_{\text{min}}$ and such words are possible to find. So for the moment we are going to assume that W is not empty. Note that the minimum distance of the words in W is not specified for now. Next, the number of words in W is doubled by increasing the words' length by one bit by affixing first a '0' and then a '1' to the rightmost position of all words in W. So now W contains words of length L_1+1. These words are then checked with the codewords in C. Those words in W which satisfy the minimum converging distance required are retained, the others are discarded. So at the end of this operation, we are left with a set of words which when compared to the codewords of C satisfy the required minimum diverging and converging distances. The only other requirement left to

$$\begin{bmatrix} 1 & 0 & 0 & 0 & 0 & 1 & 0 & 1 & 0 & 0 & 1 & 1 & 0 \\ 0 & 1 & 0 & 0 & 0 & 0 & 0 & 1 & 1 & 1 & 0 & 1 \\ 0 & 0 & 1 & 0 & 0 & 1 & 1 & 1 & 0 & 0 & 0 & 1 \\ 0 & 0 & 0 & 1 & 0 & 0 & 1 & 1 & 1 & 0 & 1 & 0 & 0 \\ 0 & 0 & 0 & 0 & 1 & 1 & 1 & 0 & 0 & 1 & 1 & 0 & 0 \end{bmatrix}$$

Fig. 1. Generator matrix for (13,5,5) fixed-length linear block code

Anticode
↓

$$\begin{bmatrix} 1 & 0 & 0 & 1 & 0 & 1 & 0 & 0 & 1 & 1 & \vdots & 0 & 0 & 0 \\ 0 & 1 & 0 & 0 & 1 & 1 & 1 & 0 & 1 & 0 & \vdots & 0 & 0 & 0 \\ 0 & 0 & 1 & 1 & 1 & 0 & 0 & 1 & 1 & 0 & \vdots & 0 & 0 & 0 \\ 0 & 0 & 0 & 0 & 0 & 0 & 1 & 1 & 1 & 0 & \vdots & 1 & 1 & 0 \\ 0 & 0 & 0 & 1 & 1 & 1 & 0 & 0 & 0 & 0 & \vdots & 0 & 1 & 1 \end{bmatrix}$$

Fig. 2. Rearranged generator matrix for the (13,5,5) fixed-length linear block code with (3,5,2) anticode in the rightmost position

```
0 0 0 0 0 0 0 0 0 0 0 0 0
1 0 0 1 0 1 0 0 1 1 0 0 0
0 1 0 0 1 1 1 0 1 0 0 0 0
1 1 0 1 1 0 1 0 0 1 0 0 0
0 0 1 1 1 0 0 1 1 0 0 0 0
1 0 1 0 1 1 0 1 0 1 0 0 0
0 1 1 1 0 1 1 1 0 0 0 0 0
1 1 1 0 0 0 1 1 1 1 0 0 0
0 0 0 0 0 0 1 1 1 0 1 1 0
1 0 0 1 0 1 1 1 0 1 1 1 0
0 1 0 0 1 1 0 1 0 0 1 1 0
1 1 0 1 1 0 0 1 1 1 1 1 0
0 0 1 1 1 0 1 0 0 0 1 1 0
1 0 1 0 1 1 1 0 1 1 1 1 0
0 1 1 1 0 1 0 0 1 0 1 1 0
1 1 1 0 0 0 0 0 1 1 1 1 0
0 0 0 1 1 1 0 0 0 0 0 1 1
1 0 0 0 1 0 0 0 1 1 0 1 1
0 1 0 1 0 0 1 0 1 0 0 1 1
1 1 0 0 0 1 1 0 0 1 0 1 1
0 0 1 0 0 1 0 1 1 0 0 1 1
1 0 1 1 0 0 0 1 0 1 0 1 1
0 1 1 0 1 0 1 1 0 0 0 1 1
1 1 1 1 1 1 1 1 1 0 1 1 1
0 0 0 1 1 1 1 1 0 1 0 0 1
1 0 0 0 1 0 1 1 0 1 1 0 1
0 1 0 1 0 0 0 1 0 0 1 0 1
1 1 0 0 0 1 0 1 1 1 1 0 1
0 0 1 0 0 1 1 0 0 0 1 0 1
1 0 1 1 0 0 1 0 1 1 1 0 1
0 1 1 0 1 0 0 0 1 0 1 0 1
1 1 1 1 1 1 0 0 0 1 1 0 1
```

Fig. 3. Codebook for (13,5,5) code and the derived VLEC code, C_1

satisfy is that these words, being of the same length, must have minimum distance at least equal to b_{min}. So here again we must choose the maximum number of words (s_2) from what is left in W such that these words form a fixed-length code with minimum distance at least b_{min} [6]. These words are then added to the codewords already in C to form a VLEC code ($s_1@L_1$, b_{min}; $s_2@(L_1+1)$, b_{min}; d_{min}, c_{min}). This whole procedure is then repeated for longer length codewords until either there are no further possible words to be found, or else when the required number of codewords is reached.

The problem with this basic construction algorithm is that if L_1 is chosen to be too small, not enough codewords may be found, whereas if L_1 is chosen to be too large, the average codeword length would be non-optimal. In addition, the codeword lengths are not matched to the source statistics. In order to have more control of the code construction, we must alter the basic algorithm in two important aspects. First, when finding the maximum number of words which satisfy the minimum block, diverging and converging distances with some given length, we must allow the possibility of eventually dropping some of these words. This may enable us to find more codewords of longer length than otherwise would be possible, hence increasing the possibility of finding the required number of codewords. The second alteration that is required is that some codeword lengths may be allowed to be skipped, i.e. we may allow the affixing of more than one bit at a time to the set W of words which satisfy the diverging distance.

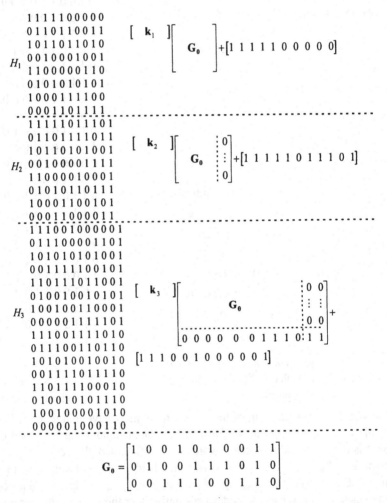

$$G_0 = \begin{bmatrix} 1 & 0 & 0 & 1 & 0 & 1 & 0 & 0 & 1 & 1 \\ 0 & 1 & 0 & 0 & 1 & 1 & 1 & 0 & 1 & 0 \\ 0 & 0 & 1 & 1 & 1 & 0 & 0 & 1 & 1 & 0 \end{bmatrix}$$

Fig. 4. VLEC code C_2 (8@10,5; 8@11,5; 16@12,5; 3,2) with $d_{free} = 5$ and its horizontal linear sub-codes

These alterations to the basic algorithm will also make it possible to produce more than one VLEC code with the specified parameters and number of codewords, each one with a different codeword length distribution. In addition, different values for L_1 may be used to give yet more VLEC codes. In order to limit this search, some maximum codeword length is also specified. The generation of this possible set of VLEC codes will enable us to match the codeword lengths to the source statistics in a rather brute force way by calculating the average codeword length for each code constructed and choosing that one which gives the minimum average codeword length for the given source. This procedure can be considerably sped up by incorporating the code selection process in the same construction algorithm, because for most of the codes constructed with this algorithm, the average codeword length becomes larger than for codes found earlier (or some specified value) even after the first few codewords are found, in which case the construction for that particular code is stopped and the next one considered immediately.

This algorithm is general enough to construct both non-linear and horizontally linear VLEC codes.

Using this algorithm, a (1@6,-; 1@7,-; 4@8,5; 5@9,5; 5@10,5; 6@11,5; 4@12,5; 3,2) non-linear VLEC code may be constructed for the 26-symbol English source [6] (the source statistics are the same as those given in Table 2 of [3]) giving an average codeword length of 8.47 bits or a code rate $R = 0.591$. The average codeword length for the horizontally linear VLEC code given in Fig. 4 when used to encode the 26-symbol English source is 10.46 bits. A horizontally linear (1@6,-; 1@7,-; 4@8,5; 4@9,5; 4@10,5; 2@11,5; 2@12,8; 2@13,10; 2@14,7; 2@15,6; 2@16,6; 3,2) VLEC code with average codeword length equal to 8.73 bits may also be constructed using the Heuristic algorithm. This indicates that the code-anticode construction is not optimal.

Fig. 5 shows a performance comparison, over an AWGN channel with BPSK modulation, between the (1@6,-; 1@7,-; 4@8,5; 5@9,5; 5@10,5; 6@11,5; 4@12,5; 3,2) non-linear VLEC code and a standard cascade consisting of a Huffman and a BCH code and also with a rate half, constraint length (K) two, convolutional code. All the schemes used have similar rate and the same minimum/free distance. All codes are hard decision decoded. Clearly, the VLEC code shows a coding gain over the other schemes.

5 Conclusions

One of the disadvantages of combined source and channel coding is that the resultant code must be matched both to the source and channel statistics. The code-anticode construction ignores the source statistics and hence the resultant code may not be efficient, as the example codes given here show. On the other hand the heuristic construction algorithm gives good VLEC codes but is computationally expensive and it would be difficult to construct codes with $s > 128$. Construction of good VLEC codes is not easy due to the multitude of parameters that must be simultaneously optimised.

Another construction for VLEC codes is discussed in [9]. This involves the concatenation of standard fixed-length block codes. However this construction has two disadvantages. As for the case of the code-anticode construction, the codeword length distribution is not well matched to the source statistics. In addition, with this construction, d_{min} must be the same as b_{min}. This usually results in codes having a larger average codeword length than is otherwise possible for a given d_{free} [6].

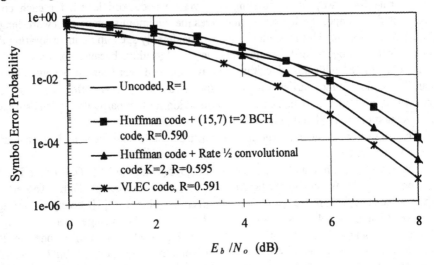

Fig. 5. Free/Minimum distance five codes used to encode the 26-symbol English source

References

1. D.A. Huffman, "A method for the construction of minimum redundancy codes", Proc. IRE, Vol. 40, pp. 1098-1101, Sep. 1952.
2. J.C. Maxted & J.P. Robinson, "Error recovery for variable length codes", *IEEE Trans. Inform. Theory*, Vol. IT-31, No. 6, pp. 794-801, Nov. 1985.
3. V. Buttigieg & P.G. Farrell, "A maximum a-posteriori (MAP) decoding algorithm for variable-length error-correcting codes", *Codes and cyphers: Cryptography and coding IV*, Essex, England, The Institute of Mathematics and its Applications, pp. 103-119, 1995.
4. V. Buttigieg & P.G. Farrell, "On variable-length error-correcting codes", *Proc. 1994 IEEE ISIT*, Trondheim, Norway, p. 507, 27 Jun. - 1 Jul. 1994b.
5. V. Buttigieg & P.G. Farrell, "Sequential decoding of variable-length error-correcting codes", *Proc. Eurocode 94*, Côte d'Or, France, pp. 93-98, 24-28 Oct. 1994c.
6. V. Buttigieg, "Variable-length error-correcting codes", *Ph.D. Thesis*, University of Manchester, England, 1995.
7. P.G. Farrell & A. Farrag, "Further properties of linear binary anticodes", *Electron. Lett.*, Vol. 10, No. 16, p. 340, Aug. 1974.
8. P.G. Farrell, "An introduction to anticodes", Internal Report, Kent, England, The University of Kent at Canterbury, 1977.
9. M.A. Bernard & B.D. Sharma, "Variable length perfect codes", *J. Inform. & Optimization Sciences*, Vol. 13, No. 1, pp. 143-151, 1992.

Lecture Notes in Computer Science

For information about Vols. 1–945

please contact your bookseller or Springer-Verlag

Vol. 981: I. Wachsmuth, C.-R. Rollinger, W. Brauer (Eds.), KI-95: Advances in Artificial Intelligence. Proceedings, 1995. XII, 269 pages. (Subseries LNAI).

Vol. 982: S. Doaitse Swierstra, M. Hermenegildo (Eds.), Programming Languages: Implementations, Logics and Programs. Proceedings, 1995. XI, 467 pages. 1995.

Vol. 983: A. Mycroft (Ed.), Static Analysis. Proceedings, 1995. VIII, 423 pages. 1995.

Vol. 984: J.-M. Haton, M. Keane, M. Manago (Eds.), Advances in Case-Based Reasoning. Proceedings, 1994. VIII, 307 pages. 1995.

Vol. 985: T. Sellis (Ed.), Rules in Database Systems. Proceedings, 1995. VIII, 373 pages. 1995.

Vol. 986: Henry G. Baker (Ed.), Memory Management. Proceedings, 1995. XII, 417 pages. 1995.

Vol. 987: P.E. Camurati, H. Eveking (Eds.), Correct Hardware Design and Verification Methods. Proceedings, 1995. VIII, 342 pages. 1995.

Vol. 988: A.U. Frank, W. Kuhn (Eds.), Spatial Information Theory. Proceedings, 1995. XIII, 571 pages. 1995.

Vol. 989: W. Schäfer, P. Botella (Eds.), Software Engineering — ESEC '95. Proceedings, 1995. XII, 519 pages. 1995.

Vol. 990: C. Pinto-Ferreira, N.J. Mamede (Eds.), Progress in Artificial Intelligence. Proceedings, 1995. XIV, 487 pages. 1995. (Subseries LNAI).

Vol. 991: J. Wainer, A. Carvalho (Eds.), Advances in Artificial Intelligence. Proceedings, 1995. XII, 342 pages. 1995. (Subseries LNAI).

Vol. 992: M. Gori, G. Soda (Eds.), Topics in Artificial Intelligence. Proceedings, 1995. XII, 451 pages. 1995. (Subseries LNAI).

Vol. 993: T.C. Fogarty (Ed.), Evolutionary Computing. Proceedings, 1995. VIII, 264 pages. 1995.

Vol. 994: M. Hebert, J. Ponce, T. Boult, A. Gross (Eds.), Object Representation in Computer Vision. Proceedings, 1994. VIII, 359 pages. 1995.

Vol. 995: S.M. Müller, W.J. Paul, The Complexity of Simple Computer Architectures. XII, 270 pages. 1995.

Vol. 996: P. Dybjer, B. Nordström, J. Smith (Eds.), Types for Proofs and Programs. Proceedings, 1994. X, 202 pages. 1995.

Vol. 997: K.P. Jantke, T. Shinohara, T. Zeugmann (Eds.), Algorithmic Learning Theory. Proceedings, 1995. XV, 319 pages. 1995.

Vol. 998: A. Clarke, M. Campolargo, N. Karatzas (Eds.), Bringing Telecommunication Services to the People – IS&N '95. Proceedings, 1995. XII, 510 pages. 1995.

Vol. 999: P. Antsaklis, W. Kohn, A. Nerode, S. Sastry (Eds.), Hybrid Systems II. VIII, 569 pages. 1995.

Vol. 1000: J. van Leeuwen (Ed.), Computer Science Today. XIV, 643 pages. 1995.

Vol. 1002: J.J. Kistler, Disconnected Operation in a Distributed File System. XIX, 249 pages. 1995.

Vol. 1004: J. Staples, P. Eades, N. Katoh, A. Moffat (Eds.), Algorithms and Computation. Proceedings, 1995. XV, 440 pages. 1995.

Vol. 1005: J. Estublier (Ed.), Software Configuration Management. Proceedings, 1995. IX, 311 pages. 1995.

Vol. 1006: S. Bhalla (Ed.), Information Systems and Data Management. Proceedings, 1995. IX, 321 pages. 1995.

Vol. 1007: A. Bosselaers, B. Preneel (Eds.), Integrity Primitives for Secure Information Systems. VII, 239 pages. 1995.

Vol. 1008: B. Preneel (Ed.), Fast Software Encryption. Proceedings, 1994. VIII, 367 pages. 1995.

Vol. 1009: M. Broy, S. Jähnichen (Eds.), KORSO: Methods, Languages, and Tools for the Construction of Correct Software. X, 449 pages. 1995. Vol.

Vol. 1010: M. Veloso, A. Aamodt (Eds.), Case-Based Reasoning Research and Development. Proceedings, 1995. X, 576 pages. 1995. (Subseries LNAI).

Vol. 1011: T. Furuhashi (Ed.), Advances in Fuzzy Logic, Neural Networks and Genetic Algorithms. Proceedings, 1994. (Subseries LNAI).

Vol. 1012: M. Bartos̆ek, J. Staudek, J. Wiedermann (Eds.), SOFSEM '95: Theory and Practice of Informatics. Proceedings, 1995. XI, 499 pages. 1995.

Vol. 1013: T.W. Ling, A.O. Mendelzon, L. Vieille (Eds.), Deductive and Object-Oriented Databases. Proceedings, 1995. XIV, 557 pages. 1995.

Vol. 1014: A.P. del Pobil, M.A. Serna, Spatial Representation and Motion Planning. XII, 242 pages. 1995.

Vol. 1015: B. Blumenthal, J. Gornostaev, C. Unger (Eds.), Human-Computer Interaction. Proceedings, 1995. VIII, 203 pages. 1995.

Vol. 1017: M. Nagl (Ed.), Graph-Theoretic Concepts in Computer Science. Proceedings, 1995. XI, 406 pages. 1995.

Vol. 1018: T.D.C. Little, R. Gusella (Eds.), Network and Operating Systems Support for Digital Audio and Video. Proceedings, 1995. XI, 357 pages. 1995.

Vol. 1019: E. Brinksma, W.R. Cleaveland, K.G. Larsen, T. Margaria, B. Steffen (Eds.), Tools and Algorithms for the Construction and Analysis of Systems. Selected Papers, 1995. VII, 291 pages. 1995.

Vol. 1020: I.D. Watson (Ed.), Progress in Case-Based Reasoning. Proceedings, 1995. VIII, 209 pages. 1995. (Subseries LNAI).

Vol. 1021: M.P. Papazoglou (Ed.), OOER '95: Object-Oriented and Entity-Relationship Modeling. Proceedings, 1995. XVII, 451 pages. 1995.

Vol. 1022: P.H. Hartel, R. Plasmeijer (Eds.), Functional Programming Languages in Education. Proceedings, 1995. X, 309 pages. 1995.

Vol. 1023: K. Kanchanasut, J.-J. Lévy (Eds.), Algorithms, Concurrency and Knowlwdge. Proceedings, 1995. X, 410 pages. 1995.

Vol. 1024: R.T. Chin, H.H.S. Ip, A.C. Naiman, T.-C. Pong (Eds.), Image Analysis Applications and Computer Graphics. Proceedings, 1995. XVI, 533 pages. 1995.

Vol. 1025: C. Boyd (Ed.), Cryptography and Coding. Proceedings, 1995. IX, 291 pages. 1995.